TMS and tDCS for Psychiatric Disorders

Wei Zheng • Yuping Ning
Editors

TMS and tDCS for Psychiatric Disorders

Non-invasive Brain Stimulation

Springer

Editors
Wei Zheng
Department of Psychiatry
The Affiliated Brain Hospital,
Guangzhou Medical University
Guangzhou, China

Key Laboratory of Neurogenetics and
Channelopathies of Guangdong Province
and the Ministry of Education of China
Guangzhou Medical University
Guangzhou, China

Yuping Ning
Department of Psychiatry
The Affiliated Brain Hospital,
Guangzhou Medical University
Guangzhou, China

Key Laboratory of Neurogenetics and
Channelopathies of Guangdong Province
and the Ministry of Education of China
Guangzhou Medical University
Guangzhou, China

ISBN 978-981-96-8503-5 ISBN 978-981-96-8504-2 (eBook)
https://doi.org/10.1007/978-981-96-8504-2

This Springer imprint is published by the registered company Springer Nature Singapore Pte Ltd.
The registered company address is: 152 Beach Road, #21-01/04 Gateway East, Singapore 189721, Singapore

Foreword

It is an honor to introduce *TMS and tDCS for Psychiatric Disorders*. This comprehensive book explores the application of noninvasive brain stimulation (NIBS) techniques in the management of psychiatric conditions. NIBS, encompassing methods such as transcranial magnetic stimulation (TMS) and transcranial direct current stimulation (tDCS), modulates cortical excitability and brain function through targeted stimulation. In recent years, these techniques have undergone significant advancements and are widely implemented in psychiatric treatments. TMS, for instance, modulates cortical excitability by delivering stimulation at frequencies from 1 to 50 Hz, inducing either excitatory or inhibitory effects depending on the protocol. Similarly, anodal tDCS is generally considered to enhance cortical activity in targeted regions, whereas cathodal tDCS is associated with inhibitory effects. Numerous meta-analyses and randomized controlled trials have consistently validated the efficacy of these modalities, particularly in treating major depressive disorder.

Co-editors Dr. Wei Zheng and Dr. Yuping Ning, along with their collaborators, have developed this detailed and authoritative resource. It comprises 13 chapters authored by experts from more than 15 tertiary hospitals. This book provides a thorough examination of advanced strategies for leveraging TMS and tDCS in psychiatric care. Each chapter is meticulously structured to enhance understanding and practical application, addressing fundamental principles and complex methodologies. Designed to offer concise and accessible insights, this text serves as an indispensable reference for general practitioners, psychiatrists, and researchers specializing in NIBS technologies. Its comprehensive coverage positions it as a foundational work in advancing the global adoption of NIBS for psychiatric disorders.

This publication is particularly timely, offering up-to-date knowledge and actionable guidance on the principles, mechanisms, and clinical applications of TMS and tDCS in psychiatry. The editors and contributors are to be commended for assembling a work of significant value, poised to improve treatment efficacy and safety for patients with psychiatric disorders.

Lin Lu

Academician, Chinese Academy of Sciences
Beijing, China

President, Shandong First Medical University
Taian, China

Boya Chair Professor, Peking University
Beijing, China

Director, National Clinical Research Center
for Mental Disorders, National Center for Mental
Disorders (Peking University Sixth Hospital)
Beijing, China

Preface

Our understanding of nervous system plasticity has significantly advanced in recent years, encompassing both healthy individuals and those experiencing neurological and neuropsychiatric dysfunctions. This progress has laid a solid foundation for the application of neuromodulation techniques in managing neurodegenerative diseases. Since the introduction of galvanic vestibular stimulation (GVS) in the 1950s, neuromodulation has undergone exponential growth, evolving from basic electrical stimulation to a diverse array of physical stimulation modalities, including electricity, magnetism, and beyond. Advances in noninvasive brain stimulation (NIBS), such as the development of improved coils, refined stimulus paradigms, and the integration of neuronavigation, have significantly enhanced precision, expanded research insights, and elevated clinical applicability. While pharmacological and psychotherapeutic interventions often fail to meet the therapeutic expectations of patients with psychiatric disorders due to limited efficacy, NIBS emerges as a promising complementary treatment modality, effectively bridging this therapeutic gap.

The clinical efficacy and underlying mechanisms of NIBS in treating psychiatric disorders exhibit considerable variability. This book focuses exclusively on the two most widely utilized NIBS techniques in clinical practice: transcranial magnetic stimulation (TMS) and transcranial direct current stimulation (tDCS). To facilitate a comprehensive understanding of the therapeutic potential and mechanisms of these techniques across various psychiatric disorders, the book integrates the latest clinical guidelines and addresses key questions. What are TMS and tDCS, and how have these methods evolved? What novel approaches are being adopted in their application? How are these techniques employed in different psychiatric disorders, and what are the associated treatment protocols and patterns? What do current findings reveal about their efficacy and safety? How have clinical experiences shaped the use of TMS and tDCS, and how might recent discoveries contribute to managing psychiatric episodes? Finally, what clinical recommendations guide the application of these techniques in treating specific psychiatric conditions?

Despite the wealth of clinical essays and books on NIBS and its applications in psychiatric disorders, a critical challenge persists: the need to deliver a comprehensive synthesis of foundational principles and encyclopedic knowledge tailored for clinical practice. To address this, the editors propose a dual-layered classification approach, using specific psychiatric disorders as the primary framework and treatment regimens and patterns of TMS and tDCS as the secondary structure. This approach aims to guide the design of NIBS paradigms that target distinct symptoms across various mental disorders, enhancing clinical diagnosis and treatment.

This book represents a thorough and systematic exploration of NIBS in psychiatric disorders, a subject to which I have dedicated years of research and profound interest. I extend my heartfelt gratitude to everyone who provided invaluable assistance and guidance throughout this endeavor; without their support, the completion of this project would not have been possible. As the Latin phrase per aspera ad astra ("through hardship to the stars") reminds us, the journey of compiling and presenting this material has been both challenging and rewarding. The process has remained steadfastly rooted in rigorous inquiry, intellectual curiosity, and an enduring passion for medical research. While every effort has been made to ensure the accuracy and completeness of this book, areas for improvement undoubtedly remain. Future editions will address these aspects, incorporating revisions and updates as the field continues to evolve.

To the readers, it is my hope that this work offers both answers and inspiration, guiding you through an intellectual odyssey of discovery and understanding in the transformative domain of NIBS for psychiatric disorders.

Guangzhou, China Wei Zheng
Guangzhou, China Yu-Ping Ning

Contents

About the Editors

Wei Zheng is an Associate Professor in the Department of Psychiatry, the Affiliated Brain Hospital, Guangzhou Medical University, Guangzhou, China. He completed resident psychiatrist training in the Affiliated Brain Hospital, Guangzhou Medical University, China, and obtained PhD degree in Guangzhou Medical University. His research focuses on neuroregulatory techniques related research. He has authored or co-authored over 190 papers in international (SCI) journals. He is serving as one of the Editor-in-Chief of *Alpha Psychiatry* (IF=3.5) at present. He was awarded the title of Top 2% Global Scientists in 2023 and 2024. His team has made some advances in antipsychotic-induced hyperproteinemia and weight gain, ketamine, and neuroregulatory techniques including TMS, tDCS, tACS, and ECT for treatment-resistant depression.

Yuping Ning is a Professor at the Department of Psychiatry, the Affiliated Brain Hospital, Guangzhou Medical University, Guangzhou, China. She is the director of the Affiliated Brain Hospital, Guangzhou Medical University, PRC. She is a Professor in neuropsychiatry and supervisor of PhD students. She holds some positions such as member of Chinese Society of Psychiatry, Chairman of Guangdong Society of Psychiatry, member of standing committee of Guangdong Society of Neurology, etc. From 1986 to 1992, she educated in XiangYa Medical College and got MD degree there. From 1992 to 1997, she practiced psychiatry in Guangzhou Brain Hospital. From 1997 to 2003, she practiced and pursued her PhD in the Department of Neurology, Sun Yat-Sen Medical

University. Her research interests are in neurodegenerative diseases (AD and PD) and depression. As a visiting scholar, she performed genetic study in Parkinson disease in Juntendo University, Japan. She has published 200 papers so far.

Transcranial Magnetic Stimulation

1

Yong Liu and Shaohua Hu

Abstract

Psychiatric disorders have become a major global health challenge, and traditional treatments often have limitations. Transcranial magnetic stimulation (TMS), a noninvasive neuromodulation therapy, has shown significant potential in clinical settings. This chapter discusses the fundamental principles of TMS, common modalities such as repetitive TMS (rTMS), intermittent theta burst stimulation (iTBS), and deep TMS, as well as treatment procedures, precautions, and underlying mechanisms. TMS modulates neuronal excitability, synaptic plasticity, and the functional connectivity of brain circuits, with variations in stimulation frequency, intensity, and pulse parameters significantly influencing its outcomes. While TMS has proven effective in treating resistant psychiatric conditions like depression with fewer side effects, further research is needed to optimize its mechanisms, personalize treatment protocols, and enhance equipment. This chapter aims to provide psychiatrists, neuroscientists, and other medical professionals with a comprehensive understanding of TMS, promoting its advancement and broader application in psychiatric treatment.

Keywords

Transcranial magnetic stimulation · Noninvasive neuromodulation · Psychiatric disorders

Y. Liu · S. Hu (✉)
Department of Psychiatry, the First Affiliated Hospital, Zhejiang University School of Medicine, Hangzhou, China

The Zhejiang Key Laboratory of Precision Psychiatry, Hangzhou, China
e-mail: drliuyong@zju.edu.cn; dorhushaohua@zju.edu.cn

W. Zheng, Y. Ning (eds.), *TMS and tDCS for Psychiatric Disorders*,
https://doi.org/10.1007/978-981-96-8504-2_1

Abbreviations

AMPA	α-amino-3-hydroxy-5-methyl-4-isoxazolepropionic acid receptor
aTMS	accelerated repetitive transcranial magnetic stimulation
cTBS	continuous theta burst stimulation
CaMKII	calcium/calmodulin-dependent protein kinase II
dTMS	deep transcranial magnetic stimulation
DLPFC	dorsolateral prefrontal cortex
ECT	electroconvulsive therapy
GABA	γ-aminobutyric acid
iTBS	intermittent theta burst stimulation
LTP	long-term potentiation
LTD	long-term depression
MST	magnetic seizure therapy
NMDA	N-methyl-D-aspartate receptor
piTBS	prolonged intermittent theta burst stimulation
pTMS	priming (guiding) transcranial magnetic stimulation
rTMS	repetitive transcranial magnetic stimulation
sTMS	synchronized transcranial magnetic stimulation
SAINT	Stanford Accelerated Intelligent Neuromodulation Therapy
TMS	transcranial magnetic stimulation
TBS	theta burst stimulation
Nav	voltage-gated sodium channels

1.1 Introduction

Psychiatric disorders have become a significant global health challenge, with profound implications for public health. According to the World Health Organization (WHO), approximately 970 million people worldwide were living with a mental disorder in 2019. The prevalence of psychiatric conditions has been steadily increasing, with an estimated 25% rise in the number of affected individuals between 2000 and 2019 [1]. This growing burden not only affects the well-being of patients but also places immense strain on families and healthcare systems, highlighting the societal and economic impact of mental health disorders.

Traditional treatment methods for psychiatric disorders, including pharmacotherapy and psychotherapy, offer symptomatic relief but have notable limitations. Pharmacological treatments are often associated with side effects, and their efficacy may be insufficient for some patients. While psychotherapy can be beneficial, it tends to be time-consuming and requires a high level of patient adherence. In this context, transcranial magnetic stimulation (TMS), an emerging noninvasive neuromodulation therapy, has gained significant attention as a promising alternative.

TMS operates on the principles of electromagnetic induction, using a coil placed on the scalp to generate a magnetic field that penetrates the skull and induces weak electrical currents in the cerebral cortex. This electrical activity modulates neuronal activity in specific brain regions, offering a means to influence dysfunctional neural circuits. Recent studies have demonstrated the significant potential of TMS in treating psychiatric disorders [2–4]. Its noninvasive nature and ability to precisely target brain regions make it an attractive alternative to traditional pharmacological treatments, especially for patients who are resistant to medication or experience intolerable side effects. With ongoing advancements in TMS technology and a growing body of clinical evidence, TMS is emerging as a valuable tool in the management of psychiatric disorders, offering hope for patients with conditions that are difficult to treat by conventional means.

In this chapter, our goal is to provide a comprehensive exploration of TMS in psychiatric treatment. This includes an overview of the fundamental principles and common parameters of TMS, as well as advanced techniques and approaches used in clinical practice. We will also discuss treatment procedures, key precautions, underlying mechanisms, and factors influencing the efficacy of TMS. Through this examination, we aim to offer a thorough understanding of how TMS is utilized to address various psychiatric conditions and optimize therapeutic outcomes.

Additionally, the chapter explores the latest evidence-based findings and future directions for TMS research. By integrating a wealth of clinical data and cutting-edge research, this work serves as an authoritative and practical resource for psychiatrists, neuroscientists, and other medical professionals. It is intended to promote the further development of TMS in psychiatry and improve the quality of life and treatment outcomes for patients suffering from mental disorders.

1.1.1 Fundamentals of TMS

TMS modulates neural activity in the brain based on the principles of electromagnetic induction. When a pulsed current is passed through a coil placed on the scalp, it generates a time-varying magnetic field in accordance with Ampère's law. This magnetic field can penetrate the skull, despite its electrical impedance and magnetic permeability. The magnetic field strength of TMS, typically in the range of 1–2 Tesla, is sufficient to reach the cerebral cortex.

According to Faraday's law of electromagnetic induction, the changing magnetic field induces an electric field in the conductive tissue of the cerebral cortex. This electric field, in turn, generates an induced current in the neurons. When the induced current affects the neurons, it alters their membrane potential. If the membrane potential reaches the depolarization threshold, voltage-gated sodium channels open, allowing sodium ions to influx and triggering an action potential, thereby increasing neuronal excitability. Conversely, if the membrane potential is hyperpolarized, neuronal excitability is reduced [5].

1.1.2 TMS and Its Advanced Approaches

In recent years, TMS has given rise to a variety of advanced application modalities, which mainly include the following.

1.1.2.1 Repetitive Transcranial Magnetic Stimulation (rTMS)

Repetitive transcranial magnetic stimulation is a common derivative form of TMS. It exerts regulatory effects on brain neural activities by repetitively applying magnetic stimulation to the brain at specific frequencies and intensities. The frequency and intensity settings of rTMS are crucial in determining its impact on neuronal excitability. For example, the frequency and intensity can be adjusted according to the specific treatment needs and the patient's condition [6].

1.1.2.2 Prolonged Intermittent Theta Burst Stimulation (piTBS)

Prolonged intermittent theta burst stimulation represents a special stimulation paradigm. It is based on the principle of theta burst stimulation (TBS), which involves pulse-train stimulation at a theta wave frequency (in the range of 4–7 Hz). Moreover, piTBS further extends the stimulation duration based on intermittent TBS. From a neurophysiological perspective, this stimulation modality exhibits potential advantages in inducing brain plasticity. It can more effectively modulate the connection strength between brain neurons [7].

1.1.2.3 Accelerated Repetitive Transcranial Magnetic Stimulation (aTMS)

Accelerated repetitive transcranial magnetic stimulation is achieved by optimizing relevant parameters such as stimulation frequency. Through this optimization, a greater number of stimulation pulses can be delivered within a relatively shorter time frame. This accelerated stimulation paradigm can effectively reduce the treatment course while ensuring the attainment of the expected therapeutic efficacy [8]. For patient populations who urgently need rapid symptom improvement or have difficulties in complying with long-term treatment, aTMS represents a more appropriate option. However, it should be noted that this stimulation method requires relatively high-performance equipment and precise control of stimulation parameters.

1.1.2.4 Deep Transcranial Magnetic Stimulation (dTMS)

Deep transcranial magnetic stimulation has the capacity to generate a stronger magnetic field, enabling it to penetrate deeper into brain tissues. The enhanced magnetic field strength of dTMS is achieved through specific coil designs and engineering optimizations [9]. These modifications allow for a more focused and intense magnetic field that can overcome the attenuation and scattering effects as the magnetic field penetrates the overlying brain layers and skull. Studies have demonstrated that dTMS can have differential effects on neural circuits within deeper brain regions, potentially modulating neurotransmitter release and synaptic plasticity in a distinct manner compared to superficial TMS [10].

1.1.2.5 Priming (Guiding) Transcranial Magnetic Stimulation (pTMS)

Priming (guiding) transcranial magnetic stimulation likely utilizes an initial application of a weak or specific patterned magnetic stimulation to "prime" the neural response mechanisms in the brain, thereby creating favorable conditions for subsequent more effective stimulation or neuromodulation procedures.

Mechanistically, it serves as a pretreatment approach. By altering the state of neural excitability in specific local brain regions, it enables the subsequent therapeutic magnetic stimulation to act more efficiently. It is analogous to providing a "guiding signal" to the brain, guiding it into a state that is more receptive to the main stimulation. Furthermore, research has suggested that pTMS-induced priming can lead to changes in synaptic efficacy and neuronal firing patterns, which may enhance the overall therapeutic efficacy in neurological and psychiatric disorders [11, 12].

1.1.2.6 Synchronized Transcranial Magnetic Stimulation (sTMS)

Synchronized transcranial magnetic stimulation (sTMS) involves the synchronization of magnetic stimulation with the natural rhythms of the brain (such as brainwaves). From the perspective of the precision of neuromodulation, this synchronized stimulation modality can more precisely target the temporal windows of brain neural activity, thereby enhancing the stimulation effect [13]. Additionally, it may potentially enhance the treatment efficacy for cognitive dysfunction. Evidence has shown that sTMS synchronized with specific brain rhythms can modulate the synchronization and coherence of neural networks, leading to improved cognitive performance and potentially therapeutic benefits in patients with cognitive impairments [14].

1.1.2.7 Various Forms of Theta Burst Stimulation (TBS)

1. *Intermittent TBS (iTBS)*: This stimulation modality intermittently applies pulse train stimulation at theta wave frequencies. From the perspective of neuroplasticity, iTBS can modulate long-term potentiation (LTP) or long-term depression (LTD) effects in brain neurons and holds significant application value in research and treatment related to neuroplasticity such as learning and memory. In a recent study, patients who received Stanford Accelerated Intelligent Neuromodulation Therapy (SAINT), potentially with an iTBS-based protocol, exhibited significant improvements in depressive symptoms in a shorter time frame compared to traditional rTMS protocols [15].
2. *Continuous TBS (cTBS)*: Unlike intermittent TBS, cTBS continuously applies pulse train stimulation at theta wave frequencies. In terms of its neuromodulatory effects, it generally exerts an inhibitory influence on brain activity and may hold potential for treating conditions such as treatment-resistant depression and suicidal ideation [16].
3. *Bilateral TBS*: Bilateral theta burst stimulation is an advanced form of TMS that applies theta burst protocols to both the left and right dorsolateral prefrontal cortex (DLPFC). It involves the simultaneous or sequential delivery of intermittent or continuous theta burst stimulation, which modulates neuronal excitability and functional connectivity in these key brain regions. Bilateral TBS has shown promising therapeutic potential in treating mood disorders, such as depression, by enhancing neuroplasticity and restoring balance in dysregulated neural circuits. As a noninvasive neuromodulation technique, it offers a novel alternative to traditional pharmacological interventions, particularly for patients with treatment-resistant conditions [17, 18].

1.1.2.8 Magnetic Seizure Therapy (MST)

Magnetic seizure therapy is a relatively high-intensity magnetic stimulation method that can induce a seizure-like response in the brain similar to that in electroconvulsive therapy (ECT). However, compared with ECT, MST has better spatial selectivity. That is, by adjusting the location and parameters of magnetic stimulation, the brain area of seizure onset can be more precisely controlled. Therefore, in treating severe mental diseases such as treatment-resistant depression, MST can be used as an alternative to ECT to improve patients' symptoms by inducing large-scale changes in brain neural activity [19].

1.1.3 Common Parameters of TMS

TMS involves several key parameters that determine its effects on neural activity. Stimulation frequency is a crucial factor. High-frequency TMS (>1 Hz) typically increases neuronal excitability and may induce long-term potentiation (LTP)-like effects, while low-frequency TMS (<1 Hz) usually decreases excitability, similar to long-term depression (LTD) mechanisms. Intensity, measured in percentage of the maximum stimulator output, influences the depth and extent of neural activation. Pulse width, which is the duration of each individual magnetic pulse, affects the energy delivered and the resulting electric field in the brain. The number of pulses per session and the total number of sessions over a treatment course also play significant roles in modulating neural plasticity and achieving therapeutic outcomes. Additionally, the coil design, such as figure-of-eight or circular, determines the focality and distribution of the magnetic field, thereby impacting the specific brain regions targeted. Precise control and optimization of these common parameters are essential for both research applications and clinical effectiveness of TMS.

1.2 Treatment Procedures and Precautions

1.2.1 Treatment Procedures

Pretreatment preparation includes gathering the patient's medical history, conducting physical and neurological exams, and obtaining informed consent after explaining the TMS process, risks, and discomforts. Patient positioning is essential, with the patient sitting comfortably, ensuring proper head and neck support, and adjusting the head for accurate coil placement on the target area.

Treatment process begins with initial titration, starting at a lower intensity and gradually increasing while monitoring the patient's responses (e.g., motor evoked potentials, sensations). Stimulation is then applied according to the protocol (continuous or intermittent). Continuous monitoring of vital signs, neurological status, and patient comfort is essential throughout the session.

Posttreatment procedure involves reevaluating the patient to assess immediate effects or side effects, including a brief neurological examination and recording subjective experiences. Detailed treatment records should be maintained, documenting patient information, treatment parameters (e.g., coil type, intensity, frequency), responses during treatment, and posttreatment evaluations for future reference.

1.2.2 TMS Treatment Precautions

1.2.2.1 Seizure Risk

To minimize seizure risk during TMS, patients with a history of epilepsy or seizure disorders must undergo thorough screening, including seizure type, frequency, and current antiepileptic treatment. Stimulation parameters should be carefully managed, limiting frequency (usually <20–30 Hz), titrating intensity, and controlling pulse count and duration to reduce seizure risk.

1.2.2.2 Scalp and Head Discomfort

Scalp and head discomfort are common during TMS. Mild scalp pain may result from coil pressure, magnetic field effects, or muscle contractions, and can be relieved by adjusting the coil or reducing intensity. Headaches or migraines may occur, potentially due to changes in cerebral blood flow, and should be managed with medical guidelines.

1.2.2.3 Other Precautions

Other precautions include screening for metallic implants, particularly ferromagnetic materials, as they may interact with the magnetic field. TMS during pregnancy and lactation is generally avoided due to insufficient safety data, and should only be considered with a multidisciplinary assessment if benefits outweigh risks.

1.3 Principle and Mechanism

1.3.1 Alteration of Membrane Potential and Cortical Excitability

1.3.1.1 Membrane Potential Alterations

TMS-induced currents modulate neuronal membrane potential [20]. Depolarization elevates the membrane potential to threshold, initiating an action potential and enhancing neuronal excitability. Conversely, hyperpolarization lowers the membrane potential, reducing the likelihood of reaching threshold and decreasing neuronal excitability.

1.3.1.2 Role of Ion Channels

Ion channels on the neuronal membrane are pivotal in modulating excitability changes induced by TMS. Voltage-gated sodium channels (Nav) facilitate action potential generation upon depolarization, allowing sodium influx [21]. In GABAergic neurons, TMS may influence chloride channels, such as the GABAA receptor complex, altering inhibitory functions and potentially modulating excitatory neuronal activity through GABAergic regulation [22].

1.3.2 Altering Synaptic Plasticity and Cortical Functional Connectivity

1.3.2.1 Long-Term Potentiation (LTP) and Long-Term Depression (LTD)

TMS can induce long-term potentiation (LTP) or long-term depression (LTD) at synapses between cortical neurons, depending on the stimulation parameters. High-frequency TMS enhances glutamate release, activating ionotropic glutamate receptors (e.g., AMPA, NMDA), leading to calcium influx and the activation of intracellular signaling pathways, such as CaMKII. This results in the insertion of AMPA receptors, enhancing synaptic transmission and generating LTP [23, 24]. Conversely, low-frequency TMS may reduce glutamate release or alter its timing, modulate GABA release, and induce hyperpolarization or AMPA receptor internalization, reducing synaptic efficiency and promoting LTD [25, 26]. TMS may modulate the morphology and density of neuronal dendritic spines, which are integral to synaptic plasticity. These structural changes can lead to an increased complexity of neuronal connectivity, thereby promoting enhanced brain plasticity, as demonstrated in various studies [27, 28].

1.3.2.2 Regulation of Local Neural Circuits and Long-Distance Brain Connectivity

TMS can regulate local neural circuits within the cerebral cortex by modulating the strength and synchrony of neuronal connections [29]. Through adjustments in the excitatory and inhibitory balance, TMS alters the information processing capacity of these circuits. Additionally, TMS can impact long-distance brain connectivity, influencing distant brain regions by transmitting signals via white matter tracts, such as the corpus callosum, thereby affecting global brain network activity [30].

1.4 Effect Factors

1.4.1 Stimulation Frequency

1.4.1.1 Frequency Classification Related to Physiological Effects

Stimulation frequency in TMS refers to the number of magnetic stimulation pulses applied per unit time (usually per second), with the unit being hertz (Hz). It is a crucial parameter characterizing the rhythm of TMS stimulation and reflects the temporal sequence properties of magnetic stimulation. Low-frequency stimulation (usually <1 Hz) inhibits neuronal activity by hyperpolarizing the membrane, reducing excitatory neurotransmitter release and promoting inhibitory neurotransmitter release, akin to long-term depression (LTD). In contrast, high-frequency stimulation

(>1 Hz) depolarizes neurons, enhancing excitability and synaptic connections, resembling long-term potentiation (LTP).

1.4.2 Stimulation Intensity

1.4.2.1 Definition and Quantification Method

Stimulation intensity is a parameter measuring the degree of impact of the magnetic field generated by TMS on brain neural tissues. In practical applications, it is usually expressed as a percentage of the maximum output power of the transcranial magnetic stimulator. This quantification method facilitates comparison and standardization of stimulation intensity under different equipment and experimental/treatment conditions.

1.4.2.2 Impact on Neural Tissues and Safety Considerations

Stimulation intensity directly correlates with neuronal activation [31, 32]. Higher intensity increases the magnetic field strength, inducing a larger current in brain tissues, which may trigger neuronal action potentials. However, excessive intensity can lead to discomfort, including scalp pain and headaches, and may elevate the risk of seizures. Therefore, careful titration is essential to determine the optimal intensity for effective neuronal stimulation while minimizing adverse effects. For motor cortex stimulation, the intensity is typically adjusted to achieve a stable minimum motor evoked potential (MEP), serving as a reference for safe and effective stimulation.

1.4.3 Pulse Duration

1.4.3.1 Definition and Physical Significance

Pulse duration refers to the length of time a single magnetic stimulation pulse lasts, which is the temporal dimension parameter of the TMS pulse. From a physical perspective, it determines the time span during which the magnetic field acts on brain neural tissues.

1.4.3.2 Impact on Stimulation Effects

Pulse duration significantly affects the spatial characteristics of TMS stimulation [33]. A shorter pulse duration tends to produce a more focal stimulation effect because the shorter action time concentrates the magnetic field energy in a smaller area, allowing for more precise stimulation of specific target regions in the cerebral cortex. Conversely, a longer pulse duration distributes the magnetic field energy more widely over time, which may expand the stimulation range but could reduce the spatial precision. In neuroscience research, especially when precisely locating and studying brain functional areas, careful control of pulse duration is essential to accurately explore the functions of specific brain regions and minimize unnecessary interference with surrounding nontarget areas.

1.4.4 Number of Pulses and Stimulation Time

1.4.4.1 Definition and Interrelationship

The number of pulses refers to the total number of magnetic stimulation pulses applied during a single TMS treatment session or experimental operation. Stimulation time refers to the duration from the start to the end of the entire TMS stimulation process. There is a close relationship between them, and the stimulation time can be calculated by the number of pulses and the pulse frequency (Stimulation time = Number of pulses/Pulse frequency).

1.4.4.2 Impact on Neuromodulation Effects

The number of pulses and stimulation time jointly determine the cumulative effect of TMS on brain neuron activities. A larger number of pulses and a longer stimulation time usually subject neurons to more continuous stimulation, resulting in more persistent neuromodulation effects. For example, accelerated TBS protocols present an opportunity to deliver higher pulse doses in shorter periods of time, thus resulting in faster and potentially more clinically effective treatment [34].

1.4.5 Interstimulus Interval

1.4.5.1 Definition and Mechanism of Action

The interstimulus interval refers to the time interval between two adjacent magnetic stimulation pulses during the process of repeatedly applying magnetic stimulation. It is another important parameter for regulating the rhythm of TMS stimulation and plays a crucial role in simulating the natural neural activity rhythm of the brain and regulating neuronal plasticity.

1.4.5.2 Application in Neuroplasticity Regulation

An appropriate interstimulus interval can simulate the rhythm of the brain's natural neural activity, promoting synchronized activities between neurons [35, 36]. For example, in the theta burst stimulation (TBS) mode, the interstimulus interval is set according to the rhythm of the theta wave (4–7 Hz). This interstimulus interval that matches the brain's natural rhythm can effectively regulate the brain's plasticity, enhancing or inhibiting synaptic plasticity through mechanisms such as affecting postsynaptic potentials, neurotransmitter release, and intracellular signal transduction. In research and treatment related to diseases associated with neuroplasticity (such as memory disorders, neurorehabilitation, etc.), reasonably setting the interstimulus interval is an important strategy for optimizing TMS treatment effects.

1.5 Conclusion

TMS is a noninvasive neuromodulation method. It functions via electromagnetic induction, affecting neuronal activity in the cerebral cortex. There are various advanced TMS approaches such as rTMS, piTBS, and so on, each with distinct applications. Treatment involves pretreatment steps, actual stimulation with monitoring, and posttreatment evaluation. Precautions are needed regarding seizures, auditory/vestibular issues, discomfort, and for specific patient groups. TMS impacts membrane potential, synaptic plasticity, and neural circuits through mechanisms related to ion channels and different stimulation parameters. Future TMS research should clarify its mechanisms better. Clinical applications may expand with personalized protocols and integration with other treatments. Technological improvements in coil design and monitoring could enhance precision. Wider patient access and better acceptance are crucial for its full potential to be realized, promising greater impact on treating neuropsychiatric disorders in the future.

Conflict of Interest The authors declare no competing interests.

References

1. World Health Organization. World mental health report: transforming mental health for all. Geneva: World Health Organization; 2022.
2. Sippel LM, Hamblen JL, Kelmendi B, Alpert JE, Carpenter LL, Grzenda A, et al. Novel pharmacologic and other somatic treatment approaches for posttraumatic stress disorder in adults: state of the evidence; 2024. p. 1045–58. https://doi.org/10.1176/appi.ajp.20230950. [cited 2024/Dec/01]. Available from: https://pubmed.ncbi.nlm.nih.gov/39616450
3. Cash RFH, Cocchi L, Lv J, Fitzgerald PB, Zalesky A. Functional magnetic resonance imaging-guided personalization of transcranial magnetic stimulation treatment for depression; 2021. p. 337–9. https://doi.org/10.1001/jamapsychiatry.2020.3794. [cited 2021/Mar/01]. Available from: https://pubmed.ncbi.nlm.nih.gov/33237320
4. Hua Q, Wang L, He K, Sun J, Xu W, Zhang L, et al. Repetitive transcranial magnetic stimulation for auditory verbal hallucinations in schizophrenia: a randomized clinical trial; 2024. p. e2444215. https://doi.org/10.1001/jamanetworkopen.2024.44215. [cited 2024/Nov/04]. Available from: https://pubmed.ncbi.nlm.nih.gov/39527055
5. Peterchev AV, Wagner TA, Miranda PC, Nitsche MA, Paulus W, Lisanby SH, et al. Fundamentals of transcranial electric and magnetic stimulation dose: definition, selection, and reporting practices; 2012. p. 435–53. https://doi.org/10.1016/j.brs.2011.10.001. [cited 2012/Oct]. Available from: https://pubmed.ncbi.nlm.nih.gov/22305345
6. Morriss R, Briley PM, Webster L, Abdelghani M, Barber S, Bates P, et al. Connectivity-guided intermittent theta burst versus repetitive transcranial magnetic stimulation for treatment-resistant depression: a randomized controlled trial; 2024. p. 403–13. https://doi.org/10.1038/s41591-023-02764-z. [cited 2024/Feb]. Available from: https://pubmed.ncbi.nlm.nih.gov/38228914
7. Huang Y, Edwards MJ, Rounis E, Bhatia KP, Rothwell JC. Theta burst stimulation of the human motor cortex; 2005. p. 201–6. [cited 2005/Jan/20]. Available from: https://pubmed.ncbi.nlm.nih.gov/15664172
8. Sonmez AI, Camsari DD, Nandakumar AL, Voort JLV, Kung S, Lewis CP, et al. Accelerated TMS for depression: a systematic review and meta-analysis; 2019. p. 770–81. [cited

2019/Mar]. Available from: https://pubmed.ncbi.nlm.nih.gov/31207865 doi: 10.1016/j.psychres.2018.12.041
9. Guadagnin V, Parazzini M, Fiocchi S, Liorni I, Ravazzani P. Deep transcranial magnetic stimulation: modeling of different coil configurations; 2016. p. 1543–50. https://doi.org/10.1109/TBME.2015.2498646. [cited 2016/Jul]. Available from: https://pubmed.ncbi.nlm.nih.gov/26560868
10. Albert GC, Cook CM, Prato FS, Thomas AW. Deep brain stimulation, vagal nerve stimulation and transcranial stimulation: an overview of stimulation parameters and neurotransmitter release; 2009. p. 1042–60. https://doi.org/10.1016/j.neubiorev.2009.04.006. [cited 2009/July]. Available from: https://pubmed.ncbi.nlm.nih.gov/19433105
11. Fitzgerald PB, Hoy K, McQueen S, Herring S, Segrave R, Been G, et al. Priming stimulation enhances the effectiveness of low-frequency right prefrontal cortex transcranial magnetic stimulation in major depression; 2008. p. 52–8. https://doi.org/10.1097/jcp.0b013e3181603f7c. [cited 2008/Feb]. Available from: https://pubmed.ncbi.nlm.nih.gov/18204341
12. Hassanzahraee M, Zoghi M, Jaberzadeh S. How different priming stimulations affect the corticospinal excitability induced by noninvasive brain stimulation techniques: a systematic review and meta-analysis; 2018. p. 883–99. https://doi.org/10.1515/revneuro-2017-0111. [cited 2018/Nov/27]. Available from: https://pubmed.ncbi.nlm.nih.gov/29604209
13. Zrenner B, Zrenner C, Gordon PC, Belardinelli P, McDermott EJ, Soekadar SR, et al. Brain oscillation-synchronized stimulation of the left dorsolateral prefrontal cortex in depression using real-time EEG-triggered TMS; 2020. p. 197–205. https://doi.org/10.1016/j.brs.2019.10.007. [cited 2020]. Available from: https://pubmed.ncbi.nlm.nih.gov/31631058
14. Thut G, Miniussi C. New insights into rhythmic brain activity from TMS-EEG studies; 2009. p. 182–9. [cited 2009/Apr]. Available from: https://pubmed.ncbi.nlm.nih.gov/19286414. https://doi.org/10.1016/j.tics.2009.01.004.
15. Cole EJ, Stimpson KH, Bentzley BS, Gulser M, Cherian K, Tischler C, et al. Stanford Accelerated intelligent neuromodulation therapy for treatment-resistant depression; 2020. p. 716–26. https://doi.org/10.1176/appi.ajp.2019.19070720. [cited 2020/Aug/01]. Available from: https://pubmed.ncbi.nlm.nih.gov/32252538
16. Zhao H, Jiang C, Zhao M, Ye Y, Yu L, Li Y, et al. Comparisons of accelerated continuous and intermittent theta burst stimulation for treatment-resistant depression and suicidal ideation; 2024. p. 26–33. https://doi.org/10.1016/j.biopsych.2023.12.013. [cited 2024/Jul/01]. Available from: https://pubmed.ncbi.nlm.nih.gov/38142717
17. Blumberger DM, Mulsant BH, Thorpe KE, McClintock SM, Konstantinou GN, Lee HH, et al. Effectiveness of standard sequential bilateral repetitive transcranial magnetic stimulation vs bilateral theta burst stimulation in older adults with depression: the FOUR-D randomized non-inferiority clinical trial; 2022. p. 1065–73. https://doi.org/10.1001/jamapsychiatry.2022.2862. [cited 2022/Nov/01]. Available from: https://pubmed.ncbi.nlm.nih.gov/36129719
18. Stubbeman WF, Zarrabi B, Bastea S, Ragland V, Khairkhah R. Bilateral neuronavigated 20Hz theta burst TMS for treatment refractory depression: an open label study; 2018. p. 953–5. https://doi.org/10.1016/j.brs.2018.04.012. [cited 2018]. Available from: https://pubmed.ncbi.nlm.nih.gov/29724677
19. Daskalakis ZJ, Dimitrova J, McClintock SM, Sun Y, Voineskos D, Rajji TK, et al. Magnetic seizure therapy (MST) for major depressive disorder; 2020. p. 276–82. [cited 2020/Jan]. Available from: https://pubmed.ncbi.nlm.nih.gov/31486777. https://doi.org/10.1038/s41386-019-0515-4.
20. Kozyrev V, Eysel UT, Jancke D. Voltage-sensitive dye imaging of transcranial magnetic stimulation-induced intracortical dynamics; 2014. p. 13553–8. https://doi.org/10.1073/pnas.1405508111. [cited 2014/Sep/16]. Available from: https://pubmed.ncbi.nlm.nih.gov/25187557
21. Addolorato G, Leggio L, Hopf FW, Diana M, Bonci A. Novel therapeutic strategies for alcohol and drug addiction: focus on GABA, ion channels and transcranial magnetic stimulation; 2012. p. 163–77. https://doi.org/10.1038/npp.2011.216. [cited 2012/Jan]. Available from: https://pubmed.ncbi.nlm.nih.gov/22030714

22. Sohn MN, Brown JC, Sharma P, Ziemann U, McGirr A. Pharmacological adjuncts and transcranial magnetic stimulation-induced synaptic plasticity: a systematic review; 2024. p. E59–76. https://doi.org/10.1503/jpn.230090. [cited 2024]. Available from: https://pubmed.ncbi.nlm.nih.gov/38359933
23. O'Donnell P, Dijkstra FM, Damar U, Quanhong L, de Goede AA, Xu L, et al. Transcranial magnetic stimulation as a translational biomarker for AMPA receptor modulation; 2021. p. 325. https://doi.org/10.1038/s41398-021-01451-2. [cited 2021/May/27]. Available from: https://pubmed.ncbi.nlm.nih.gov/34045439
24. Richter MA, de Jesus DR, Hoppenbrouwers S, Daigle M, Deluce J, Ravindran LN, et al. Evidence for cortical inhibitory and excitatory dysfunction in obsessive compulsive disorder; 2012. p. 1144–51. https://doi.org/10.1038/npp.2011.300. [cited 2012/Apr]. Available from: https://pubmed.ncbi.nlm.nih.gov/22169948
25. Baur D, Galevska D, Hussain S, Cohen LG, Ziemann U, Zrenner C. Induction of LTD-like corticospinal plasticity by low-frequency rTMS depends on pre-stimulus phase of sensorimotor μ-rhythm; 2020. p. 1580–7. https://doi.org/10.1016/j.brs.2020.09.005. [cited 2020]. Available from: https://pubmed.ncbi.nlm.nih.gov/32949780
26. Gersner R, Kravetz E, Feil J, Pell G, Zangen A. Long-term effects of repetitive transcranial magnetic stimulation on markers for neuroplasticity: differential outcomes in anesthetized and awake animals; 2011. p. 7521–6. https://doi.org/10.1523/JNEUROSCI.6751-10.2011. [cited 2011/May/18]. Available from: https://pubmed.ncbi.nlm.nih.gov/21593336
27. Meneses-San Juan D, Lamas M, Ramírez-Rodríguez GB. Repetitive transcranial magnetic stimulation reduces depressive-like behaviors, modifies dendritic plasticity, and generates global epigenetic changes in the frontal cortex and hippocampus in a rodent model of chronic stress; 2023. https://doi.org/10.3390/cells12162062. [cited 2023/Aug/14]. Available from: https://pubmed.ncbi.nlm.nih.gov/37626872
28. Cambiaghi M, Cherchi L, Masin L, Infortuna C, Briski N, Caviasco C, et al. High-frequency repetitive transcranial magnetic stimulation enhances layer II/III morphological dendritic plasticity in mouse primary motor cortex; 2021. p. 113352. https://doi.org/10.1016/j.bbr.2021.113352. [cited 2021/Jul/23]. Available from: https://pubmed.ncbi.nlm.nih.gov/33979657
29. Ferrarelli F, Phillips ML. Examining and modulating neural circuits in psychiatric disorders with transcranial magnetic stimulation and electroencephalography: present practices and future developments; 2021. p. 400–13. https://doi.org/10.1176/appi.ajp.2020.20071050. [cited 2021/May/01]. Available from: https://pubmed.ncbi.nlm.nih.gov/33653120
30. Fox MD, Buckner RL, White MP, Greicius MD, Pascual-Leone A. Efficacy of transcranial magnetic stimulation targets for depression is related to intrinsic functional connectivity with the subgenual cingulate; 2012. p. 595–603. https://doi.org/10.1016/j.biopsych.2012.04.028. [cited 2012/Oct/01]. Available from: https://pubmed.ncbi.nlm.nih.gov/22658708
31. Thordstein M, Saar K, Pegenius G, Elam M. Individual effects of varying stimulation intensity and response criteria on area of activation for different muscles in humans. A study using navigated transcranial magnetic stimulation; 2013. p. 49–53. https://doi.org/10.1016/j.brs.2012.01.004. [cited 2013/Jan]. Available from: https://pubmed.ncbi.nlm.nih.gov/22410475
32. Nahas Z, Lomarev M, Roberts DR, Shastri A, Lorberbaum JP, Teneback C, et al. Unilateral left prefrontal transcranial magnetic stimulation (TMS) produces intensity-dependent bilateral effects as measured by interleaved BOLD fMRI; 2001. p. 712–20. [cited 2001/Nov/01]. Available from: https://pubmed.ncbi.nlm.nih.gov/11704079
33. Shirota Y, Sommer M, Paulus W. Strength-duration relationship in paired-pulse transcranial magnetic stimulation (TMS) and its implications for repetitive TMS; 2016. p. 755–61. https://doi.org/10.1016/j.brs.2016.04.019. [cited 2016]. Available from: https://pubmed.ncbi.nlm.nih.gov/27234142
34. Cole E, O'Sullivan SJ, Tik M, Williams NR. Accelerated theta burst stimulation: safety, efficacy, and future advancements; 2024. p. 523–35. https://doi.org/10.1016/j.biopsych.2023.12.004. [cited 2024/Mar/15]. Available from: https://pubmed.ncbi.nlm.nih.gov/38383091

35. George MS, Huffman S, Doose J, Sun X, Dancy M, Faller J, et al. EEG synchronized left prefrontal transcranial magnetic stimulation (TMS) for treatment resistant depression is feasible and produces an entrainment dependent clinical response: a randomized controlled double blind clinical trial; 2023. p. 1753–63. [cited 2023]. Available from: https://pubmed.ncbi.nlm.nih.gov/38043646. https://doi.org/10.1016/j.brs.2023.11.010.
36. Pantazatos SP, Mclntosh JR, Saber GT, Sun X, Doose J, Faller J, et al. The timing of transcranial magnetic stimulation relative to the phase of prefrontal alpha EEG modulates downstream target engagement; 2023. p. 830–9. https://doi.org/10.1016/j.brs.2023.05.007. [cited 2023]. Available from: https://pubmed.ncbi.nlm.nih.gov/37187457

Transcranial Direct Current Stimulation

2

Jiaxin Fu, Lihao Zheng, Hongwen Song,
and Xiaochu Zhang

Abstract

Transcranial direct current stimulation (tDCS) is a noninvasive brain stimulation technique that exhibits promising therapeutic potential in the treatment of chronic pain, neurological disorders, and psychiatric disorders. Despite its widespread application, there remains a lack of standardized guidelines regarding the impact of various tDCS parameters (such as polarity, electrode size and placement, stimulation duration) on the induction of neurophysiological and behavioral effects. In this chapter, we present a comprehensive overview of the current understanding of tDCS encompassing its fundamental principles and common parameters, treatment protocols and precautionary measures, potential mechanisms underlying its effects, as well as influential factors. Future studies are warranted to further elucidate its mechanism of action and explore more efficacious stimulus programs.

Keywords

Neuromodulation · Noninvasive brain stimulation · Transcranial direct current stimulation (tDCS) · High-definition tDCS (HD-tDCS) · Personalized tDCS · Home-based tDCS

J. Fu · L. Zheng · H. Song (✉) · X. Zhang (✉)
Division of Life Science and Medicine, Department of Radiology, The First Affiliated Hospital of USTC, School of Life Science, University of Science & Technology of China, Hefei, China
e-mail: shw0519@ustc.edu.cn; zxcustc@ustc.edu.cn

W. Zheng, Y. Ning (eds.), *TMS and tDCS for Psychiatric Disorders*,
https://doi.org/10.1007/978-981-96-8504-2_2

Abbreviations

^{1}H MRS	proton magnetic resonance spectroscopy
A	electric current
AEs	adverse events
AMPA	α-amino-3-hydroxy-5-methyl-4-isoxazolepropionic acid
ASL	arterial spin labeling
C	stimulation charge
C3, Fp2, F3, F4	the standard electroencephalography (EEG) electrode position
CBF	cerebral blood flow
CBZ	channel blocker carbamazepine
DA	dopamine
DC	direct current
DLPFC	dorsolateral prefrontal cortex
DMN	the default mode network
DMO	dextromethorphan
E	excitatory neurotransmitters
EEG	electroencephalogram
FLU	channel blocker flunarizine
fMRI	functional magnetic resonance imaging
FPNs	frontoparietal networks
GABA	gamma-aminobutyric acid
GABAA	gamma-aminobutyric acid type A
HD-tDCS	high-definition transcranial direct current stimulation
I	inhibitory neurotransmitters
J	stimulation energy, with the unit being joules
LOR	Lorazepam
LTD	long-term depression
LTP	long-term potentiation
MEP	motor evoked potential
MHC I	major histocompatibility complex I
MR	magnetic resonance
MRI	magnetic resonance imaging
M1	the primary motor cortex
M1-SO	an electrode site connected to the sensorimotor cortex
NA	noradrenaline
NMDA	N-methyl-D-aspartate
PET	positron emission computed tomography
rCBF	regional cerebral blood flow
RSN	resting-state network
rTMS	repetitive transcranial magnetic stimulation
SHAM	sham stimulation
t	time
tDCS	transcranial direct current stimulation
tNAA	N-acetylaspartate

TMS transcranial magnetic stimulation
V voltage
VMR vasomotor reactivity
W stimulation power, the unit is watts

2.1 Introduction

2.1.1 Fundamentals of tDCS

Transcranial direct current stimulation (tDCS) is a noninvasive brain stimulation technique, in which electrodes placed on the scalp deliver a weak direct current (typically 1–2 mA) to modulate neuronal excitability in specific targeted brain regions [1–3]. The systematic investigation of tDCS can be traced back to the 1960s; however, most research groups at that time employed current intensities and stimulation durations that were insufficient to induce profound changes in cerebral excitability. It was not until 2000 when tDCS was definitively demonstrated to polarize the cerebral cortex through the application of weak constant currents, thereby modulating neuronal excitability. This discovery led to a resurgence of research interest in tDCS [1]. Concurrently, advancements in signal processing and imaging technologies, such as functional magnetic resonance imaging (fMRI), positron emission computed tomography (PET), and electroencephalogram (EEG), have facilitated the integration of tDCS with these techniques, propelling tDCS into the domain of robust brain function analysis and neurophysiology research, thus reestablishing it as a prominent subject of investigation [4, 5].

The tDCS device utilized is a battery-powered current generator capable of delivering a constant electrical current flow of up to 2 mA. Typically, the device employs two large sponge-electrodes soaked in saline solution or a ring array for targeted delivery of the current [4, 6]. The two sponges, soaked in saline solution, typically have an area ranging from 20 to 35 cm^2 and are equipped with slits for the placement of electrodes (an anode and cathode), thus forming an electrode-sponge unit [7, 8]. The purpose of electrodes in tDCS is to facilitate the efficient transmission of current from the stimulation device to the scalp [5]. Standard tDCS electrodes (pads) are typically square 5 × 5 cm or 5 × 7 cm [9], 9 cm^2, and smaller electrodes (pads) are also used. Standard tDCS electrode assemblies employ either metallic or conductive rubber electrodes, which serve as the site of electrochemical reactions and should not come into direct contact with the skin [5, 10]. Electrolytes are commonly isotonic saline, saturated in a sponge that envelops the electrode, serving as a buffer between the electrode and the skin. Sufficient electrolyte volume is essential to prevent chemicals generated at the electrode from reaching the skin; however, conductive gels and/or creams have also been employed [5, 10]. Due to their size and material composition, the number of pad electrodes is typically restricted to a maximum of 3–4 [10]. The shape and size of the sponges are intentionally designed to facilitate a homogeneous distribution of current across the stimulation area, thereby mitigating the risk of skin burns resulting from localized electric concentrations (referred to as "hot spots") at the interface between the sponge and skin [5, 7].

tDCS involves the application of either anodal or cathodal stimulation, which typically elicit contrasting effects [4]. The positioning of the anode and cathode electrodes determines the direction of electrical current flow across the brain [6]. The electrode through which the current is directed into the brain is commonly referred to as the anode, while the electrode through which the current exits the body is known as the cathode [11]. Precise placement of these electrodes plays a crucial role in tDCS application, and any deviations from predetermined scalp locations may significantly impact its outcomes [6]. Typically, electrode positioning on the scalp follows guidelines derived from the International EEG 10–20 System [5, 7]. One electrode-sponge unit is positioned on the scalp, while the second can be placed at a different cephalic location (referred to as bipolar or bicephalic montage) or an extracephalic location (known as unipolar or monocephalic montage), typically on the shoulder or upper arm [7]. The most commonly utilized electrode montages in tDCS research include the left anodal M1-SO (C3-right supraorbital region [Fp2]) montage and the bilateral frontal (F3-F4) montage [6]. Once desired locations are identified, elastic straps are utilized to securely fasten the electrode assembly onto the scalp for precise current delivery. Inadequate or excessive tightening of these straps may result in electrode displacement during the course of a tDCS session. Moreover, excessive tightening of electrode straps can lead to an elevated likelihood of saline leakage from the electrode sponges [5].

Conventional tDCS, utilizing large electrode configurations, leads to the activation of relatively extensive cortical regions. A fraction of the electrical current is diverted by the scalp, while the remaining portion effectively reaches the brain tissue and elicits diffuse alterations in cortical excitability. To address the limited spatial resolution of conventional tDCS, several methods have been proposed to enhance stimulation focus. In 2009, a novel electrode configuration known as high-definition tDCS (HD-tDCS) was developed with the aim of improving stimulation focality and enabling precise targeting of specific brain regions [12]. In contrast to conventional tDCS, which typically employs two larger electrodes, HD-tDCS utilizes multiple smaller electrodes (approximately 1 cm^2) for enhanced spatial targeting. Moreover, the position and current intensity at each electrode can be optimized to improve precision or targeting [5, 7, 13]. Smaller electrodes offer the advantage of accommodating a larger number of electrodes and/or positioning them in closer proximity [14]. A commonly employed electrode configuration for HD tDCS is a 4×1 ring, consisting of a central electrode surrounded by four polarity-opposite peripheral electrodes [13]. In comparison to conventional tDCS, the 4×1 ring-shaped HD-tDCS elicits a more localized brain stimulation, confining the induced modulation of cortical excitability within the boundaries of the ring electrode. The concentration of the electric field's peak within the central electrode area enhances stimulation precision and current intensity, ensuring a high current density in the target region while minimizing undesired effects on nontarget brain areas [13, 15]. In addition to enhanced focality, the effects of HD-tDCS on cortical excitability persist for a minimum duration of 30 min longer than those observed with conventional tDCS, thereby further augmenting interest in this technology [12, 13]. However, the focal nature of the stimulation leads to an increase in interindividual variability in electric fields, which presents a significant challenge for the practical implementation of HD-tDCS [13].

2.1.2 Common Parameters of tDCS

Despite its limited current clinical application, extensive research conducted abroad over the past decade has demonstrated the efficacy of tDCS in stimulating the human cerebral cortex and established its optimal stimulation patterns. Key parameters for tDCS include the delivered current intensity (in mA), stimulation duration (in min), and electrode size and placement [7]. By calculating the current density at the electrode, which is obtained by dividing the delivered current by the size of the electrode, one can determine it based on the electrode size and applied current [7, 10]. Typically used for tDCS are current intensities ranging from 0.5 to 2 mA with corresponding current densities between 0.029 and 0.08 mA/cm^2. In standard tDCS protocols in humans, a current density of approximately 0.05 mA/cm^2 is typically generated [9]. The duration of the stimulation commonly ranges between 20 and 40 min [7]. In addition to the key parameters mentioned above, the parameters of tDCS also include stimulation charge (in coulombs, C), stimulation charge density ($A \times t/m^2 = C/m^2$), stimulation power (in watts, $W = V \times A$), and stimulation energy (in joules, $J = V \times A \times t$). Stimulation charge is determined by multiplying current by duration. Stimulation charge density is charge divided by electrode area, and is also an average metric. Stimulation power is voltage multiplied by current. Stimulation energy is power multiplied by duration [10].

2.2 Treatment Procedures and Precautions

As a noninvasive and well-established intervention, tDCS has been extensively investigated for its therapeutic potential in various conditions, including chronic pain and neuropsychiatric disorders such as stroke, Alzheimer's disease, refractory epilepsy, Parkinson's disease, major depressive disorder, substance addiction, and fibromyalgia [7, 16].

In terms of safety and tolerability, numerous systematic reviews and meta-analyses have consistently demonstrated that the application of tDCS to both motor and non-motor areas, in accordance with current tDCS safety guidelines, is associated with minimal adverse effects in healthy individuals as well as patients with various neurological disorders [14, 17–19]. In the subsample reporting AEs, the most common were, for active vs. sham tDCS group, itching, tingling, headache, burning sensation, and discomfort [18]. Several empirical studies have also corroborated the safety of tDCS, for example, Prateek C. Gandiga and colleagues evaluated the effects of tDCS and sham stimulation (SHAM) on healthy subjects and stroke patients' self-report measures of attention, fatigue, duration of elicited sensations, and discomfort, they found that tDCS and SHAM elicited comparably minimal discomfort and duration of sensations [20]. The study conducted by M.A. Nitsche and colleagues demonstrated that the application of anodal and cathodal tDCS protocols to the human brain did not elicit any structural changes in brain tissue or induce alterations in the blood–brain barrier, as confirmed by magnetic resonance imaging (MRI) [21].

The treatment procedures of tDCS includes the following seven points:

1. *Physical examination:* Prior to commencing tDCS treatment, it is imperative for the physician to conduct a comprehensive safety screening in order to ascertain the patient's suitability for undergoing tDCS therapy.
2. *Patient history and assessment:* Gather comprehensive patient history, including clinical presentation and current medication regimen.
3. *Informed consent:* Obtain informed consent from the patient before initiating treatment.
4. *Electrode placement:* Typically, electrode placement is determined using the International 10–20 System [7]. Precise localization can be achieved using standardized positioning electrode caps compatible with the tDCS device, or alternative methods such as neuro-navigation systems (e.g., magnetic resonance imaging-guided localization) or physiologically-based methods (e.g., transcranial magnetic stimulation-induced motor evoked potentials) [8, 22].
5. *Stimulation parameter determination:* Currently, most clinical tDCS studies utilize a continuous stimulation mode, where a constant current (typically 1–2 mA) is applied for a set duration (typically 20–30 min). The current can be ramped up and down gradually. Parameters requiring control include current amplitude and stimulation duration. Medical technicians can titrate individual treatment parameters for different patients based on their needs [23–25]. In clinical practice, it is recommended to start with the minimum stimulation current and duration recommended by consensus guidelines, and adjust these parameters based on patient response and the severity of adverse events. Ultimately, the optimal therapeutic stimulation parameters should be determined. Twenty minutes of 2.0 mA tDCS is a typical starting point to assess tolerability and efficacy of the dose [6].
6. *Treatment administration:* Initiate tDCS treatment according to the established parameters.
7. *Posttreatment monitoring and adverse event reporting:* Monitor the patient after treatment and document any adverse reactions.

The precautions of tDCS includes the following seven points:

1. *Training and qualification:* Prior to treating patients with tDCS, medical personnel should undergo training and pass competency assessments in a qualified facility where this technology is implemented [5]. A treatment team comprised of licensed physicians, nurses, and technicians should be involved, with a licensed physician responsible for prescribing tDCS therapy. Other trained personnel can administer tDCS under the physician's supervision. Training should encompass, but is not limited to, safety protocols, treatment principles, equip-

ment operation, and implementation of commonly used evidence-based prescriptions.

2. *Safety screening:* Ensure that patients do not have any skull defects or fractures, and are not wearing metal implants, pacemakers, or other devices that could increase local current density and pose safety risks [5].
3. *Electrical safety:* Maintain electrical safety during treatment. As tDCS devices are electrical equipment, caution must be exercised to prevent safety hazards.
4. *Electrode fixation:* Utilize nonconductive headwear, such as elastic bands, to secure electrodes in their designated positions. Improper fixation can alter current distribution [5].
5. *Patient monitoring and response assessment:* If the patient experiences discomfort, excessive distraction, pain, or any other adverse reaction, reduce the stimulation intensity and duration as necessary. Continuously and objectively assess patient response and any adverse events to determine if parameter optimization or cessation of tDCS treatment is required [6].
6. *Stimulation dosage limits:* Daily tDCS sessions should not exceed the maximum stimulation dosage (total duration of 60 min at a stimulation intensity of 4 mA, with electrodes ranging from 1 to 100 cm^2) [26].
7. *Posttreatment observation:* Following treatment, the patient should be monitored for a period of time to ensure no adverse reactions occur before discharge.

2.3 Principle and Mechanism

The prevailing perspective on the fundamental mechanisms of tDCS is that it initially modifies neuronal activity by altering membrane potentials, thereby inducing changes in action potential rate, while exerting no significant impact on synaptic plasticity. Longer-lasting effects of tDCS may arise from alterations in neurotransmitter release, modifications in the functioning of membrane-bound receptors and ion channels, or other neurobiological and neurochemical changes [27]. Despite the absence of a fully elucidated mechanism for tDCS, significant advancements have been made in comprehending its underlying mechanisms over the past two decades through the integration of techniques such as transcranial magnetic stimulation (TMS), pharmacological investigations, and neuroimaging.

2.3.1 Alteration of Membrane Potential and Cortical Excitability

The application of tDCS has demonstrated its reliability as a cortical excitability modulation technique [7]. During tDCS, low-amplitude direct currents permeate the skull and enter the brain to modify the transmembrane neuronal potential, thereby influencing excitability levels and modulating individual neuron firing rates [4]. Priori and colleagues investigated the functional effects of scalp direct current (DC) stimulation with very weak intensity (<0.5 mA, 7 s) on the human motor cortex by evaluating alterations in motor potentials elicited through transcranial magnetic brain stimulation. Their findings demonstrate that such low-intensity anodal scalp DC, combined with cathodal DC, significantly reduces the excitability of the human motor cortex [28]. Nitsche and Paulus have demonstrated that the application of weak direct current through the human scalp can modulate motor cortex excitability, with effects lasting for several minutes after stimulation cessation [1]. Furthermore, they have shown that transcranial direct current stimulation in humans is capable of inducing sustained elevations in cortical excitability. The application of transcranial magnetic stimulation revealed a significant increase in motor cortical excitability, reaching approximately 150% above baseline levels for up to 90 min following the cessation of stimulation [29]. Importantly, it should be noted that the subthreshold current employed in tDCS does not elicit action potentials; rather, it modulates cortical excitability levels in a polarity-dependent manner. In general, anodal stimulation typically enhances excitability while cathodal stimulation generally diminishes excitability; however, the polarization direction is strictly contingent upon the orientation of axons and dendrites within the induced electrical field [1, 4, 5, 7, 8, 30]. The effects of tDCS on cortical excitability, both anodal and cathodal, are believed to be attributed to alterations in the resting membrane potential (depolarization and hyperpolarization) [30]. Anodal stimulation enhances spontaneous neuronal firing and depolarizes the resting membrane potential, thereby augmenting neuronal excitability; conversely, cathodal stimulation hyperpolarizes the resting membrane potential, suppressing neuronal excitability. Additionally, while the nominal targets of tDCS are often positioned beneath the electrodes, the resulting current flow generated by conventional tDCS extends across all cortical regions between and surrounding the electrodes, thereby not being confined solely to the localized area underneath [7]. Consequently, the effects of tDCS can directly or indirectly modulate remote neural networks [4, 8]. EEG studies provide further support for these findings by demonstrating that stimulation of a specific region (e.g., frontal cortex) elicits synchronous changes in oscillatory activity across the entire brain [4]. In terms of cellular mechanisms, the application of tDCS induces alterations in local pH levels (dependent on changes in electrolytic hydrogen ion concentration) and ion concentrations (e.g., intracellular calcium ion concentration) [31]. A prolonged and constant electric field can induce migration of transmembrane proteins, leading to changes in their steric and conformational properties as well as the local alteration of tissue acid–base balance. This, in turn, directly and indirectly affects membrane function outside the synapse by locally modifying ionic concentrations and properties of ion channels [4, 31]. In summary, a potential mechanism of action for tDCS involves modulating the membrane potential of superficial excitatory interneurons in the human motor cortex [5, 30]. By manipulating ion channels or altering electrical gradients that affect the balance of ions inside and outside neural membranes [28, 32], tDCS can induce enduring changes in cortical excitability and neural cell membrane potential.

2.3.2 Altering Synaptic Plasticity and Cortical Functional Connectivity

The mechanisms underlying changes in cortical excitability induced by tDCS differ between the effects observed during stimulation and those occurring poststimulation, despite similar neurophysiological effects [9]. Prolonged effects of tDCS can be attributed to modifications in the efficacy of N-methyl-D-aspartate (NMDA) receptors, alterations in gamma-aminobutyric acid-ergic activity, and modulation of interneurons, resulting in enduring changes in synaptic efficacy [7, 30, 31]. Long-term potentiation (LTP) and long-term depression (LTD) have been proposed as the likely mechanisms underlying the persistent effects of tDCS [28, 32–35]. These effects are contingent upon the levels of postsynaptic calcium and involve N-methyl-D-aspartate (NMDA) and α-amino-3-hydroxy-5-methyl-4-isoxazolepropionic acid (AMPA) receptors, metabotropic glutamate receptors, as well as gamma-aminobutyric acid GABA-A and GABA-B receptors [8, 9]. They exert persistent facilitatory/inhibitory modulation on synaptic connections. The modulation of neocortical LTP-like and LTD-like plasticity involves the influence of various neuromodulators, including noradrenaline (NA), dopamine (DA), and serotonin. These neuromodulators possess the capacity to facilitate both LTP and LTD at synaptic levels [9]. Therefore, it has been proposed that tDCS could potentially modulate neurotransmitter concentrations, thereby modifying the balance between excitatory and inhibitory (E/I) neurotransmitters and subsequently eliciting long-term potentiation/depression (LTP/LTD) processes that may facilitate cortical reorganization. Previous studies have demonstrated that anodal tDCS applied over the primary motor cortex (M1) results in a reduction of GABA levels, whereas cathodal stimulation leads to a significant decrease in both glutamate and GABA concentrations. This finding is consistent with the notion that tDCS-induced plasticity resembling LTP in the neocortex, which is believed to be modulated by GABA, critically depends on GABA modulation [35]. Vincent P Clark and colleagues employed proton magnetic resonance spectroscopy (^{1}H MRS) to quantitatively assess changes in brain metabolite concentrations following tDCS stimulation.

The right parietal cortex, beneath the stimulating electrode, exhibited significantly elevated levels of combined glutamate and glutamine, indicating a disruption in the Glutamine (Gln)—Glutamate (Glu) cycle with tDCS. Nonsignificant increases were also observed in corresponding regions of the contralateral hemisphere. The application of tDCS resulted in alterations in levels of Glu and Gln, which can reasonably be inferred to indicate changes in glutamatergic neurotransmission, thereby affecting neural activation and metabolism. Furthermore, tDCS significantly increased N-acetylaspartate (tNAA) in the right parietal cortex beneath the stimulating electrode, while showing a trend toward the opposite direction in the corresponding region of the left hemisphere [27].

The advancement of brain imaging technologies, such as PET, EEG, and fMRI, has facilitated a paradigm shift in our understanding of the brain from a simplistic and isolated structure–function relationship to the domain of functional connectivity networks. The application of tDCS stimulation, which modulates synaptic connections, holds promise for enhancing the functional connectivity between distinct cortical regions. Rafael Polania and colleagues investigate the impact of tDCS on cortical network function, revealing that tDCS applied to the primary motor cortex enhances connectivity activity within the cortico-cortical and cortico-subcortical motor neural network [36] components, including premotor cortex, parietal lobe, thalamus, and caudate nucleus. In another study conducted by the researchers, EEG recordings were utilized to investigate alterations in functional cortical networks induced by excitatory anodal tDCS over the primary motor cortex during voluntary hand movements. The findings revealed a significant increase in functional connectivity patterns within premotor, motor, and sensorimotor areas of the stimulated hemisphere in the 60–90 Hz frequency range, which were specifically associated with changes in the topological functional organization of the entire brain. The functional connectivity in the entire dynamic brain is shifted and modulated by the combined effects of motor activity and excitatory transcranial stimulation [37]. Furthermore, Charlotte J. Stagg and colleagues employed whole-brain arterial spin labeling (ASL) to investigate alterations in brain perfusion and observed an increase in perfusion within regions anatomically connected to the DLPFC during anodal tDCS. Additionally, they found a decrease in functional coupling between the left dorsolateral prefrontal cortex (DLPFC) and bilateral thalami, suggesting that tDCS may play a role in modulating the functional connectivity between the DLPFC and thalamus [16]. In addition to its direct effects, tDCS can also induce changes in distal cortical and subcortical brain regions through functional connectivity between different areas of the brain. Specifically, prefrontal stimulation with tDCS has been shown to modulate resting-state functional connectivity [7]. Keeser and colleagues investigated the impact of tDCS on resting-state network (RSN) connectivity, as measured by fMRI, and observed significant alterations in regional brain connectivity within the default mode network (DMN) and frontoparietal networks (FPNs). These changes were observed both near the primary stimulation site and in interconnected brain regions following real tDCS [38]. Furthermore, tDCS elicited polarity-specific modifications in cerebral blood flow (CBF).

2.3.3 Nonneuronal Mechanisms

In fact, tDCS not only modulates neuronal activity but also exerts an influence on nonneuronal cells, such as vascular endothelial cells to induce changes in blood vessel diameter or impact nonneuronal cell types like microglia and astrocytes [39]. For instance, tDCS has been shown to directly alter blood vessel diameter, resulting in localized vasodilation at the stimulation site and subsequent skin reddening. This effect can also be observed within the cerebrovascular system [40, 41]. These responses can be quantified by assessing CBF and vasomotor reactivity (VMR). Studies have demonstrated widespread increases and decreases in regional CBF within cortical and subcortical areas following anodal and cathodal tDCS interventions [42–44]. Bragina and colleagues reported that the application of 0.1 mA anodal tDCS over the right barrel cortex or motor cortex, using epicranial electrodes with a contact area of 3.5 mm^2, resulted in vasodilation under the anode, significantly enhancing CBF, along with elevated levels of blood oxygenation [44]. In an experimental rat model, tDCS has been demonstrated to induce polarity-specific and long-lasting changes in CBF. Specifically, anodal tDCS leads to increased CBF while cathodal tDCS results in decreased CBF [43]. Moreover, Shinde and colleagues found that a current intensity of 4 mA during anodal tDCS produced a larger increase in CBF compared to 2 mA, and Jamil and colleagues showed that both 2 mA anodal and cathodal tDCS induced greater polarity-specific changes in CBF compared to intensities of 0.5 and 1 mA [39, 44, 45]. These studies, conducted on both animal models and human subjects, suggest that the impact of tDCS on CBF is contingent upon the polarity and dosage of the stimulation. However, since CBF measurements were limited to the cortical area targeted by tDCS, it remains unclear whether tDCS-induced changes in CBF are confined to the ECA or if they extend beyond this site [39]. The impact of tDCS on the VMR is characterized by a complex and intricate mechanism involving the autonomic nervous system. Anodal tDCS has been shown to decrease VMR in a clinical experiment, whereas cathodal tDCS has been found to increase it [40]. However, another clinical study reported that anodal tDCS did not induce any significant alterations in VMR among both healthy controls and patients [46].

In addition to the aforementioned factors, studies in mouse models have revealed the potential effects of this technique on glial cell function [47–49]. Cathodal tDCS applied over the motor cortex has been found to downregulate astrocyte activity, while anodal tDCS administered via epicranial electrodes over the visual and somatosensory cortices enhances intracellular calcium ion (Ca^{2+}) concentration within astrocytes and activates microglia in these regions. However, the electric field intensities employed in these investigations typically exceed the standard dosages utilized in human tDCS treatments. Furthermore, the impact of tDCS on glial cells may arise from their pivotal roles in neuronal cell polarization and signal propagation. In combination, these effects are somewhat distinct from its impact on neuronal mechanisms.

2.3.4 Others

Recently conducted studies have also indicated that the mechanisms underlying tDCS are intricately linked to immune and inflammatory processes. Notably, in rats, anodal tDCS administered at a current density of 14.29 mA/cm^2 has been demonstrated to upregulate the gene expression of major histocompatibility complex I (MHC I), while cathodal tDCS at the same current density enhances the expression of osteopontin [39], an immunomodulatory protein. In mouse/rat brain disease models, cathodal tDCS using skull electrodes at a current intensity ranging from 200 to 500 μA and a current density between 14.29 and 57.14 mA/cm^2 promotes the migration of oligodendrocyte precursors toward the lesion site [50]. The administration of this treatment leads to a significant reduction in pro-inflammatory factors, promotes white matter myelination, mitigates autophagy [51], and suppresses phagocytic anti-inflammatory activity at the site of stimulation [49, 52]. However, in human tDCS studies, a relatively low current density is typically employed (approximately 0.03–0.08 mA/cm^2) [53]. Animal models have shown no long-term effects of tDCS on inflammatory markers beyond 30 min [54] when the current density is as low as 0.3 mA/cm^2.

Similarly, recent animal model studies have investigated the direct effects of tDCS on the blood–brain barrier. In these cases, in vivo rat brain studies have demonstrated that the application of 1 mA tDCS through skull electrodes with a current density of 8 mA/cm^2 enhances the blood–brain barrier permeability and facilitates the transport of large molecules into the extracellular matrix of brain tissue [54]. Additionally, it reduces the density of the extracellular matrix, thereby promoting increased diffusion of solutes through the extracellular space irrespective of their charge [54]. However, when a current of 1 mA tDCS (with a current density of approximately 0.03 mA/cm^2) was administered over the motor cortex in healthy volunteers, magnetic resonance imaging (MRI) revealed no evidence of lesions or disruption to the blood–brain barrier following 13 min of anodal stimulation and 9 min of cathodal stimulation [21].

In general, lower current densities may not be adequate to elicit immune and inflammatory responses; however, higher current densities in tDCS exhibit a modulatory influence on these processes, thereby potentially paving the way for novel nonneuronal mechanisms in future research.

2.4 Effect Factors

The effects of tDCS critically depend on the specific parameters of stimulation employed. While these parameters may vary, it has been demonstrated that polarity, duration, current density (i.e., intensity relative to electrode size), and stimulation location significantly influence the neuromodulatory outcomes of tDCS [4, 8–10, 16, 30]. Subsequent sections will provide a comprehensive discussion on the factors that impact tDCS efficacy.

2.4.1 Stimulus Polarity

The effects of tDCS on cortical excitability are polarity specific [9, 30]. Anodal stimulation enhances cortical excitability during and after stimulation, provided that the duration of stimulation is sufficient. Conversely, cathodal stimulation decreases excitability within the cortex [5, 8, 30]. Nitsche and colleagues reported that opposite effects were observed in individual cortical layers, highlighting the dependence of DC stimulation effects on the interplay between electric flow direction and neuronal geometry [1]. The impact of anodal tDCS during stimulation appears to be exclusively reliant on alterations in membrane potential. The effects of anodal stimulation were reduced by the Ca^{2+} channel blocker flunarizine (FLU) and abolished by the Na^+ channel blocker carbamazepine (CBZ). Neither dextromethorphan

(DMO), an NMDA receptor antagonist, nor Lorazepam (LOR), a GABA-A receptor agonist, exerted any modulatory effect on the intrastimulation response. In contrast to anodal stimulation, blockade of voltage-dependent Ca^{2+} or Na^{+} channels did not influence excitability shifts during cathodal stimulation. Moreover, the effects of intrastimulation cathodal tDCS were not modulated by either NMDA or GABA blockade [9]. Additionally, a study investigated the impact of tDCS pretreatment on repetitive transcranial magnetic stimulation (rTMS) action and revealed that preconditioning with cathodal tDCS could reverse the suppressive effect typically observed with 1 Hz rTMS, leading to a seemingly paradoxical facilitation in the human corticospinal motor system [55]. Conversely, anodal tDCS exerted opposite effects on subsequent conditioning rTMS trains.

2.4.2 Duration

The duration of tDCS after-effects is contingent upon the stimulation duration [21]; if the stimulation duration is sufficiently prolonged, the modified cortical excitability can persist for up to 1 h poststimulation. Nitsche reported that by varying the current intensity and duration, one can regulate both the strength and duration of these after-effects [1]. Notably, acute excitability alterations induced by tDCS for 4 s do not result in subsequent after-effects (at an intensity of 1 mA with an electrode size of 35 cm^2) [5]. Short-lasting after-effects lasting up to 10 min can be achieved with 5 and 7-min stimulation, while stimulation for 9 min under the cathode electrode on the motor cortex and for 13 min under the anode electrode result in after-effects lasting approximately 1 h (using electrodes measuring 5 × 7 cm and a motor cortex-contralateral supraorbital ridge electrode arrangement) [5]. However, several studies have indicated that prolonged duration of stimulation does not necessarily correspond to improved therapeutic outcomes. Katia Monte-Silva reported that applying anodal tDCS at 1 mA for 26 min with a surface area of 35 cm^2 resulted in decreased excitability compared to a shorter stimulation period of 13 min, suggesting the presence of a nonlinear relationship between stimulation parameters and excitability [56].

2.4.3 Current Intensity

Similarly, there is not a strictly linear relationship between the intensity of stimulation and the effects of tDCS. In a study conducted by Koen Cuypers and colleagues, three groups of right-handed subjects received 20 min of anodal, cathodal, or sham tDCS applied to the left prefrontal cortex. The results showed that at 1 mA tDCS, there were no significant effects on subjects' performance. However, when the intensity was increased to 2 mA, verbal fluency significantly improved with anodal tDCS while mildly decreased with cathodal tDCS [24]. Vincent P Clark conducted a study to investigate the impact of tDCS on learning and performance for a challenging visual perceptual task. The results showed that anodal tDCS over parietal cortex led to significant improvements, with the amount of learning being proportional to current strength. Subjects who received 2.0 mA of anodal tDCS current over right inferior frontal or right parietal cortex demonstrated more than twice the performance improvement compared to those who received 0.1 mA sham tDCS current [27]. Furthermore, in a tDCS study conducted by Koen Cuypers and colleagues, a significant enhancement in motor performance at retention was observed for 1.5 mA tDCS compared to sham tDCS, while no statistically significant differences were reported between 1 mA tDCS and sham tDCS. Conversely, other studies have also documented more pronounced effects with lower current densities. Andisheh Bastani and colleagues [57] reported that a current density of 0.013 mA/cm^2 elicited greater cortical excitability compared to the commonly employed 0.029 mA/cm^2, with no statistically significant distinctions observed between this density and higher densities of 0.058 and 0.083 mA/cm^2.

2.4.4 Others

In addition to the aforementioned stimulation parameters, there exist other factors that exert an influence on the efficacy of tDCS. Primarily, a crucial aspect in tDCS lies in determining the optimal placement of electrodes on the cranium. The neurochemical response to tDCS may exhibit variations based on anatomical location, thereby accounting for divergent outcomes observed across different studies [27]. The amount of current that reaches neuronal tissue, however, is contingent upon several uncontrollable factors, encompassing the resistance posed by various cephalic structures such as the integumentary system, cranium, vasculature, and cerebral matter [30]. Accurately quantifying the trajectory of current flow across the human subjects' cerebral cortex presents inherent challenges, necessitating our reliance on indirect observations from human and animal studies as well as mathematical models to ascertain the path and density of current within specific cortical regions [9]. Studies investigating physiological changes following tDCS and computational modeling studies of predicted current flow have demonstrated that the relative positioning of electrodes significantly influences both the spatial distribution and magnitude of delivered current to the brain [5]. Additionally, the effects of tDCS can be influenced by the placement of electrodes. Caroline Paquette and colleagues employed positron emission tomography (PET) to quantify the impact on activation-induced changes in regional cerebral blood flow (ΔrCBF) during bilateral tDCS. The study revealed that the implementation of bilateral tDCS electrode mount can effectively induce alterations in brain activation during movement, with the cathode exhibiting stronger effects compared to the anodal effect observed in unilateral stimulation [58].

Accumulating evidence suggests that the effects of transcranial direct current stimulation (tDCS) exhibit a cumulative pattern, necessitating consecutive daily applications to achieve clinically meaningful outcomes [6]. However, it is noteworthy that the behavioral effects of single tDCS sessions are relatively transient, lasting for a maximum duration of a few tens of minutes. Conversely, employing multiple spaced sessions may extend the duration of these behavioral effects up to several weeks [9]. The time interval between stimulation sessions has been demonstrated to be a crucial determinant of the outcomes during the second period of stimulation. For instance, Monte-Silva and colleagues reported that cathodal tDCS exhibited enhanced lasting inhibitory effects (measured 1–2 h after the second stimulation period) when there was a 24-h separation between the two periods of stimulation, whereas no such enhancement was observed with a 3-h interval. Angelo Alonzo and colleagues investigated the impact of daily or alternate-day tDCS sessions on motor cortical excitability. Their findings revealed that administering tDCS on a daily basis resulted in significantly greater increases in motor evoked potential (MEP) amplitude compared to an alternate-day schedule over a fixed 5-day period. These results suggest that a time interval of 1 day falls within the critical window for consolidative and cumulative excitatory effects, while a 2-day interval lies outside this crucial period [2].

2.5 Summary and Outlook

In conclusion, tDCS demonstrates immense potential in both basic and clinical applications as a noninvasive neuromodulation technique and promising therapeutic tool due to its user-friendly nature, cost-effectiveness, high tolerability, and minimal disruption to subjects [1]. By modulating membrane potential, cortical excitability, synaptic plasticity, and cortical functional connectivity, tDCS exerts a significant impact on brain function. Moreover, the regulatory effects of tDCS on nonneuronal cells have further enhanced our understanding of its mechanisms of action within the brain.

When applying tDCS, it is crucial to consider the impact of stimulation polarity, duration, and current intensity, as well as electrode placement, in order to meticulously design studies, optimize the efficacy of tDCS applications, and accurately predict potential responses of clinical populations to this treatment modality. Additionally, careful consideration must be given to the poststimulation effects of tDCS when utilizing it.

The future research can prioritize the following two aspects for conducting long-term studies:

1. *Personalized tDCS treatments:* Variations in brain anatomy and genetic polymorphisms have an impact on the efficacy of tDCS treatments [39]. MRI-based head models offer a limited means to investigate how individual anatomical factors affect the neurophysiological outcomes of tDCS [59].
2. *tDCS treatment at home:* The therapeutic effects of tDCS necessitate consecutive daily applications in order to achieve clinically significant outcomes. To address the practical challenge of patients having to make repeated visits for treatment, the exploration of home-based tDCS represents a crucial avenue for future research. The provision of tDCS treatment at home obviates the necessity for patients to travel to medical or research facilities for each treatment session [6]. Moreover, tDCS treatment at home can enhance treatment compliance and facilitate access to tDCS for patients residing in geographically remote areas [60, 61].

Acknowledgments H.S. and X.Z. conceived of the project. J.F., L.Z., and H.S. drafted the manuscript. All authors approved the final version of the manuscript for submission.

Disclosure/Conflicts of Interest The authors declare no competing interests.

Financial Support This work was supported by grants from the National Natural Science Foundation of China (Grant Nos. 71942003, 32171080, 32100886, 32161143022, and 32200914); the Chinese National Programs for Brain Science and Brain-like Intelligence Technology (Grant No. 2021ZD0202101); the Institute of Health and Medicine, HefeiComprehensive National Science Center, Global Select Project (Grant No. DJK-LX-2022008); the Open Foundation of the Institute of Linguistics and Applied Linguistics, Anhui Jianzhu University (YYX2024KF01); the Open Fund of Key Laboratory of Philosophy and Social Science of Anhui Province on Adolescent Mental Health and Crisis Intelligence Intervention (SYS2024B01); and also supported by Shanghai Key Laboratory of Brain-Machine Intelligence for Information Behavior.

References

1. Nitsche MA, Paulus W. Excitability changes induced in the human motor cortex by weak transcranial direct current stimulation. J Physiol. 2000;527(Pt 3):633–9.
2. Alonzo A, Brassil J, Taylor JL, Martin D, Loo CK. Daily transcranial direct current stimulation (tDCS) leads to greater increases in cortical excitability than second daily transcranial direct current stimulation. Brain Stimul. 2012;5:208–13.
3. Grundmann L, et al. Effects of transcranial direct current stimulation of the primary sensory cortex on somatosensory perception. Brain Stimul. 2011;4:253–60.
4. Zaghi S, Acar M, Hultgren B, Boggio PS, Fregni F. Noninvasive brain stimulation with low-intensity electrical currents: putative mechanisms of action for direct and alternating current stimulation. Neuroscientist. 2010;16:285–307.
5. Woods AJ, et al. A technical guide to tDCS, and related non-invasive brain stimulation tools. Clin Neurophysiol. 2016;127:1031–48.
6. Charvet LE, Shaw MT, Bikson M, Woods AJ, Knotkova H. Supervised transcranial direct current stimulation (tDCS) at home: a guide for clinical research and practice. Brain Stimul. 2020;13:686–93.
7. RC Kadosh. The stimulated brain: cognitive enhancement using non-invasive brain stimulation. Elsevier. 2014.
8. Fregni F, et al. Evidence-based guidelines and secondary meta-analysis for the use of transcranial direct current stimulation in neurological and psychiatric disorders. Int J Neuropsychopharmacol. 2021;24:256–313.
9. Stagg CJ, Nitsche MA. Physiological basis of transcranial direct current stimulation. Neuroscientist. 2011;17:37–53.
10. Bikson M, et al. Safety of transcranial direct current stimulation: evidence based update 2016. Brain Stimul. 2016;9:641–61.
11. Frohlich F, Townsend L. Closed-loop transcranial alternating current stimulation: towards personalized non-invasive brain stimulation for the treatment of psychiatric illnesses. Curr Behav Neurosci Rep. 2021;8:51–7.
12. Elyssa Kok T, Schaette R, Shekhawat GS. Impact of tDCS and HD-tDCS on tinnitus perception: a scoping review. Prog Brain Res. 2021;262:225–44.
13. Müller D, Habel U, Brodkin ES, Clemens B, Weidler C. HD-tDCS induced changes in resting-state functional connectivity: Insights from EF modeling. Brain Stimul. 2023;16:1722–32.
14. Aparício LVM, et al. A systematic review on the acceptability and tolerability of transcranial direct current stimulation treatment in neuropsychiatry trials. Brain Stimul. 2016;9:671–81.
15. Parlikar R, et al. High definition transcranial direct current stimulation (HD-tDCS): a systematic review on the treatment of neuropsychiatric disorders. Asian J Psychiatr. 2021;56:102542.

16. Stagg CJ, et al. Widespread modulation of cerebral perfusion induced during and after transcranial direct current stimulation applied to the left dorsolateral prefrontal cortex. J Neurosci. 2013;33:11425–31.
17. Poreisz C, Boros K, Antal A, Paulus W. Safety aspects of transcranial direct current stimulation concerning healthy subjects and patients. Brain Res Bull. 2007;72:208–14.
18. Bikson M, et al. Transcranial electrical stimulation nomenclature. Brain Stimul. 2019;12:1349–66.
19. Godinho MM, et al. Safety of transcranial direct current stimulation: evidence based update 2016. Brain Stimul. 2017;10:983–5.
20. Gandiga PC, Hummel FC, Cohen LG. Transcranial DC stimulation (tDCS): a tool for double-blind sham-controlled clinical studies in brain stimulation. Clin Neurophysiol. 2006;117:845–50.
21. Nitsche MA, et al. MRI study of human brain exposed to weak direct current stimulation of the frontal cortex. Clin Neurophysiol. 2004;115:2419–23.
22. Lefaucheur J-P, et al. Evidence-based guidelines on the therapeutic use of repetitive transcranial magnetic stimulation (rTMS): an update (2014-2018). Clin Neurophysiol. 2020;131:474–528.
23. Boggio PS, et al. Effects of transcranial direct current stimulation on working memory in patients with Parkinson's disease. J Neurol Sci. 2006;249:31–8.
24. Iyer MB, et al. Safety and cognitive effect of frontal DC brain polarization in healthy individuals. Neurology. 2005;64:872–5.
25. Cuypers K, et al. Is motor learning mediated by tDCS intensity? PLoS One. 2013;8:e67344.
26. Antal A, et al. Low intensity transcranial electric stimulation: safety, ethical, legal regulatory and application guidelines. Clin Neurophysiol. 2017;128:1774–809.
27. Clark VP, Coffman BA, Trumbo MC, Gasparovic C. Transcranial direct current stimulation (tDCS) produces localized and specific alterations in neurochemistry: a ^{1}H magnetic resonance spectroscopy study. Neurosci Lett. 2011;500:67–71.
28. Priori A, Berardelli A, Rona S, Accornero N, Manfredi M. Polarization of the human motor cortex through the scalp. Neuroreport. 1998;9:2257–60.
29. Nitsche MA, Paulus W. Sustained excitability elevations induced by transcranial DC motor cortex stimulation in humans. Neurology. 2001;57:1899–901.
30. Koops S, Van Den Brink H, Sommer IEC. Transcranial direct current stimulation as a treatment for auditory hallucinations. Front Psychol. 2015;6:244.
31. Ardolino G, Bossi B, Barbieri S, Priori A. Non-synaptic mechanisms underlie the after-effects of cathodal transcutaneous direct current stimulation of the human brain. J Physiol. 2005;568:653–63.
32. Notturno F, et al. Neuroprotective effect of cathodal transcranial direct current stimulation in a rat stroke model. J Neurol Sci. 2014;342:146–51.
33. Hummel F, et al. Effects of non-invasive cortical stimulation on skilled motor function in chronic stroke. Brain. 2005;128:490–9.
34. Nitsche MA, et al. Modulating parameters of excitability during and after transcranial direct current stimulation of the human motor cortex. J Physiol. 2005;568:291–303.
35. Polanía R, Nitsche MA, Ruff CC. Studying and modifying brain function with non-invasive brain stimulation. Nat Neurosci. 2018;21:174–87.
36. Polanía R, Paulus W, Antal A, Nitsche MA. Introducing graph theory to track for neuroplastic alterations in the resting human brain: a transcranial direct current stimulation study. NeuroImage. 2011;54:2287–96.
37. Polanía R, Nitsche MA, Paulus W. Modulating functional connectivity patterns and topological functional organization of the human brain with transcranial direct current stimulation. Hum Brain Mapp. 2011;32:1236–49.
38. Keeser D, et al. Prefrontal transcranial direct current stimulation changes connectivity of resting-state networks during fMRI. J Neurosci. 2011;31:15284–93.
39. Yavari F, Chhabra H, Polania R, Nitsche MA. Mechanisms of action of transcranial direct current stimulation. In: Interventional psychiatry. Elsevier; 2024. p. 149–86. https://doi.org/10.1016/B978-0-443-18496-3.00006-9.

40. Giorli E, et al. Transcranial direct current stimulation and cerebral vasomotor reserve: a study in healthy subjects. J Neuroimaging. 2015;25:571–4.
41. Gagne C, Zika O, Dayan P, Bishop SJ. Impaired adaptation of learning to contingency volatility in internalizing psychopathology. elife. 2020;9:e61387.
42. Lang N, et al. How does transcranial DC stimulation of the primary motor cortex alter regional neuronal activity in the human brain? Eur J Neurosci. 2005;22:495–504.
43. Mielke D, et al. Cathodal transcranial direct current stimulation induces regional, long-lasting reductions of cortical blood flow in rats. Neurol Res. 2013;35:1029–37.
44. Jamil A, et al. Current intensity- and polarity-specific online and aftereffects of transcranial direct current stimulation: an fMRI study. Hum Brain Mapp. 2020;41:1644–66.
45. Shinde AB, Lerud KD, Munsch F, Alsop DC, Schlaug G. Effects of tDCS dose and electrode montage on regional cerebral blood flow and motor behavior. NeuroImage. 2021;237:118144.
46. List J, et al. Impact of tDCS on cerebral autoregulation in aging and in patients with cerebrovascular diseases. Neurology. 2015;84:626–8.
47. Zhang K-Y, et al. Cathodal tDCS exerts neuroprotective effect in rat brain after acute ischemic stroke. BMC Neurosci. 2020;21:21.
48. Takeda I, et al. Controlled activation of cortical astrocytes modulates neuropathic pain-like behaviour. Nat Commun. 2022;13:4100.
49. Cherchi L, Anni D, Buffelli M, Cambiaghi M. Early application of ipsilateral cathodal-tDCS in a mouse model of brain ischemia results in functional improvement and perilesional microglia modulation. Biomol Ther. 2022;12:588.
50. Rueger MA, et al. Multi-session transcranial direct current stimulation (tDCS) elicits inflammatory and regenerative processes in the rat brain. PLoS One. 2012;7:e43776.
51. Guo T, Fang J, Tong ZY, He S, Luo Y. Transcranial direct current stimulation ameliorates cognitive impairment via modulating oxidative stress, inflammation, and autophagy in a rat model of vascular dementia. Front Neurosci. 2020;14:28.
52. Nitsche MA, Paulus W. Transcranial direct current stimulation—update 2011. Restor Neurol Neurosci. 2011;29:463–92.
53. Callai EMM, et al. Evaluation of the immediate effects of a single transcranial direct current stimulation session on astrocyte activation, inflammatory response, and pain threshold in naïve rats. Behav Brain Res. 2022;428:113880.
54. Xia Y, et al. Modulation of solute diffusivity in brain tissue as a novel mechanism of transcranial direct current stimulation (tDCS). Sci Rep. 2020;10:18488.
55. Siebner HR, et al. Preconditioning of low-frequency repetitive transcranial magnetic stimulation with transcranial direct current stimulation: evidence for homeostatic plasticity in the human motor cortex. J Neurosci. 2004;24:3379–85.
56. Monte-Silva K, et al. Induction of late LTP-like plasticity in the human motor cortex by repeated non-invasive brain stimulation. Brain Stimul. 2013;6:424–32.
57. Bastani A, Jaberzadeh S. Differential modulation of corticospinal excitability by different current densities of anodal transcranial direct current stimulation. PLoS One. 2013;8:e72254.
58. Paquette C, Sidel M, Radinska BA, Soucy J-P, Thiel A. Bilateral transcranial direct current stimulation modulates activation-induced regional blood flow changes during voluntary movement. J Cereb Blood Flow Metab. 2011;31:2086–95.
59. Mosayebi-Samani M, et al. The impact of individual electrical fields and anatomical factors on the neurophysiological outcomes of tDCS: a TMS-MEP and MRI study. Brain Stimul. 2021;14:316–26.
60. Pilloni G, et al. Long term at-home treatment with transcranial direct current stimulation (tDCS) improves symptoms of cerebellar ataxia: a case report. J Neuroeng Rehabil. 2019;16:41.
61. Sandran N, Hillier S, Hordacre B. Strategies to implement and monitor in-home transcranial electrical stimulation in neurological and psychiatric patient populations: a systematic review. J Neuroeng Rehabil. 2019;16:58.

Major Depressive Disorder

3

TianHong Zhang, ZhanMing Shi, LingYun Zeng, and JiJun Wang

Abstract

Major depressive disorder (MDD) presents a significant challenge due to its severity and the limited effectiveness of conventional pharmacological treatments and psychotherapies for many patients. This underscores the need for alternative therapeutic approaches. Transcranial magnetic stimulation (TMS) and transcranial direct current stimulation (tDCS) have emerged as promising noninvasive brain stimulation techniques that address these needs. This chapter systematically reviews the applications of various TMS and tDCS paradigms, including bilateral repeat TMS (rTMS), accelerated rTMS, deep TMS, priming rTMS, synchronized rTMS, intermittent theta-burst stimulation (iTBS), accelerated iTBS, continuous theta-burst stimulation (cTBS), bilateral TBS, magnetic seizure therapy, high-definition tDCS, and a comparison between rTMS and tDCS. By exploring these physical treatment modalities, the chapter highlights their potential in effectively managing MDD and offers insights into their respective roles in enhancing patient care.

Keywords

rTMS · Deep TMS · Priming rTMS · Synchronized rTMS · iTBS · Accelerated iTBS · cTBS · Bilateral TBS · MST · tDCS · HD-tDCS · Depression

T. Zhang (✉) · J. Wang
Shanghai Mental Health Center, Shanghai Jiaotong University School of Medicine, Shanghai Engineering Research Center of Intelligent Psychological Evaluation and Intervention, Shanghai Key Laboratory of Psychotic Disorders, Shanghai, People's Republic of China

Z. Shi
Chongqing Jiangbei Mental Health Center, Chongqing, People's Republic of China

L. Zeng
Department of Psychiatric Rehabilitation, Shenzhen Kangning Hospital, ShenZhen, GuangDong, China

W. Zheng, Y. Ning (eds.), *TMS and tDCS for Psychiatric Disorders*,
https://doi.org/10.1007/978-981-96-8504-2_3

Abbreviations

aMST	accelerated MST
aTBS	accelerated theta burst stimulation
CBT	cognitive behavioral therapy
cTBS	continuous theta-burst stimulation
DLPFC	dorsolateral prefrontal cortex
ECT	electroconvulsive therapy
EEG	electroencephalogram
FDA	Food and Drug Administration
HAMA	Hamilton Rating Scale for Anxiety
HAMD	Hamilton Rating Scale for Depression
HD-tDCS	high-definition tDCS
HF	high frequency
iTBS	intermittent theta-burst stimulation
LF	low frequency
MADRS	Montgomery–Asberg Depression Rating Scale
MDD	major depressive disorder
MRI	magnetic resonance imaging
MST	magnetic seizure therapy
MT	motor threshold
NNT	number needed to treat
OR	odds ratio
QEEG	quantitative EEG
RCT	randomized controlled trial
rTMS	repetitive transcranial magnetic stimulation
SNT	Stanford Neuromodulation Therapy
tDCS	transcranial direct current stimulation
TMS	transcranial magnetic stimulation
TRD	treatment-resistant depression
RBANS	Repeatable Battery for the Assessment of Neuropsychological Status
RUL	right unilateral
WHO	World Health Organization

3.1 Introduction

Major depressive disorder (MDD) is a group of mental health conditions characterized by persistent low mood, loss of interest, and impaired daily functioning. Symptoms may include feelings of sadness, fatigue, sleep disturbances, changes in appetite, difficulty concentrating, and lowered self-esteem. These disorders not only affect emotional well-being but can also impact cognitive processes, behavior, and physical health. The prevalence of MDD is relatively high, with approximately 3–8% of the global population experiencing major depression at some point in their lives [1]. Statistical data show that women are generally at a higher risk than men

[2], and both adolescents and the elderly are particularly vulnerable to these conditions [3–5]. In recent years, the incidence of MDD has been rising, highlighting the need for increased awareness and intervention. The severity of MDD is significant and can lead to serious consequences if left untreated. Individuals often experience debilitating emotional pain, which can result in social isolation, decreased work and academic performance, and an increased risk of suicide [6]. According to the World Health Organization (WHO), depression is a leading cause of disability worldwide, severely impacting individuals' daily functioning and social capabilities. Research indicates that individuals with depression may experience a reduction in work performance [7], and the lost productivity was estimated at approximately 28–30 lost days per year [8]. This impairment significantly affects their ability to engage in daily activities and social interactions.

The effectiveness of conventional pharmacological treatments and psychotherapies for MDD is often limited, with many patients experiencing inadequate responses. For instance, a systematic review by Cipriani et al. [9] found that only about 50% of patients achieve remission after the first course of antidepressant therapy, and many do not respond to multiple treatment attempts. Furthermore, a study indicated that approximately 30% of patients with MDD remain treatment-resistant, highlighting the urgent need for alternative therapeutic options [9]. Additionally, a meta-analysis revealed that cognitive behavioral therapy (CBT), while effective for some, has a response rate of only about 40–60% in clinical settings, suggesting that a significant proportion of patients may not benefit from standard psychological interventions [10]. These limitations underscore the necessity for innovative treatments such as transcranial magnetic stimulation (TMS) and transcranial direct current stimulation (tDCS) to address the unmet needs of individuals suffering from MDD.

Repetitive TMS (rTMS) is a technique approved by the U.S. Food and Drug Administration (FDA) for the treatment-resistant depression (TRD). Conventional rTMS involves high frequency (HF) to the left dorsolateral prefrontal cortex (DLPFC) and low frequency (LF) to the right DLPFC. However, not all depressed patients could benefit from standard rTMS protocols. A meta-analysis showed that patients who received HF rTMS to the left DLPFC had an average response rate of 29.3% [11]. LF rTMS to the right DLPFC had similar therapeutic effects to HF DLPFC [12]. Therefore, newer forms of rTMS paradigms are warranted to improve antidepressant response and remission rates in patients with depression, especially those who are refractory to adequate antidepressant trials. In this chapter, we review the efficacy and safety of conventional and newer forms of rTMS and tDCS paradigms, including bilateral rTMS, accelerated rTMS, deep TMS, priming TMS, synchronized TMS, intermittent theta burst stimulation (iTBS), accelerated iTBS, continuation TBS, bilateral TBS, magnetic seizure therapy (MST), high-definition tDCS (HD-tDCS), and a comparison between rTMS and tDCS. Bilateral rTMS is defined as LF stimulation of the right DLPFC in addition to HF stimulation of the left DLPFC. Accelerated rTMS and accelerated iTBS is defined as more than one rTMS or iTBS session per day. Deep TMS is defined as a specialized (H) coil that is more precise and allows deeper stimulation through brief magnetic pulses that induce targeted neuronal depolarization in the cerebral cortex. Priming rTMS is defined as a specific protocol in which a brief period of low-intensity high-frequency TMS ("priming stimulation") is performed first, followed by the main treatment of low-frequency right DLPFC rTMS. Synchronized rTMS is defined as the use of rotating spherical neodymium magnets placed along the midline of the scalp to deliver low-field sinusoidal waveform stimulation that is synchronized to an individual's alpha electroencephalogram (EEG) frequency. Continuous TBS (cTBS) is defined as a 40-s train of uninterrupted TBS. Bilateral TBS was administered with right-side continuation followed by FDA-approved left-side iTBS. MST is defined as the use of magnetic fields to induce a generalized seizure. HD-tDCS was delivered using the 1 × 1 tDCS low-intensity stimulator with a 4 × 1 adapter. In the context of depression treatment, several studies have aimed to replace LF-rTMS with continuous TBS delivered to the right DLPFC or HF-rTMS with iTBS delivered to the left DLPFC [13]. Currently, the FDA has approved the deep rTMS and iTBS device for the treatment of TRD patients.

3.2 TMS

3.2.1 rTMS

3.2.1.1 Unilateral rTMS

Efficacy

MDD presents a complex landscape of symptoms and treatment responses, making it a challenging condition to manage. A significant portion of patients, estimated at 30–50%, fail to achieve sufficient relief from first-line treatments, which typically include a combination of antidepressant medications and CBT [14]. Conventional antidepressants primarily target neurotransmitter systems such as serotonin, norepinephrine, and dopamine, but their effectiveness can vary greatly among individuals. While large-scale meta-analyses indicate that these medications are generally more effective than placebos, the overall impact is relatively modest, with an effect size around 0.3 [9].

TRD, characterized by a lack of response to at least two consecutive antidepressant trials at appropriate doses over a period of 4–6 weeks, poses a significant public health challenge. Patients with TRD often experience greater functional impairments and an elevated risk of suicide. Although recent FDA approvals of treatments like esketamine provide new options for TRD, there remains a pressing need for effective treatments that act quickly. TMS offers a noninvasive alternative that has shown promise in addressing these gaps, providing a viable option for patients who have not responded to traditional therapies.

Conventional rTMS therapy, particularly at 10 and 1 Hz, has demonstrated significant efficacy in treating depression. HF (10 Hz) stimulation is applied to the left DLPFC, enhancing neural activity and improving mood states. Numerous studies indicate that patients often experience marked reductions in depressive symptoms and improvements in overall functioning following treatment. Conversely, LF (1 Hz) stimulation targets the right prefrontal cortex to inhibit excessive neural activity, making it suitable for certain TRD cases.

The effectiveness of conventional rTMS for treating MDD has been thoroughly investigated, particularly regarding its acute antidepressant effects when applied to the DLPFC. A systematic review and meta-analysis [11] encompassing 29 randomized controlled trials (RCTs) with 1371 participants demonstrated a significant pooled odds ratio (OR) of 3.3 for treatment response. Additionally, the analysis revealed an OR of 3.3 for remission, indicating a robust efficacy of rTMS in alleviating depressive symptoms.

Further validation comes from another meta-analysis of 16 double-Blind parallel-design RCTs [15], which compared HF rTMS to sham treatments, yielding a significant effect size (Cohen's d) of −0.55 (95% CI, −0.75 to −0.35) for its antidepressant effects. These findings underscore the potential of rTMS as an effective intervention for patients with depression, particularly those who have not responded adequately to antidepressants.

Two multicenter RCTs have specifically focused on rTMS treatment in patients with MDD who were not on antidepressant medications. The first study, sponsored by industry, reported a 24% response rate and a 17% remission rate with active rTMS, compared to 15% and 8%, respectively, for sham rTMS [16]. The second study, funded by the National Institute of Mental Health, found a 15% response rate and 14% remission rate with active treatment, contrasting with a mere 5% response and remission rate for the sham group [17].

The recent systematic review and network meta-analysis [18] provides compelling evidence for the efficacy of HF-rTMS in treating TRD. Their analysis, which encompasses numerous prospective RCTs, found that HF-rTMS significantly increases the odds of response compared to sham treatments, with an OR of 3.07 (95% CI: 2.24–4.21). This substantial effect size indicates that the benefits of HF-rTMS are unlikely to be influenced by confounding factors or biases. Several critical aspects of this meta-analysis counter the criticisms regarding HF-rTMS's efficacy. First, the studies included were all prospective RCTs, which typically ensure a balanced representation of both observed and unobserved covariates. Second, the narrow confidence intervals around the OR indicate a high level of precision in the estimated treatment effect. Additionally, multiple sensitivity analyses conducted within the review consistently supported the primary findings, reinforcing the robustness of the results. Although not every rTMS trial demonstrated statistically significant superiority over sham treatments, this may be attributed to the complexities of treating highly comorbid populations or instances of Type II error. Overall, these findings advocate for the consideration of HF-rTMS as a viable and effective option for individuals suffering from TRD.

Overall, while rTMS demonstrates medium-range efficacy, there is ongoing research into optimizing treatment parameters and combining rTMS with other therapeutic approaches to enhance its effectiveness further. However, individual variability in treatment response is noteworthy, as some patients may not achieve the desired improvements. This underscores the need for personalized treatment plans and ongoing research to optimize the application of rTMS in depression management.

Our previous study involved a naturalistic approach to assess the efficacy of add-on rTMS in treating depressive and anxiety symptoms, focusing on a sample of 117 patients, including 42 adolescents and 75 adults [19]. All participants had mood or anxiety disorders and received at least 10 sessions of rTMS targeting the left DLPFC at a frequency of 10 Hz. The outcomes were measured using the Hamilton Rating Scale for Depression (HAMD) and the Hamilton Rating Scale for Anxiety (HAMA) at baseline, after 10 sessions, and during follow-up at 2 and 4 weeks. The results demonstrated a significantly higher response rate in adolescents, with a notable percentage of patients achieving remission compared to their adult counterparts.

This study highlights the significant efficacy of add-on rTMS in treating depressive and anxiety symptoms specifically in adolescents, suggesting that this population experiences greater clinical improvement compared to adults. This underscores the potential of rTMS as a valuable therapeutic option for young patients, particularly when conventional pharmacological treatments may be less effective or carry greater risks. The results support the notion that rTMS could serve as a promising alternative or adjunctive treatment for adolescents facing mood and anxiety disorders. Given the limitations of current pharmacological approaches, including low remission rates and concerns about adverse effects, rTMS offers a safe and effective option that can be integrated into treatment plans for young patients. The demonstrated benefits in symptom reduction and overall improvement in mental health underscore the importance of considering rTMS as a viable treatment strategy in clinical practice for this vulnerable population.

Another study investigated the efficacy of adjunct rTMS in elderly patients with acute depressive episodes. A total of 114 inpatients were included, with 54 in the elderly group (aged ≥60) and 60 in the adult group (aged 18–59) [20, 21]. All participants had a baseline HAMD score of ≥14 and were drug-free for at least 2 weeks prior to enrollment. After 4 weeks of rTMS treatment, the remission rate in the elderly group was 36%, compared to 68% in the adult group. Although the remission rate was lower in the elderly, the response rates were similar, at 77% for the elderly and 91% for adults. These findings suggest that rTMS is effective in improving depressive symptoms in elderly patients.

The results highlight the potential value of rTMS for treating depression in older adults, particularly given the increased risk of adverse effects and drug interactions associated with traditional pharmacotherapy in this population. As a noninvasive treatment option, rTMS demonstrated good tolerability and significantly reduced depressive symptoms without increasing the risk of serious adverse events. This suggests that rTMS could serve as a valuable adjunctive treatment for elderly patients experiencing acute depressive episodes, offering a promising alternative for rapid intervention. Future research should further explore its long-term effectiveness and adaptability across diverse patient backgrounds.

Safety

The rTMS is generally considered a safe treatment modality for MDD, with a well-documented safety profile derived from extensive research and clinical trials. The most common adverse effects experienced during rTMS sessions include transient discomfort at the site of stimulation, which may extend to nearby regions of the face, such as the eye, ear, and jaw. Patients might also experience twitching or muscle contractions in these areas due to the excitation of superficial nerve branches. Headaches can occur, particularly in the initial treatment sessions as individuals adjust to the HF stimulation sensation. However, these adverse effects often diminish over time as patients habituate to the procedure.

Importantly, rTMS does not increase the risk of migraine headaches in healthy individuals or those with a history of migraines. In fact, a specific device, Spring TMS, has been cleared by the FDA for treating acute migraine. Strategies to mitigate discomfort include the use of over-the-counter analgesics, such as acetaminophen or ibuprofen, and topical anesthetics applied to the scalp prior to treatment. Adjusting the amplitude of TMS pulses may also enhance tolerability, although the antidepressant efficacy of lower intensities remains uncertain. Overall, while rTMS is associated with some minor adverse effects, there is no evidence of harmful changes to brain tissue when delivered within established safety parameters, making it a viable option for individuals with TRD.

Treatment Regimen

The treatment regimen for conventional rTMS involves several key components, including the stimulation target, intensity, frequency, duration of each session, and the total number of treatment sessions.

- *Stimulation target*: For HF (10 Hz) rTMS, the primary stimulation target is the left DLPFC, typically identified at the F3 electrode site using the 10–20 EEG system. Conversely, LF (1 Hz) rTMS targets the right DLPFC, generally at the F4 electrode site.
- *Stimulation intensity*: The intensity of rTMS is often set at a percentage of the patient's motor threshold (MT), usually around 110% for 10 Hz stimulation to ensure effective neural activation. For 1 Hz stimulation, the intensity is typically lower, often around 90% of the MT, to achieve the desired inhibitory effects without causing discomfort.
- *Stimulation frequency*: HF rTMS is delivered at 10 Hz, while LF rTMS is delivered at 1 Hz. This difference in frequency is strategically employed to either enhance or inhibit neural activity in the targeted brain areas.
- *Stimulation time*: Each rTMS session usually lasts between 20 and 40 min, depending on the specific protocol and patient tolerance.
- *Number of treatment sequences*: The overall treatment regimen typically consists of daily sessions over a period of 4–6 weeks, amounting to a total of approximately 20–30 treatment sessions. Adjustments may be made based on individual patient responses and clinical considerations to optimize therapeutic outcomes.

Clinical Recommendations

Conventional rTMS should be considered for broader clinical use in treating MDD, whether as an adjunct to other therapies, as an augmentation strategy, or as a standalone treatment. Its ease of administration and the substantial body of clinical evidence supporting its efficacy make rTMS a practical option for clinicians. Given its favorable safety profile and the flexibility in treatment protocols, rTMS can be integrated into comprehensive treatment plans tailored to individual patient needs. Clinicians are encouraged to utilize rTMS alongside pharmacotherapy and psychotherapy to enhance overall treatment outcomes. Regular monitoring and follow-up assessments will ensure optimal patient care and allow for adjustments in treatment strategies as necessary, ultimately improving the management of depression in diverse patient populations. Notably, rTMS also offers distinct advantages in special populations, such as adolescents and the elderly, where traditional treatments may pose greater risks or be less effective.

Overall, level A evidence (definite efficacy) was applied to HF-rTMS of the left DLPFC and level B evidence (probable efficacy) was applied to LF-rTMS of the right DLPFC for MDD [22].

3.2.1.2 Bilateral rTMS

Efficacy

Bilateral rTMS is a new and promising way to treat MDD. A review of seven randomized controlled trials (RCTs) on using bilateral rTMS to treat MDD shows that it is more effective than sham rTMS [23]. Bilateral rTMS was as effective as left and right unilateral rTMS [24], demonstrating that bilateral rTMS is an effective treatment.

Safety

The adverse effects associated with bilateral rTMS for MDD include a variety of sensations such as anxiety, cognitive complaints, difficulty sleeping, dizziness, fatigue, feeling nauseous, headache, hypomania, lactation, lightheadedness, metallic taste, nightmares, pain, racing thoughts, scalp discomfort, tinnitus, worsening mood, flu [25–36]. The majority of these adverse effects are transient and typically resolve after discontinuation of rTMS treatment. rTMS administration is contraindicated in patients with metallic implants or intracranial lead devices. Caution should also be exercised when administering rTMS to patients taking medications known to lower seizure thresholds.

Treatment Regimen

The bilateral rTMS approach involves the administration of LF stimulation to the right DLPFC and HF stimulation to the left DLPFC within the same treatment session. Specifically, 600 pulses at 1 Hz to the right DLPFC followed by 1500 pulses at 10 Hz to the left DLPFC: 5 times per week for 3 weeks for a total of 15 treatments according to the study by Blumberger et al. [26].

Clinical Recommendations
In Blumberger et al.'s study, only bilateral rTMS produced significantly greater antidepressant effects than sham [26]. It is therefore recommended that bilateral rTMS protocols over the DLPFC (LF on the right side and HF on the left side) be considered as a probable antidepressant treatment (level B evidence) for patients with MDD [22].

3.2.1.3 Accelerated rTMS

Efficacy
To enhance the antidepressant response and reduce the number of stimulation days, it has been suggested that increasing the number of rTMS sessions per day (more than one daily session) might be more effective [37]. The main goal of accelerated rTMS protocols is to reduce the patient and operator burden of repeated sessions over several weeks. A systematic review and meta-analysis suggested that accelerated rTMS could improve the severity of depressive symptoms [38].

Safety
Accelerated rTMS protocols appear to be safe and well tolerated in depressed patients [38], even in the elderly [39]. Common adverse effects include headache, malaise, nausea, and dizziness. A small number of patients may experience moderate to severe adverse effects, including functional impairment [39].

Treatment Regimen
In the accelerated rTMS protocol of Fitzgerald et al. [40], patients received three treatments per day for 3 days in week 1. In week 2, three treatments were given over 2 days, and in week 3, three treatments were given on a single day. In the three daily treatment sessions, 83, 83, and 84 trains of 10 Hz rTMS were delivered to the left DLPFC, respectively. 4.2-s trains were delivered at 120% of resting MT with a 15-s intertrain interval (10,500 pulses per day across the three sessions for a total of 63,000 pulses). Sessions were 15–30 min apart. And in the study by Loo et al. [37], two treatment sessions were given on each day of the week, separated by 2 h, stimulation parameters per session were 10 Hz, 30 trains of 5 s duration, 25 s between trains, at 110% of the subject's MT.

Clinical Recommendations
Although evidence supports similar efficacy of accelerated rTMS protocols and rTMS protocols with only one stimulation session per day [40] and in favor of sham stimulation [37], it is premature to recommend the use of accelerated rTMS protocols for the treatment of MDD.

3.2.2 Deep TMS

3.2.2.1 Efficacy

Deep TMS presents a compelling rationale for treating depression by targeting deeper brain structures that are often involved in mood regulation. Unlike traditional TMS, which primarily affects superficial cortical areas, deep TMS is designed to reach deeper nuclei, such as the subcallosal cingulate and ventral striatum. This ability to penetrate deeper into the brain allows for a more comprehensive modulation of the neural circuits associated with MDD.

One of the key mechanisms by which deep TMS operates is through its impact on neurotransmitter systems situated within these deeper nuclei. By stimulating these areas, deep TMS can significantly enhance the release of serotonin, dopamine, and norepinephrine, neurotransmitters that are essential for mood stabilization. This targeted stimulation of deeper structures facilitates a more profound therapeutic effect, potentially leading to greater symptom relief than superficial stimulation alone [41].

Recent findings from a study [42] utilizing resting-state quantitative EEG (QEEG) highlight the neurophysiological changes resulting from deep TMS. The study demonstrated a reduction in slow-frequency brain activity (delta and theta waves) in the prefrontal cortex following 36 treatment sessions. This attenuation of slow-wave brain activity may be a critical factor in how deep TMS alleviates depressive symptoms. Notably, baseline QEEG measures predicted treatment response with 93% accuracy, suggesting a strong link between neurophysiological markers and therapeutic outcomes.

Moreover, deep TMS is particularly effective in promoting neuroplasticity within the deeper brain regions. This is crucial for patients with depression, as they often exhibit reduced neuroplasticity. By directly stimulating deeper nuclei, deep TMS encourages the formation of new neural connections and enhances synaptic strength. This process of neurogenesis is vital for long-term improvements in mood and cognitive function, making deep TMS a powerful tool in the treatment of MDD.

Finally, the noninvasive nature of deep TMS enhances its attractiveness as a treatment option for those who may not respond to traditional therapies. Its capability to reach deeper brain structures while minimizing effects on surrounding tissues underscores its therapeutic potential, establishing a solid rationale for its use in managing depression effectively.

The deep TMS demonstrates significant efficacy in the treatment of depression. The systematic review [43] indicates that deep TMS shows strong clinical efficacy in patients with MDD, particularly among those who are resistant to traditional treatment. This therapeutic method effectively modulates neural networks associated with mood by stimulating deeper regions of the brain.

Multiple clinical trials comparing deep TMS to sham treatment have shown that patients receiving deep TMS experience significantly better outcomes in terms of symptom relief. For instance, studies report marked improvements in remission and response rates among deep TMS-treated patients. These findings highlight deep

TMS as a promising alternative therapy, especially for individuals who do not respond well to conventional pharmacotherapy [44, 45].

The deep TMS has emerged as a promising treatment for TRD, demonstrating significant efficacy in various clinical studies. Another systematic review by Berlim et al. [11, 46] highlighted that deep TMS can lead to substantial improvements in depressive symptoms, particularly in patients who have not responded to conventional therapies [46]. The review synthesized data from multiple studies, indicating that deep TMS is effective in reducing depression scores measured by validated scales such as the HAMD and the Montgomery–Asberg Depression Rating Scale (MADRS).

In an RCT conducted by Holtzheimer et al. [47], patients undergoing deep TMS showed notable reductions in depressive symptoms compared to those receiving sham treatment. The study reported that a significant percentage of participants experienced a 50% or greater improvement in their MADRS scores after several weeks of deep TMS therapy. This finding underscores the potential of deep TMS as an effective intervention for individuals with chronic depression [47].

Additional research by Fitzgerald and Segrave [48] emphasized the durable effects of deep TMS, with many patients maintaining improvements in depressive symptoms well after the treatment period. Long-term follow-up studies showed that the benefits of deep TMS could persist, making it a viable option for ongoing management of TRD [48].

A recent study [49] aimed to evaluate the efficacy of deep TMS in patients who did not respond to the initial acute treatment phase. In this RCT, 33 medication-free participants who completed 20 sessions of deep TMS over 4 weeks were identified as nonresponders. Following this, they continued with active deep TMS treatment twice weekly for up to 12 additional weeks. The findings were promising: 72.7% of nonresponders eventually achieved responder status during the continuation phase, with 60.6% responding within the first 4 weeks of extended treatment. Notably, 63.6% of participants reached remission at some point during the study. These results suggest that continuing deep TMS treatment beyond the initial acute course may significantly benefit patients who initially do not respond, offering hope for long-term symptom relief in those with TRD.

An RCT [50] explored the efficacy and safety of accelerated deep TMS protocols using the H1-coil in patients with TRD. A total of 28 participants underwent accelerated deep TMS sessions lasting 20 min, delivered twice daily for either 10 or 15 days. The primary outcomes assessed included changes in HAMD scores and any discontinuation of treatment due to adverse events. Results showed significant reductions in HAMD scores, with decreases of 13 points (59% improvement) for the 10-day protocol and 13 points (62% improvement) for the 15-day protocol, indicating substantial clinical benefit. Remission rates were also notable, with 38% achieving remission after the 10-day protocol and 42% after the 15-day protocol. Additionally, significant and clinically relevant reductions in Beck Depression Inventory-II scores were observed, particularly during the first week. These promising findings suggest that twice-daily accelerated deep TMS may be a safe and effective intervention for TRD, highlighting the need for further research through larger RCTs to compare accelerated deep TMS with standard deep TMS protocols.

Finally, the practical applications of deep TMS in clinical practice have been supported by its positive outcomes in real-world settings. Studies conducted in outpatient clinics have consistently reported high rates of response and remission among patients receiving deep TMS, reinforcing its role as an effective treatment modality for those suffering from depression that does not adequately respond to traditional therapies [51].

3.2.2.2 Safety

The deep TMS is generally considered a safe intervention, characterized by a profile of adverse effects that are mostly mild and manageable. The systematic review of clinical trials highlights that the most frequently reported adverse effects include headaches and localized pain or discomfort at the stimulation site. These effects are typically transient and resolve shortly after treatment, making them relatively easy to address.

While serious adverse events are rare, they have been documented, including instances of seizures and suicidal ideation or attempts. Such occurrences underscore the importance of careful screening and monitoring of patients undergoing deep TMS, as the potential for serious adverse effects, although low, necessitates vigilance in clinical practice.

The review also assessed the risk of bias across studies, finding a collectively low risk according to the Grading of Recommendations, Assessment, Development, and Evaluations checklist. This suggests that the reported safety outcomes are reliable and support the overall safety profile of deep TMS as an intervention.

Available literature indicates that deep TMS can serve as a safe alternative or adjunctive treatment for individuals who do not respond adequately to traditional therapies. Its safety profile particularly supports its use in treating conditions such as MDD, obsessive–compulsive disorder, and substance use disorders, where treatment-resistant patients may benefit the most.

Looking ahead, ongoing research is necessary to further understand the long-term safety and potential adverse effects of deep TMS, especially in diverse patient populations. This continued investigation will help to ensure that deep TMS remains a viable option in the therapeutic landscape for psychiatric and cognitive disorders.

3.2.2.3 Treatment Regimen

The deep TMS employs specialized H-coils, which are designed to stimulate deeper brain regions compared to traditional TMS coils. These coils can reach brain areas approximately 4 cm beneath the scalp, enabling more effective treatment for various psychiatric and cognitive disorders.

The treatment typically involves HF stimulation, usually at 18–20 Hz, delivered at an intensity of 120% of the resting motor threshold (MT). Each session lasts about 20–40 min, and the regimen often consists of daily treatments over a period of 2–4 weeks, tailored to the specific disorder and individual response.

Depending on the condition being treated, different H-coil designs may target various brain structures. For example, the prefrontal cortex is often targeted for depression, while the anterior cingulate cortex may be targeted for anxiety disorders. This specificity helps maximize treatment efficacy.

Throughout the treatment, patients are closely monitored to assess tolerability and effectiveness. Common adverse effects include mild headaches and discomfort at the site of stimulation, which are usually manageable.

After completing the treatment regimen, patients undergo evaluations to measure changes in symptoms and overall functioning. Follow-up sessions may be scheduled to help maintain the benefits of the treatment, particularly for those who have not responded well to traditional therapies.

3.2.2.4 Clinical Recommendations

The deep TMS has demonstrated significant efficacy in treating MDD, particularly in patients who are TRD or have not responded adequately to traditional pharmacological therapies. Clinical trials indicate that deep TMS can lead to marked improvements in depressive symptoms, making it a viable alternative or adjunct treatment option. Clinicians should consider deep TMS for patients with a confirmed diagnosis of MDD who exhibit persistent symptoms despite prior treatments.

The recommended treatment regimen for deep TMS in MDD typically involves HF stimulation (18–20 Hz) delivered at an intensity of 120% of the resting MT over a series of daily sessions lasting 20–40 min, often conducted over 2–4 weeks. Continuous monitoring of patient responses and adverse effects is crucial, as mild headaches and discomfort at the stimulation site are common. Regular follow-up assessments should be conducted to evaluate treatment efficacy and make necessary adjustments to the regimen, ensuring optimal outcomes for patients.

Based on two high-quality studies [45, 52], therefore, level A evidence (definite efficacy) is recommended for deep HF-rTMS of the left DLPFC in patients with MDD, including the elderly [22].

3.2.3 Priming rTMS

3.2.3.1 Efficacy

In Fitzgerald et al.'s study of 60 MDDs (30 real, 30 sham), the priming group showed a significantly greater reduction in depression scores on the MADRS than the sham group [53]. In another study comparing a priming LF rTMS protocol to a bilateral rTMS protocol in a large series of 179 patients, a significant mean reduction of >50% in HDRS-17 score (with a response rate of 56% and a remission rate of 40%) was found in both treatment groups, but no difference between groups at the end of the 4-week protocol [31]. Priming rTMS can also produce similar antidepressant effects in late-life depression [54].

3.2.3.2 Safety

Priming rTMS is well tolerated for the treatment of depressed adults and the elderly. No major adverse events have been reported with priming rTMS. Common adverse events included scalp discomfort, lacrimation, malaise, headache, and nausea. All of these adverse effects were mild in intensity, resolved within 1 h of completion of treatment, and did not require medical intervention.

3.2.3.3 Treatment Regimen

Twenty trains of 5 s duration at 6 Hz and 90% of the resting MT followed by 1 Hz stimulation (110% of MT, one continuous train of 900 pulses). Each treatment was administered for up to 4 weeks, 5 days per week (total of 20 treatment sessions) for adult MDDs [31] and 6 Hz rTMS over the right DLPFC at 80% resting MT for 10 min (600 stimulations; 20 trains of 5 s each) followed by 1 Hz rTMS over the right DLPFC at 100% resting MT for 21 min (1200 stimulations; 60 pulses per train, with 20 trains, and a 5 s intertrain interval), total of 10 daily sessions, for 2 weeks for late-life depression.

3.2.3.4 Clinical Recommendations

Although evidence supports a similar efficacy of priming rTMS protocols and bilateral rTMS protocols [31] and in favor of sham stimulation [54], it is premature to recommend the use of accelerated rTMS protocols for the treatment of MDD.

3.2.4 Synchronized rTMS

3.2.4.1 Efficacy

Synchronizing a TMS pulse with a person's underlying EEG rhythm can modify the brain's response. Compared with unsynchronized rTMS, patients with MDD who received synchronized rTMS showed no difference in overall clinical response [55]. No significant difference was found between fixed and randomly synchronized rTMS groups [56]. Either fixed or random frequency active synchronized rTMS had a statistically significantly greater percentage reduction in depression severity compared to sham [56]. However, two other studies found no difference in efficacy between active and sham synchronized rTMS [57, 58].

3.2.4.2 Safety

Synchronized rTMS is well tolerated for the treatment of MDD. No major adverse events have been reported with synchronized rTMS. Common adverse events included headache, dizziness, visual disturbance, dyschezia, back pain, insomnia, and respiratory tract infection. Common adverse effects resolved shortly after treatment.

3.2.4.3 Treatment Regimen

According to the study by George et al. [55], for each rTMS session, subjects had EEG setups, received synchronized rTMS, and received the following dose of rTMS delivered over the left prefrontal cortex: 6–13 Hz (at individual alpha frequency), 120% MT, 40 pulses per train, average 4-s pulse train, 3000 pulses per session, one session per weekday, as the train length varied by frequency (e.g., it takes 4 s at 10 Hz to get 40 pulses, but 5 s at 8 Hz). The intertrain off time (8–12 s) was adjusted so that all treatment sessions were the same length regardless of individual alpha frequency. Treatments were administered for a fixed interval of 6 weeks. Interruptions during treatment were allowed as needed for patient comfort or convenience using the "pause" selection on the device. For fixed frequency synchronized rTMS, set to the subject's average individual alpha frequency ± 0.1 Hz; for random frequency synchronized rTMS, with a random stimulus frequency that varied between 8 and 13 Hz [56].

3.2.4.4 Clinical Recommendations

Synchronized rTMS may be an effective treatment for MDD. It is premature to recommend the use of synchronized rTMS protocols for the treatment of MDD.

3.2.5 TBS

3.2.5.1 iTBS

Efficacy

TBS represents an innovative advancement in TMS techniques, tailored for the treatment of MDD [13]. The rationale for employing TBS lies in its unique stimulation pattern, which delivers bursts of HF pulses (50 Hz) at intervals designed to replicate theta rhythms found in the brain. This approach has been shown to enhance synaptic plasticity—an essential mechanism for learning and memory—thereby potentially addressing the neurobiological deficits present in depression.

One of the key mechanisms by which TBS may exert its effects is through the modulation of cortical excitability. Animal studies [59] have highlighted TBS's ability to induce both long-term potentiation and long-term depression in neuronal circuits [60]. These processes are crucial for the adaptive functioning of neural networks, particularly in areas like the DLPFC, which is often implicated in mood regulation. By affecting these pathways, TBS can alter the excitability and connectivity of brain regions associated with depressive symptoms.

Additionally, TBS's design allows for a more efficient application compared to rTMS. With treatment durations of only 1–3 min, TBS is not only time-efficient but also reduces the burden on patients and clinicians. This efficiency is particularly important in clinical settings where patient throughput and comfort are critical considerations. The reduced intensity required for TBS (typically around 80% of the active MT) further ensures that the method is both safe and tolerable, minimizing the risk of adverse effects that can accompany longer or more intense stimulation protocols [61].

Finally, the application of TBS in MDD aligns with emerging insights into the neurobiological basis of depression, particularly the interhemispheric asymmetry observed in the DLPFC. By utilizing intermittent TBS on the left DLPFC and continuous TBS on the right, this technique aims to restore balance between these hemispheres, addressing the dysregulation often seen in depressed individuals. This targeted approach not only supports the theoretical underpinnings of TBS but also reflects a growing understanding of the need for personalized and precise treatment modalities in psychiatry.

In summary, the rationale for TBS–TMS in MDD is grounded in its ability to leverage principles of synaptic plasticity, enhance cortical excitability, and provide a more efficient, patient-friendly treatment option that addresses the specific neurobiological features of MDD.

In the context of the efficacy of TBS, several studies have demonstrated its significant effectiveness in treating MDD. An important study [62] designed as an RCT conducted at three Canadian university hospitals, involving participants aged 18–65 with TRD. Eligible individuals had to show a current episode of MDD with an HAMD-17 score of at least 18 and a history of insufficient response to at least two antidepressant treatments. Participants were randomly assigned to receive either HF rTMS at 10 Hz or iTBS, with both treatments administered five times a week over 4–6 weeks.

Results indicated that both treatment groups experienced significant reductions in HAMD-17 scores, with the 10 Hz rTMS group improving from a baseline score of 23.5–13.4, and the iTBS group from 23.6 to 13.4. The estimated adjusted difference in scores favored iTBS, confirming its non-inferiority to 10 Hz rTMS, with a p-value of 0.0011. This finding suggests that iTBS can achieve similar clinical outcomes in a much shorter treatment time, enhancing the feasibility of rTMS therapy. Regarding adverse effects, both treatments exhibited comparable tolerability, although the self-reported pain level was slightly higher in the iTBS group. The dropout rates were similar, with 6% in the 10 Hz rTMS group and 8% in the iTBS group, and headaches were the most commonly reported adverse effect. Overall, the study concluded that iTBS is an effective alternative to traditional rTMS, potentially increasing treatment capacity for depression without sacrificing effectiveness or safety. This RCT highlighted that patients receiving TBS exhibited a marked reduction in depressive symptoms, with many reporting substantial improvements within just a few sessions. Follow-up assessments indicated that these positive outcomes were maintained for several months after treatment, suggesting that TBS not only provides immediate relief but also long-lasting benefits.

Overall, the accumulating evidence supports the efficacy of TBS as a promising therapeutic option in mental health care. As more clinical trials are conducted, the potential for TBS to become a widely accepted treatment modality in psychiatric practice becomes increasingly evident. The combination of rapid symptom relief and minimal adverse effects positions TBS as a vital tool for improving patient outcomes in depression management.

The Stanford Neuromodulation Therapy (SNT) protocol represents the most accelerated theta burst stimulation (aTBS) approach for MDD to date [63–65]. This innovative method delivers a total of 90,000 pulses of iTBS over 5 days, equating to five times the standard dose typically administered in a conventional

6-week protocol. Each day, 10 sessions are conducted with a pulse intensity of 90% of the resting MT, utilizing personalized targeting based on functional connectivity magnetic resonance imaging (MRI) to optimize treatment outcomes.

Remarkably, the SNT protocol has demonstrated a high remission rate of 90% in open-label studies and 78.6% in its first RCT. The adverse effects reported were minimal and comparable to those of conventional TMS protocols, suggesting that this accelerated approach is not only effective but also maintains a favorable safety profile. However, due to the significant alterations in stimulation parameters compared to standard iTBS, further RCTs are necessary to evaluate the individual contributions of these changes to the overall clinical efficacy of SNT.

A recent study [66] involved 44 participants diagnosed with TRD and moderate to severe suicidal ideation, who were randomly assigned to receive either accelerated iTBS or accelerated cTBS treatment. Each participant underwent 10 sessions of TBS over 5 consecutive days, totaling 90,000 pulses. The stimulation was precisely targeted to the left DLPFC for iTBS and the right DLPFC for continuous TBS, utilizing neuronavigation techniques. Clinical assessments were conducted at baseline, week 1, week 3, and week 5 to evaluate changes in suicidal ideation, depression, and anxiety symptoms.

The results indicated that accelerated cTBS was significantly more effective than iTBS in reducing suicidal ideation and anxiety symptoms at the 1-week mark after treatment. Specifically, the cTBS group achieved a mean change of 50.91% in suicidal ideation, compared to 26.67% in the iTBS group. Both treatments showed comparable antidepressant effects over time, with response rates for depression remaining stable. Notably, the cTBS group demonstrated a higher rate of remission of suicidal ideation at the conclusion of the study, suggesting that right DLPFC cTBS may offer a potent therapeutic option for managing suicidality in patients with TRD.

Despite its promising results, there are concerns regarding the potential for rapid relapses following accelerated treatments, as well as the durability of the antidepressant effects. Current data indicate that responses from accelerated TBS may be sustained for at least 4 weeks posttreatment, but long-term studies are needed to assess the lasting impact of these protocols. Furthermore, issues related to financial accessibility and the necessity for ongoing maintenance treatment raise important considerations for patients seeking this advanced therapy, especially as aTBS protocols are not currently covered by insurance in the United States.

Safety

As with any therapeutic technique, ensuring the safety of TBS is paramount. This focus on safety is supported by a growing body of research that outlines the risk factors, adverse effects, and monitoring protocols necessary for safe application.

The safety profile of TBS is generally favorable, with the most commonly reported adverse effects being mild and transient. Patients may experience localized discomfort at the site of stimulation, temporary headaches, or fatigue. These effects are typically short-lived, often resolving within minutes to hours posttreatment. Studies have shown that serious adverse events, such as seizures, are exceedingly rare in healthy individuals undergoing TBS. However, it is essential for practitioners to conduct thorough pretreatment assessments to identify any contraindications, especially in individuals with a history of seizures or other neurological disorders.

To enhance safety during TBS administration, monitoring protocols are crucial. Clinicians are encouraged to observe patients closely throughout the session, looking for any immediate adverse effects. Pretreatment screening should include a detailed medical history and an assessment of neurological health. Informed consent is also a critical step, ensuring that patients understand the procedure, potential adverse effects, and the monitoring process. Posttreatment follow-up is recommended to identify any delayed reactions and to reinforce patient comfort and safety.

Long-term safety data for TBS is still being established, but initial findings are promising. Research indicates that repeated sessions of TBS do not lead to cumulative adverse effects or significant changes in cognitive function. Unlike some other brain stimulation techniques, TBS has not been associated with an increased risk of seizures or other severe complications in healthy populations. This stability over time is particularly important when considering TBS for chronic conditions, as the therapeutic benefits must be weighed against any potential risks.

Treatment Regimen

TBS is a form of rTMS characterized by its unique stimulation pattern. It typically operates at a frequency of 50 Hz, delivering bursts of stimulation in specific sequences. TBS is primarily categorized into two types: cTBS and iTBS, each exhibiting significant differences in their effects and duration.

In cTBS, stimulation is applied continuously for either 20 or 40 s, delivering 300 or 600 pulses, respectively. This method typically reduces cortical excitability for approximately 20 min to up to an hour following stimulation, effectively suppressing neuronal activity in the targeted area. Conversely, iTBS involves shorter stimulation trains (2 s) repeated every 10 s, resulting in a total of 600 pulses, which enhances cortical excitability and maintains this effect for at least 15 min.

One of the notable advantages of TBS is its reduced administration time compared to conventional rTMS, often requiring only a few minutes for treatment. This efficiency makes TBS more convenient for clinical applications, while also utilizing lower stimulation intensities (typically 80% of the active MT), which can enhance patient comfort and reduce the risk of adverse effects.

Research indicates that TBS holds promise in treating depression, particularly through its targeted modulation of the DLPFC. In clinical trials, iTBS is frequently applied to the left DLPFC, while cTBS targets the right DLPFC, aiming to restore functional balance within the brain.

During the procedure, the patient is seated comfortably, and the TMS coil is positioned over the targeted area of the scalp. The resting MT is first determined to establish the appropriate stimulation intensity. Once set, the TBS session commences, with ongoing monitoring of the patient's comfort and any potential adverse effects.

Posttreatment care involves monitoring patients for a brief period to observe any adverse effects, although serious complications are rare. Patients can generally resume their normal activities immediately after the session. It is important to note that treatment parameters may be adjusted based on individual responses, and clinicians should always refer to updated clinical guidelines for the latest information on TBS administration.

Clinical Recommendations

TBS represents a promising advancement in the treatment of MDD. Clinicians should consider TBS as a viable option, especially for patients who have not responded adequately to standard antidepressant therapies or conventional rTMS. Given its shorter administration time and lower intensity requirements, TBS offers a convenient and patient-friendly alternative, potentially increasing patient compliance and satisfaction.

When implementing TBS, it is crucial for clinicians to personalize treatment protocols based on individual patient needs and responses. Utilizing both iTBS and cTBS may be beneficial, depending on the specific symptoms and characteristics of the patient. Monitoring the patient's progress and adverse effects closely during the treatment course will help in adjusting stimulation parameters and ensuring optimal therapeutic outcomes. Collaborative decision-making between clinicians and patients regarding treatment goals and expectations is essential for enhancing the overall effectiveness of TBS.

In summary, while TBS shows great promise, further research is needed to establish standardized protocols, identify response predictors, and fully understand its neurobiological mechanisms. Ongoing clinical trials and studies will help refine the application of TBS in various psychiatric settings. As the body of evidence grows, TBS could become a cornerstone in the management of TRD, improving the quality of life for many patients.

3.2.5.2 Accelerated iTBS

Efficacy

A systematic review and meta-analysis of five double-blind RCTs in MDD or bipolar depression showed superior therapeutic efficacy compared with sham stimulation. Specifically, active accelerated iTBS was significantly superior to sham stimulation in terms of trial-defined response (39.8% vs. 19.0%), but not in remission (36.5% vs. 16.2%); however, improvement in depressive symptoms at the post-accelerated iTBS assessment did not show a statistical difference between the active and sham stimulation groups [67], with more pronounced anti-suicidal effects in MDD with higher suicide risk profiles [68]. Another accelerated form of TBS, SNT, RCT showed that a high-dose iTBS protocol with functional connectivity-guided targeting was more effective than sham stimulation for TRD, specifically that 85.7% in the active SNT group met the criterion for response (a reduction ≥50% in MADRS score) and 78.6% met the criterion for remission (a MADRS score ≤ 10) in at least one of the five posttreatment assessments [69].

Safety

Accelerated iTBS was safe with minimal adverse effects; no serious adverse effects were observed. Common adverse effects included headache, fatigue, discomfort, anxiety, dental problems, and scalp pain. These complaints resolved spontaneously after a short period of time or shortly after taking a common analgesic such as acetaminophen.

Treatment Regimen

In the context of TRD, the treatment protocol, consisting of a total of 20 iTBS sessions, was administered over 4 days with five sessions per day, for a total of 32,400 stimuli. In each session, patients received 1620 pulses per session in 54 triplet bursts with a train duration of 2 s and an intertrain interval of 8 s, the stimulation was set at 110% of resting MT [70]. For SNT, the standard protocol includes: (1) the adoption of an efficient rTMS form known as iTBS; (2) the delivery of multiple iTBS sessions daily at optimally spaced intervals; (3) the application of a higher cumulative pulse dose of stimulation; and (4) personalized targeting for stimulation of the left DLPFC to subgenual anterior cingulate cortex circuit. Ten sessions of SNT were delivered daily for a total of 18,000 pulses per day for 5 consecutive days, with stimulation delivered at 90% of resting MT, adjusted for the depth of the identified functional connectivity MRI target [69].

Clinical Recommendations

Accelerated iTBS may be an effective treatment for MDD. It is premature to recommend the use of accelerated iTBS protocols for the treatment of MDD.

3.2.5.3 Continuous TBS

Efficacy

A recent meta-analysis examining the treatment effects of cTBS in MDD or bipolar depression found no significant benefits over sham stimulation [71]. Specifically, for patients with MDD, there was no significant difference in the degree of clinical improvement between the active and sham cTBS groups. However, in patients whose medication status was unchanged prior to the study and in those who were off medication, active cTBS resulted in a significantly greater reduction in HAMD compared to sham cTBS [72]. For TRD, iTBS had better antidepressant responses than cTBS [73].

Safety

cTBS appears to be a safe and well-tolerated option for the treatment of MDD; common adverse effects include headache, dizziness, palpitations, nausea, soreness, and facial muscle twitching [74].

Treatment Regimen

The target of cTBS is the right DLPFC. In the study by Li et al. [73], a 120-s train of continuous bursts (1800 pulses) was delivered in each session per day for a total of 10 sessions. In the study by Chistyakov et al. [72], cTBS consisted of three-pulse

50 Hz bursts delivered at a rate of 5 Hz (i.e., 200 ms between each burst) in uninterrupted trains. Each treatment session consisted of 3600 stimuli delivered in four consecutive trains of 900 stimuli separated by at least 15 min. The stimulation intensity was 100% of the active MT. A total of 10 sessions were performed over 10 days.

Clinical Recommendations

cTBS may be an effective treatment for MDD. It is premature to recommend the use of accelerated TBS protocols for the treatment of MDD.

3.2.5.4 Bilateral TBS

Efficacy

A recent meta-analysis examining the treatment effects of bilateral TBS in MDD found significant advantages over sham stimulation, whether in improving depressive symptoms or in response and remission rates [75].

Safety

Bilateral TBS appears to be a well-tolerated form of TMS, with no significant differences observed between the active bilateral TBS and sham bilateral TBS groups for most adverse events. The most commonly reported adverse events were headache and dizziness, which were manageable and in most cases did not lead to treatment discontinuation. Serious adverse events (suicides and suicide attempts) occurred in the study by Prasser et al. [76].

Treatment Regimen

Bilateral TBS targeting the DLPFC: The bilateral TBS protocol for the treatment of MDD involves cTBS to the right DLPFC and iTBS to the left DLPFC using a figure-of-eight coil at an intensity of 80% of resting MT. The treatment schedule typically consists of 10 sessions, with each session consisting of cTBS (600 stimuli) followed by iTBS (600 stimuli) to the respective DLPFC regions. The treatment duration is typically 2 weeks, with daily sessions during this period.

Clinical Recommendations

Currently, a Level B of evidence (probable antidepressant efficacy) could be proposed for a sequential bilateral left-sided iTBS + right-sided cTBS protocol applied to the DLPFC in the context of patients with unipolar depressive disorder [22].

3.2.6 Magnetic Seizure Therapy (MST)

3.2.6.1 Efficacy

MST emerges from the need to enhance the treatment options for MDD, particularly for patients resistant to conventional therapies. By combining aspects of electroconvulsive therapy (ECT) and TMS, MST aims to maintain the therapeutic efficacy of ECT while minimizing its cognitive adverse effects. This is particularly relevant for mood disorders such as MDD, where treatment options are often limited and associated with significant morbidity [77].

One of the key rationales for MST is its ability to induce controlled seizures with greater precision compared to ECT. Traditional ECT often results in widespread cortical activation, which, while effective, can lead to significant cognitive impairments. In contrast, MST utilizes magnetic fields to focus stimulation on specific brain regions, allowing for targeted therapeutic effects without the extensive collateral stimulation seen in ECT. This targeted approach is believed to enhance the antidepressant effects while reducing adverse cognitive outcomes [78].

Furthermore, MST takes advantage of the brain's neuroplasticity—its ability to adapt and reorganize itself in response to therapeutic interventions [79]. By inducing seizures in a controlled environment, MST may promote beneficial changes in neural circuitry associated with mood regulation. This mechanism aligns with the growing understanding that depression is linked to disruptions in neural connectivity and function, suggesting that MST could effectively restore balance within these networks.

Additionally, MST's reliance on magnetic rather than electrical stimulation offers a unique advantage. The magnetic pulses can penetrate the skull with less resistance, leading to more effective stimulation of superficial cortical regions. This is crucial because many of the cognitive adverse effects associated with ECT are thought to arise from deeper brain structures being affected. By focusing on the cortex, MST may mitigate these adverse effects while still eliciting significant therapeutic responses.

Lastly, the rationale for MST is reinforced by the success of TMS in treating MDD. TMS has shown promising results in alleviating depressive symptoms by modulating activity in the DLPFC. MST builds upon this foundation, hypothesizing that by inducing seizures through magnetic stimulation, it could harness similar neural mechanisms, ultimately providing a novel and effective treatment avenue for those suffering from TRD [80].

The MST has emerged as a promising alternative to ECT for treating MDD. Research indicates that MST can lead to significant reductions in depressive symptoms. In a recent trial, patients receiving HF MST (100 Hz) achieved a remission rate of 33.3%, which is notably higher than that reported for many standard antidepressant treatments [81].

The efficacy of MST appears to be influenced by the frequency of stimulation. The study found that HF MST produced the highest rates of response and remission compared to medium and low-frequency MST. Specifically, a response rate of 41.7% was observed in the HF group, highlighting the potential for tailored frequency adjustments to optimize therapeutic outcomes.

Furthermore, MST demonstrated not only antidepressant effects but also maintained cognitive functioning in most patients. Performance on various cognitive measures remained stable, with some improvements noted in specific tasks, such as brief visuospatial memory. This suggests that MST may offer a therapeutic advantage over ECT, which is often associated with cognitive adverse effects.

A recent RCT [82] aimed to compare the antidepressant efficacy of MST with ultra brief pulse right unilateral (RUL) ECT in treating MDD and bipolar disorder. Conducted at three academic hospitals, the study included adults aged 18–90 years with a baseline HAMD-24 score of 18 or higher. Results indicated that both MST and ECT produced clinically meaningful antidepressant effects, with response rates of 51.4% for MST and 42.1% for ECT, while remission rates were 37.1% for MST and 26.3% for ECT among the intent-to-treat sample. Among treatment completers, the response rates were similar, at 58.6% for MST and 62.5% for ECT. Notably, the mean number of treatments required to achieve remission was higher for MST (9.0) compared to ECT (6.7). Despite this, both treatments demonstrated sustained benefits over a 6-month follow-up period, highlighting that MST's efficacy is comparable to that of ultrabrief pulse RUL ECT.

A pilot study [83] explored the effectiveness of accelerated MST (aMST) in MDD among 15 patients over a 6-day treatment course. The primary outcome was measured using the HAMD-17, which demonstrated significant reductions in depression and anxiety symptoms. Notably, 60% of participants achieved a clinical response, while 47% experienced remission. Additionally, improvements were observed in overall cognitive performance as measured by the Repeatable Battery for the Assessment of Neuropsychological Status (RBANS), particularly in immediate and delayed memory indices. These findings suggest that aMST may provide a rapid antidepressant effect without significant cognitive adverse effects, warranting further investigation in larger, RCTs.

A nonrandomized controlled trial investigated the effectiveness of MST in treating suicidality in patients with TRD. A total of 67 patients, with a mean age of 46.3 years, received an average of 19.5 MST treatments, with outcomes measured by the Beck Scale for Suicidal Ideation. The study found that 47.8% of participants achieved remission from suicidality, with notably higher remission rates observed in those receiving LF (55.2%) and moderate-frequency (54.5%) MST compared to HF (25.0%) MST. The results demonstrated a significant association between time and reductions in suicide ideation scores, suggesting that MST may effectively alleviate suicidality, particularly at lower frequencies. These findings highlight the potential of MST as a viable treatment option for suicidality in mental disorders, warranting further comparative studies with ECT [84].

Importantly, MST has shown promise in TRD. In the study, patients who had previously undergone multiple unsuccessful treatments with antidepressants reported significant symptom relief following MST, underscoring its potential as a viable option for those who do not respond to conventional therapies.

In conclusion, MST represents a significant advancement in the treatment of MDD, particularly for patients with TRD. The favorable remission rates and minimal cognitive impairment associated with HF MST make it a compelling alternative to traditional ECT, warranting further research to establish its long-term efficacy and optimize treatment protocols.

3.2.6.2 Safety

The MST has been investigated for its safety profile in the treatment of MDD. In the review [81], 140 patients were screened, and of those, 86 received at least eight treatments of MST. Throughout the treatment protocol, adverse events were closely monitored, allowing for an assessment of safety in a clinical setting. The review reported a total of 17 serious adverse events affecting 16 patients, representing about 11.4% of the participants. Among these, only a few were deemed related or possibly related to the MST procedure. Notable serious adverse events included the emergence of mania in one patient and a hospitalization due to a fall, indicating that while adverse events did occur, they were relatively rare and specific.

Additionally, one instance of a superficial burn on the head was reported, which was attributed to a malfunction of the stimulation device. This highlights the importance of monitoring the equipment and ensuring that safety protocols are followed during MST sessions to minimize risks associated with device malfunctions.

Despite these adverse events, the overall safety profile of MST appears favorable when compared to traditional ECT. The study suggests that MST may produce fewer cognitive adverse effects, which is a significant consideration for patients and healthcare providers concerned about the cognitive impact commonly associated with ECT.

While some adverse events were observed during the MST trials, the overall incidence was low, and the therapy was well-tolerated among the majority of participants. These findings support the potential of MST as a safer alternative for treating MDD, particularly for patients who are wary of the cognitive risks associated with ECT. Future studies are needed to further establish the long-term safety and efficacy of MST in diverse patient populations.

3.2.6.3 Treatment Regimen

MST is conducted in specialized clinical environments equipped for neuromodulation therapies. Patients are treated at centers such as the Temerty Centre for Therapeutic Brain Intervention, which are designed to ensure safety and efficacy.

- *Anesthesia*: General anesthesia is administered by trained anesthesiologists. Common agents include methohexital and succinylcholine, which help ensure muscle relaxation and facilitate seizure induction during the procedure.
- *Stimulation protocol*: MST is delivered using an MST device equipped with a dual circular coil, positioned over the prefrontal cortex. The treatment can involve different frequency settings: HF (100 Hz), medium-frequency (50–60 Hz), and LF (25 Hz). Each session starts with titration to determine the individual seizure threshold.
- *Treatment schedule*: Patients typically undergo MST sessions two to three times per week. A total of up to 24 treatment sessions may be administered, depending on clinical response and tolerability, allowing for flexibility based on individual needs.

- *Monitoring*: Throughout the procedure, patients are closely monitored for efficacy and any adverse effects, particularly regarding cognitive function and mood symptoms.
- *Assessment*: Depressive symptoms are evaluated using standardized scales, while cognitive function is assessed through a comprehensive neurocognitive battery before, during, and after treatment. This holistic approach aims to provide effective relief from depressive symptoms while minimizing cognitive adverse effects.

3.2.6.4 Clinical Recommendations

The MST should be considered as a viable treatment option for patients with MDD, particularly those who have not responded adequately to conventional antidepressant therapies or who experience significant cognitive adverse effects from ECT. Clinicians are encouraged to evaluate the patient's history, including previous treatment responses and any cognitive concerns, to determine if MST may offer a more tailored and effective approach for alleviating depressive symptoms.

It is essential for healthcare providers to monitor patients closely during MST treatment, assessing both the effectiveness of symptom reduction and any cognitive changes. Regular evaluations using standardized depression scales and neurocognitive assessments can help track progress and guide treatment adjustments. Additionally, conducting MST in a controlled clinical setting with experienced staff ensures patient safety and maximizes the therapeutic potential of this innovative treatment modality.

3.3 tDCS

3.3.1 Conventional tDCS

3.3.1.1 Efficacy

tDCS offers a promising approach for treating MDD due to its ability to modulate cortical excitability. The rationale for using tDCS stems from its mechanism of inducing regional changes in neuronal activity by applying a low electrical current to the scalp. This noninvasive technique can enhance or inhibit neuronal firing rates depending on the polarity of the stimulation, allowing for targeted modulation of brain areas associated with mood regulation [85].

One critical area of focus for tDCS in depression is the DLPFC, which is often implicated in the pathophysiology of MDD. Studies have shown that individuals with depression frequently exhibit hypometabolism in the left DLPFC, leading to decreased cortical excitability in this region. By applying anodal stimulation to the left DLPFC, tDCS aims to increase neuronal excitability and promote more balanced activation between the left and right prefrontal cortices, potentially alleviating depressive symptoms.

Additionally, tDCS has been linked to neuroplastic changes that may underlie its therapeutic effects. These changes resemble long-term potentiation and depression, processes that are crucial for learning and memory. By enhancing synaptic plasticity, tDCS may help to counteract the neurobiological deficits associated with depression, fostering adaptive neural responses and improving emotional regulation over time.

Furthermore, the use of tDCS aligns with emerging theories in neuropsychiatry that emphasize the importance of neuroplasticity in treating psychiatric disorders. Given that depression is often characterized by maladaptive neural circuits and impaired plasticity, tDCS provides a unique tool to facilitate the reorganization of these circuits. This makes tDCS a rational choice for therapy, as it not only targets specific brain regions involved in mood but also aims to restore the underlying neurobiological functions that are disrupted in patients with depression.

The tDCS has emerged as a promising non-pharmacological treatment for MDD, supported by a growing body of clinical evidence. A comprehensive individual patient data meta-analysis reviewed data from six RCTs involving 289 patients [85]. This analysis revealed that active tDCS significantly outperformed sham stimulation, with a clinical response rate of 34% and a remission rate of 23%. These findings indicate that a substantial portion of patients benefits from the treatment.

The efficacy of tDCS was measured using various outcome metrics, including response and remission rates. The ORs for response and remission were found to be 2.44 and 2.38, respectively, suggesting that patients receiving active tDCS were more than twice as likely to experience significant improvement in their depressive symptoms compared to those receiving sham treatment. This level of effectiveness is comparable to that seen with traditional antidepressant medications.

In terms of the number needed to treat (NNT), the analysis indicated an NNT of 7 for achieving a clinical response, meaning that for every seven patients treated with tDCS, one additional patient is likely to benefit compared to sham treatment. The NNT for remission was calculated at 9, further underscoring the treatment's potential effectiveness in alleviating severe depressive symptoms in a significant number of patients.

The meta-analysis also examined the acceptability of tDCS, finding no significant difference in drop-out rates between active (10.1%) and sham (12.2%) groups. This suggests that tDCS is a well-tolerated intervention, making it a viable option for patients who may be reluctant to use pharmacological treatments due to adverse effects or other concerns.

The tDCS has shown promise as an effective treatment for depression, with its antidepressant effects peaking at approximately 6 weeks [86]. The analysis of individual participant data from ten RCTs revealed that tDCS significantly reduced depressive symptoms compared to sham stimulation, and this effect continued to diverge from sham up to 10 weeks.

Predictors of mood outcomes were identified, indicating that higher baseline depression severity and treatment resistance were associated with worse responses, while participants with bipolar disorder and anxiety disorder exhibited better outcomes. This suggests that tDCS may be particularly beneficial for those with bipolar depression, potentially requiring longer treatment courses to prevent relapse.

The findings emphasize the importance of considering various factors, including treatment duration and the presence of comorbid conditions, in optimizing tDCS protocols for depression. Overall, the study underscores the efficacy of tDCS in alleviating depressive symptoms, particularly in TRD and among individuals with bipolar disorder.

Overall, the clinical trials and meta-analyses conducted to date provide robust evidence supporting the efficacy of tDCS for treating MDD. As a noninvasive treatment with favorable efficacy and acceptability profiles, tDCS represents a valuable addition to the range of therapeutic options available for individuals struggling with depression. Further research is warranted to optimize treatment protocols and explore its application in various clinical populations.

3.3.1.2 Safety

A comprehensive analysis of individual patient data from six RCTs evaluated the safety and acceptability of tDCS specifically in patients with MDD [87]. The study aimed to determine the rates of adverse events and dropouts associated with active tDCS compared to sham stimulation.

The findings indicated that the dropout rates between the active tDCS group (8.8%) and the sham group (12%) were not significantly different. This suggests that tDCS is as acceptable as sham stimulation, indicating good tolerability among patients. Importantly, the analysis showed that adverse event rates were also comparable between the active (73.5%) and sham (68.3%) groups, reinforcing the notion that tDCS is generally well-tolerated.

Moreover, the study found that higher current densities during tDCS were associated with lower rates of adverse events. This insight is crucial, as it suggests that optimizing stimulation parameters may enhance patient safety and comfort. Adverse effects reported included mild symptoms such as tingling and discomfort, which are consistent with previous findings in healthy volunteers.

The absence of significant differences in safety profiles between active and sham groups indicates that tDCS does not pose additional risks compared to placebo treatments. This is particularly relevant for patients with MDD, who often face challenges with existing pharmacological treatments that can have numerous adverse effects.

Overall, the results support the conclusion that tDCS is a safe and well-accepted treatment modality for MDD. Given the urgent need for effective and tolerable treatment options in depression, tDCS represents a valuable addition to the therapeutic landscape for managing MDD.

3.3.1.3 Treatment Regimen

The treatment regimen for tDCS in the context of antidepressant therapy typically involves the application of a weak electrical current through electrodes placed on the scalp. In most clinical settings, a common setup includes two electrodes: an anode (positive) and a cathode (negative). The anode is usually positioned over the left prefrontal cortex, an area associated with mood regulation, while the cathode is placed on the right prefrontal cortex or another area deemed appropriate based on the specific treatment protocol.

In terms of parameter settings, tDCS sessions generally utilize a current intensity ranging from 1 to 2 mA, with a typical duration of 20–30 min per session. The specific choice of current density can impact the treatment's efficacy and tolerability; studies have indicated that higher current densities (e.g., 0.8 A/m^2) may be associated with better outcomes and fewer adverse effects. Treatment frequency varies, but protocols often recommend multiple sessions per week over several weeks to achieve optimal therapeutic effects.

Before initiating treatment, practitioners assess individual patient characteristics, including the severity of depression and any comorbid conditions, to tailor the tDCS protocol. This personalized approach, combined with standardized operational guidelines, aims to enhance the effectiveness of tDCS while ensuring patient safety and comfort throughout the treatment course.

3.3.1.4 Clinical Recommendations

Clinical recommendations for the use of tDCS in the treatment of depression emphasize the importance of individualized treatment protocols. Practitioners should conduct thorough assessments of patients' clinical profiles, including the severity of depressive symptoms, history of treatment resistance, and any comorbid conditions. This tailored approach helps determine the optimal electrode placement, current intensity, and treatment frequency to maximize therapeutic outcomes.

It is generally recommended that tDCS sessions consist of 20–30 min of stimulation, with a current intensity typically set between 1 and 2 mA. Treatment regimens often involve multiple sessions per week over a duration of several weeks, usually around 6 weeks, to achieve sustained antidepressant effects. Patients should be closely monitored for any adverse effects or discomfort during and after sessions, allowing for adjustments to the protocol as needed.

Additionally, clinicians should educate patients about the tDCS procedure, including its safety profile and potential benefits. This information can enhance patient acceptance and adherence to the treatment regimen. Regular follow-ups are essential to evaluate treatment efficacy and make necessary modifications to the tDCS protocol, ensuring that it remains a viable option in the comprehensive management of depression.

3.3.2 HD-tDCS

3.3.2.1 Efficacy

HD-tDCS may have antidepressant effects. An RCT focused on TRD showed that HD-tDCS led to significant improvements in depressive symptoms compared to sham stimulation as measured by HAMD [88]. However, another RCT demonstrated that HD-tDCS can accurately target and modulate neural circuitry relevant to depression, with improvements in anhedonia as measured by the Snaith-Hamilton Pleasure Scale, but nonsignificant effects were observed for HD and conventional tDCS, respectively, compared to sham stimulation as measured by the HAMD [89].

3.3.2.2 Safety

Adverse effects of HD-tDCS were minimal, with only one case of pruritus reported, and this was not severe enough to cause the patient to discontinue treatment; this symptom did not require treatment. No adverse events related to HD-tDCS were reported.

3.3.2.3 Treatment Regimen

HD-tDCS uses the Visor Neuron Navigation System to precisely target the left DLPFC at a stimulation intensity of 2 mA for 20 min for a total of 10–12 sessions.

3.3.2.4 Clinical Recommendations

HD-tDCS may be an effective treatment for MDD. It is premature to recommend the use of HD-tDCS protocols for the treatment of MDD.

3.4 TMS Vs. tDCS

3.4.1 Efficacy

One RCT evaluated the clinical efficacy and safety of rTMS, tDCS, tDCS + rTMS, and sham tDCS + sham rTMS after 2 weeks of treatment in patients with MDD and found that those who received sham tDCS + active rTMS showed a significant reduction in total HAMD-24 score compared to those who received active tDCS + sham rTMS [90]. There was no significant difference between the active rTMS and active tDCS groups in the change in total RBANS scores [91]. It is worth noting that tDCS + rTMS was a more effective treatment option than either tDCS or rTMS alone for patients with MDD.

3.4.2 Safety

Not only either tDCS or rTMS intervention alone but also tDCS + rTMS intervention were safe treatment option for patients with MDD; no serious adverse events occurred among these treatment option; depressive patients tolerated the treatments well, with no seizures or manic symptoms; the common adverse effects for tDCS + rTMS were skin redness, headache, mild irritation, pruritus, and dizziness.

3.5 Conclusion

The conventional rTMS (10 Hz, 1 Hz), TBS (iTBS, cTBS, SNT, etc.), Deep TMS, MST, and tDCS are noninvasive brain stimulation techniques that have shown promise in the treatment of MDD (see Table 3.1).

The TMS has been established as an effective intervention, particularly in TRD, with robust evidence supporting its ability to induce long-lasting changes in neural activity and improve depressive symptoms. Its targeted approach allows for

Table 3.1 TMS in MDD

TMS	Clinical recommendations MDD	Levels
Unilateral rTMS of HF-rTMS applied to the left DLPFC	Recommend	[a][11, 16, 17], etc.
Unilateral rTMS of LF-rTMS applied to the right DLPFC	Recommend	[b][11, 16, 17], etc.
Bilateral rTMS	Recommend	[b][25, 26, 28], etc.
Accelerated rTMS	Recommend	[b][37, 38, 40], etc.
Deep TMS	Recommend	[a][11, 22, 46, 47], etc.
Priming rTMS	Unclear	[c][53, 54], etc.
Synchronized rTMS	Unclear	[c][55, 56], etc.
iTBS	Recommend	[a][13, 62], etc.
Accelerated iTBS	Unclear	[c][68, 69], etc.
cTBS	Unclear	[c][71, 72], etc.
Bilateral TBS	Recommend	[b][75, 76], etc.
MST	Recommend	[b][81, 82, 78], etc.

Abbreviations: *iTBS i*ntermittent theta burst stimulation, *MST* magnetic seizure therapy, *rTMS* repetitive transcranial magnetic stimulation, *TBS* theta burst stimulation, *TMS* transcranial magnetic stimulation

[a]Derived from global multicenter clinical cohorts/randomized placebo-controlled (RCT) studies

[b]Derived from single-center/real-world studies

[c]Data derived only from expert consensus or case reports

Table 3.2 tDCS in MDD

Optimization strategies	Clinical recommendations MDD	Levels
tDCS	Recommend	[a][85–87], etc.
HD-tDCS	Unclear	[b][88, 89], etc.)

Abbreviations: *tDCS* transcranial direct current stimulation

[a]Derived from global multicenter clinical cohorts/randomized placebo-controlled (RCT) studies

[b]Data derived only from expert consensus or case reports

[c]Derived from single-center/real-world studies

modulation of specific brain regions, making it a valuable option for personalized treatment strategies.

As summarized in Table 3.2, the tDCS offer alternative mechanisms for brain modulation, with tDCS providing a constant current to alter neuronal excitability. Technique has demonstrated efficacy in reducing depressive symptoms, albeit often with more modest effects compared to TMS. Their favorable safety profiles and ease of application make them attractive adjunctive therapies, particularly for patients who may benefit from a less invasive approach. Collectively, these modalities highlight the potential for innovative, non-pharmacological treatments to enhance recovery from depression and improve overall patient outcomes. Antidepressant treatment efficacy may be improved by tDCS combined with rTMS.

Acknowledgments None.

Disclosure/Conflicts of Interest The authors declare no conflicts of interest in conducting this study or preparing the manuscript.

Financial Support This study was supported by the Ministry of Science and Technology of China, National Key R&D Program of China (2023YFC2506800), National Natural Science Foundation of China (82,171,544, 82,371,505, 82,151,314, 82,101,623), Shenzhen Science and Technology Plan Project (JCYJ20220530165009020), and Shenzhen Medical and Health Three Project (SZSM202011014). The funders had no role in study design, data collection and analysis, decision to publish, or preparation of the manuscript.

References

1. Shorey S, Ng ED, Wong CHJ. Global prevalence of depression and elevated depressive symptoms among adolescents: a systematic review and meta-analysis. Br J Clin Psychol. 2022;61(2):287–305.
2. Lu J, Xu X, Huang Y, Li T, Ma C, Xu G, Yin H, Xu X, Ma Y, Wang L, Huang Z, Yan Y, Wang B, Xiao S, Zhou L, Li L, Zhang Y, Chen H, Zhang T, Yan J, Ding H, Yu Y, Kou C, Shen Z, Jiang L, Wang Z, Sun X, Xu Y, He Y, Guo W, Jiang L, Li S, Pan W, Wu Y, Li G, Jia F, Shi J, Shen Z, Zhang N. Prevalence of depressive disorders and treatment in China: a cross-sectional epidemiological study. Lancet Psychiatry. 2021;8(11):981–90.
3. Cheng X, Wang Q, Wang R, Wang Y, Chen X, Mi G, Chen X, Wang L, Wang C, Hu L, Gu L, Zhang J, Hou R. Prevalence of depressive disorders and associated demographic characteristics in Shandong: an epidemiological investigation. J Affect Disord. 2022;311:198–204.
4. Corneliusson L, Gustafson Y, Olofsson B. Prevalence of depressive disorders among the very old in the 21st century. J Affect Disord. 2024;362:706–15.
5. Spoelma MJ, Sicouri GL, Francis DA, Songco AD, Daniel EK, Hudson JL. Estimated prevalence of depressive disorders in children from 2004 to 2019: a systematic review and meta-analysis. JAMA Pediatr. 2023;177(10):1017–27.
6. Rotenstein LS, Ramos MA, Torre M, Segal JB, Peluso MJ, Guille C, Sen S, Mata DA. Prevalence of depression, depressive symptoms, and suicidal ideation among medical students: a systematic review and meta-analysis. JAMA. 2016;316(21):2214–36.
7. Plaisier I, Beekman AT, de Graaf R, Smit JH, van Dyck R, Penninx BW. Work functioning in persons with depressive and anxiety disorders: the role of specific psychopathological characteristics. J Affect Disord. 2010;125(1–3):198–206.
8. Tsuchiya M, Kawakami N, Ono Y, Nakane Y, Nakamura Y, Fukao A, Tachimori H, Iwata N, Uda H, Nakane H, Watanabe M, Oorui M, Naganuma Y, Furukawa TA, Kobayashi M, Ahiko T, Takeshima T, Kikkawa T. Impact of mental disorders on work performance in a community sample of workers in Japan: the world mental health Japan survey 2002-2005. Psychiatry Res. 2012;198(1):140–5.
9. Cipriani A, Furukawa TA, Salanti G, Chaimani A, Atkinson LZ, Ogawa Y, Leucht S, Ruhe HG, Turner EH, Higgins JPT, Egger M, Takeshima N, Hayasaka Y, Imai H, Shinohara K, Tajika A, Ioannidis JPA, Geddes JR. Comparative efficacy and acceptability of 21 antidepressant drugs for the acute treatment of adults with major depressive disorder: a systematic review and network meta-analysis. Lancet. 2018;391(10128):1357–66.
10. Hofmann SG, Asnaani A, Vonk IJ, Sawyer AT, Fang A. The efficacy of cognitive behavioral therapy: a review of meta-analyses. Cogn Ther Res. 2012;36(5):427–40.
11. Berlim MT, van den Eynde F, Tovar-Perdomo S, Daskalakis ZJ. Response, remission and drop-out rates following high-frequency repetitive transcranial magnetic stimulation (rTMS) for treating major depression: a systematic review and meta-analysis of randomized, double-blind and sham-controlled trials. Psychol Med. 2014b;44(2):225–39.

12. Chen J, Zhou C, Wu B, Wang Y, Li Q, Wei Y, Yang D, Mu J, Zhu D, Zou D, Xie P. Left versus right repetitive transcranial magnetic stimulation in treating major depression: a meta-analysis of randomised controlled trials. Psychiatry Res. 2013;210(3):1260–4.
13. Chung SW, Hoy KE, Fitzgerald PB. Theta-burst stimulation: a new form of TMS treatment for depression? Depress Anxiety. 2015;32(3):182–92.
14. Rush AJ, Trivedi MH, Wisniewski SR, Nierenberg AA, Stewart JW, Warden D, Niederehe G, Thase ME, Lavori PW, Lebowitz BD, McGrath PJ, Rosenbaum JF, Sackeim HA, Kupfer DJ, Luther J, Fava M. Acute and longer-term outcomes in depressed outpatients requiring one or several treatment steps: a STAR*D report. Am J Psychiatry. 2006;163(11):1905–17.
15. Kedzior KK, Reitz SK, Azorina V, Loo C. Durability of the antidepressant effect of the high-frequency repetitive transcranial magnetic stimulation (rTMS) in the absence of maintenance treatment in major depression: a systematic review and meta-analysis of 16 double-blind, randomized, sham-controlled trials. Depress Anxiety. 2015;32(3):193–203.
16. O'Reardon JP, Solvason HB, Janicak PG, Sampson S, Isenberg KE, Nahas Z, McDonald WM, Avery D, Fitzgerald PB, Loo C, Demitrack MA, George MS, Sackeim HA. Efficacy and safety of transcranial magnetic stimulation in the acute treatment of major depression: a multisite randomized controlled trial. Biol Psychiatry. 2007;62(11):1208–16.
17. George MS, Lisanby SH, Avery D, McDonald WM, Durkalski V, Pavlicova M, Anderson B, Nahas Z, Bulow P, Zarkowski P, Holtzheimer PE III, Schwartz T, Sackeim HA. Daily left prefrontal transcranial magnetic stimulation therapy for major depressive disorder: a sham-controlled randomized trial. Arch Gen Psychiatry. 2010;67(5):507–16.
18. Brunoni AR, Chaimani A, Moffa AH, Razza LB, Gattaz WF, Daskalakis ZJ, Carvalho AF. Repetitive transcranial magnetic stimulation for the acute treatment of major depressive episodes: a systematic review with network meta-analysis. JAMA Psychiatry. 2017;74(2):143–52.
19. Zhang T, Zhu J, Xu L, Tang X, Cui H, Wei Y, Wang Y, Hu Q, Qian Z, Liu X, Tang Y, Li C, Wang J. Add-on rTMS for the acute treatment of depressive symptoms is probably more effective in adolescents than in adults: evidence from real-world clinical practice. Brain Stimul. 2019b;12(1):103–9.
20. Qiao Y, Wang J, Zhu J, Cui H, Tang Y, Xu L, Tang X, Wei Y, Hui L, Li C, Jia Q, Zhu H, Wen H, Wang J, Zhang T. Antidepressant effect of adjunct repetitive transcranial magnetic stimulation in inpatients 60 years and older. J ECT. 2020;36(3):216–21.
21. Zhang T, Sun W, Zhu J, Tang Y, Hui L, Zhou L, Xu J, Wang J. Effect of adjunct repetitive transcranial magnetic stimulation in elderly patients with acute depressive episode: supporting evidence from a real-world observation. Am J Geriatr Psychiatry. 2019a;27(1):91–2.
22. Lefaucheur JP, Aleman A, Baeken C, Benninger DH, Brunelin J, Di Lazzaro V, Filipović SR, Grefkes C, Hasan A, Hummel FC, Jääskeläinen SK, Langguth B, Leocani L, Londero A, Nardone R, Nguyen JP, Nyffeler T, Oliveira-Maia AJ, Oliviero A, Padberg F, Palm U, Paulus W, Poulet E, Quartarone A, Rachid F, Rektorová I, Rossi S, Sahlsten H, Schecklmann M, Szekely D, Ziemann U. Evidence-based guidelines on the therapeutic use of repetitive transcranial magnetic stimulation (rTMS): an update (2014-2018). Clin Neurophysiol. 2020;131(2):474–528.
23. Berlim MT, Van den Eynde F, Daskalakis ZJ. A systematic review and meta-analysis on the efficacy and acceptability of bilateral repetitive transcranial magnetic stimulation (rTMS) for treating major depression. Psychol Med. 2013;43(11):2245–54.
24. Chen JJ, Liu Z, Zhu D, Li Q, Zhang H, Huang H, Wei Y, Mu J, Yang D, Xie P. Bilateral vs. unilateral repetitive transcranial magnetic stimulation in treating major depression: a meta-analysis of randomized controlled trials. Psychiatry Res. 2014;219(1):51–7.
25. Blumberger DM, Mulsant BH, Fitzgerald PB, Rajji TK, Ravindran AV, Young LT, Levinson AJ, Daskalakis ZJ. A randomized double-blind sham-controlled comparison of unilateral and bilateral repetitive transcranial magnetic stimulation for treatment-resistant major depression. World J Biol Psychiatry. 2012;13(6):423–35.

26. Blumberger DM, Maller JJ, Thomson L, Mulsant BH, Rajji TK, Maher M, Brown PE, Downar J, Vila-Rodriguez F, Fitzgerald PB, Daskalakis ZJ. Unilateral and bilateral MRI-targeted repetitive transcranial magnetic stimulation for treatment-resistant depression: a randomized controlled study. J Psychiatry Neurosci. 2016;41(4):E58–66.
27. Conca A, Di Pauli J, Beraus W, Hausmann A, Peschina W, Schneider H, König P, Hinterhuber H. Combining high and low frequencies in rTMS antidepressive treatment: preliminary results. Hum Psychopharmacol. 2002;17(7):353–6.
28. Fitzgerald PB, Benitez J, de Castella A, Daskalakis ZJ, Brown TL, Kulkarni J. A randomized, controlled trial of sequential bilateral repetitive transcranial magnetic stimulation for treatment-resistant depression. Am J Psychiatry. 2006;163(1):88–94.
29. Fitzgerald PB, Hoy K, Gunewardene R, Slack C, Ibrahim S, Bailey M, Daskalakis ZJ. A randomized trial of unilateral and bilateral prefrontal cortex transcranial magnetic stimulation in treatment-resistant major depression. Psychol Med. 2011;41(6):1187–96.
30. Fitzgerald PB, Hoy KE, Herring SE, McQueen S, Peachey AV, Segrave RA, Maller J, Hall P, Daskalakis ZJ. A double blind randomized trial of unilateral left and bilateral prefrontal cortex transcranial magnetic stimulation in treatment resistant major depression. J Affect Disord. 2012;139(2):193–8.
31. Fitzgerald PB, Hoy KE, Singh A, Gunewardene R, Slack C, Ibrahim S, Hall PJ, Daskalakis ZJ. Equivalent beneficial effects of unilateral and bilateral prefrontal cortex transcranial magnetic stimulation in a large randomized trial in treatment-resistant major depression. Int J Neuropsychopharmacol. 2013;16(9):1975–84.
32. Garcia-Toro M, Salva J, Daumal J, Andres J, Romera M, Lafau O, Echevarría M, Mestre M, Bosch C, Collado C, Ibarra O, Aguirre I. High (20-Hz) and low (1-Hz) frequency transcranial magnetic stimulation as adjuvant treatment in medication-resistant depression. Psychiatry Res. 2006;146(1):53–7.
33. Loo CK, Mitchell PB, Croker VM, Malhi GS, Wen W, Gandevia SC, Sachdev PS. Double-blind controlled investigation of bilateral prefrontal transcranial magnetic stimulation for the treatment of resistant major depression. Psychol Med. 2003;33(1):33–40.
34. McDonald WM, Easley K, Byrd EH, Holtzheimer P, Tuohy S, Woodard JL, Beyer K, Epstein CM. Combination rapid transcranial magnetic stimulation in treatment refractory depression. Neuropsychiatr Dis Treat. 2006;2(1):85–94.
35. Pallanti S, Bernardi S, Di Rollo A, Antonini S, Quercioli L. Unilateral low frequency versus sequential bilateral repetitive transcranial magnetic stimulation: is simpler better for treatment of resistant depression? Neuroscience. 2010;167(2):323–8.
36. Rybak M, Bruno R, Turnier-Shea Y, Pridmore S. An attempt to increase the rate and magnitude of the antidepressant effect of transcranial magnetic stimulation (TMS). A pilot study. German journal of. Psychiatry. 2005;8(4)
37. Loo CK, Mitchell PB, McFarquhar TF, Malhi GS, Sachdev PS. A sham-controlled trial of the efficacy and safety of twice-daily rTMS in major depression. Psychol Med. 2007;37(3):341–9.
38. Sonmez AI, Camsari DD, Nandakumar AL, Voort JLV, Kung S, Lewis CP, Croarkin PE. Accelerated TMS for depression: a systematic review and meta-analysis. Psychiatry Res. 2019;273:770–81.
39. Dardenne A, Baeken C, Crunelle CL, Bervoets C, Matthys F, Herremans SC. Accelerated HF-rTMS in the elderly depressed: a feasibility study. Brain Stimul. 2018;11(1):247–8.
40. Fitzgerald PB, Hoy KE, Elliot D, Susan McQueen RN, Wambeek LE, Daskalakis ZJ. Accelerated repetitive transcranial magnetic stimulation in the treatment of depression. Neuropsychopharmacology. 2018;43(7):1565–72.
41. Roth Y, Zibman S, Pell GS, Zangen A, Tendler A. Revisiting the rotational field TMS method for Neurostimulation. J Clin Med. 2023;12(3)
42. Shanok NA, Rodriguez S, Muzac S, Huertas Del Pino C, Brown L, Rodriguez R. Deep transcranial magnetic stimulation alters resting-state neurophysiological traits in major depressive disorder. J Affect Disord. 2023;337:104–11.

43. Di Passa AM, Prokop-Millar S, Yaya H, Dabir M, McIntyre-Wood C, Fein A, MacKillop E, MacKillop J, Duarte D. Clinical efficacy of deep transcranial magnetic stimulation (dTMS) in psychiatric and cognitive disorders: a systematic review. J Psychiatr Res. 2024;175:287–315.
44. Harmelech T, Roth Y, Tendler A. Deep TMS H7 coil: features, applications & future. Expert Rev Med Devices. 2021;18(12):1133–44.
45. Levkovitz Y, Isserles M, Padberg F, Lisanby SH, Bystritsky A, Xia G, Tendler A, Daskalakis ZJ, Winston JL, Dannon P, Hafez HM, Reti IM, Morales OG, Schlaepfer TE, Hollander E, Berman JA, Husain MM, Sofer U, Stein A, Adler S, Deutsch L, Deutsch F, Roth Y, George MS, Zangen A. Efficacy and safety of deep transcranial magnetic stimulation for major depression: a prospective multicenter randomized controlled trial. World Psychiatry. 2015;14(1):64–73.
46. Berlim MT, McGirr A, Van den Eynde F, Fleck MP, Giacobbe P. Effectiveness and acceptability of deep brain stimulation (DBS) of the subgenual cingulate cortex for treatment-resistant depression: a systematic review and exploratory meta-analysis. J Affect Disord. 2014a;159:31–8.
47. Holtzheimer PE, Husain MM, Lisanby SH, Taylor SF, Whitworth LA, McClintock S, Slavin KV, Berman J, McKhann GM, Patil PG, Rittberg BR, Abosch A, Pandurangi AK, Holloway KL, Lam RW, Honey CR, Neimat JS, Henderson JM, DeBattista C, Rothschild AJ, Pilitsis JG, Espinoza RT, Petrides G, Mogilner AY, Matthews K, Peichel D, Gross RE, Hamani C, Lozano AM, Mayberg HS. Subcallosal cingulate deep brain stimulation for treatment-resistant depression: a multisite, randomised, sham-controlled trial. Lancet Psychiatry. 2017;4(11):839–49.
48. Fitzgerald PB, Segrave RA. Deep brain stimulation in mental health: review of evidence for clinical efficacy. Aust N Z J Psychiatry. 2015;49(11):979–93.
49. Yip AG, George MS, Tendler A, Roth Y, Zangen A, Carpenter LL. 61% of unmedicated treatment resistant depression patients who did not respond to acute TMS treatment responded after four weeks of twice weekly deep TMS in the Brainsway pivotal trial. Brain Stimul. 2017;10(4):847–9.
50. Filipcic I, Simunovic Filipcic I, Sucic S, Milovac Z, Geres N, Matic K, Celic-Ruzic M, Zecevic Penic S, Orgulan I, Pozgaj V, Bajic Z. A pilot investigation of accelerated deep transcranial magnetic stimulation protocols in treatment-resistant depression. Eur Arch Psychiatry Clin Neurosci. 2021;271(1):49–59.
51. Kisely S, Li A, Warren N, Siskind D. A systematic review and meta-analysis of deep brain stimulation for depression. Depress Anxiety. 2018;35(5):468–80.
52. Kaster TS, Daskalakis ZJ, Noda Y, Knyahnytska Y, Downar J. Efficacy, tolerability, and cognitive effects of deep transcranial magnetic stimulation for late-life depression: a prospective randomized controlled trial. Neuropsychopharmacology. 2018;43(11):2231–8.
53. Fitzgerald PB, Hoy K, McQueen S, Herring S, Segrave R, Been G, Kulkarni J, Daskalakis ZJ. Priming stimulation enhances the effectiveness of low-frequency right prefrontal cortex transcranial magnetic stimulation in major depression. J Clin Psychopharmacol. 2008;28(1):52–8.
54. Vidya KL, Srivastava S, Singh B, Kar SK. Investigating the impact of adjunctive priming repetitive transcranial magnetic stimulation in late-life depression: a pilot single-blind randomized control study. CNS Spectr. 2024;29(1):76–82.
55. George MS, Huffman S, Doose J, Sun X, Dancy M, Faller J, Li X, Yuan H, Goldman RI, Sajda P, Brown TR. EEG synchronized left prefrontal transcranial magnetic stimulation (TMS) for treatment resistant depression is feasible and produces an entrainment dependent clinical response: a randomized controlled double blind clinical trial. Brain Stimul. 2023;16(6):1753–63.
56. Jin Y, Phillips B. A pilot study of the use of EEG-based synchronized transcranial magnetic stimulation (sTMS) for treatment of major depression. BMC Psychiatry. 2014;14:13.
57. Cook IA, Wilson AC, Corlier J, Leuchter AF. Brain activity and clinical outcomes in adults with depression treated with synchronized transcranial magnetic stimulation: an exploratory study. Neuromodulation. 2019;22(8):894–7.

58. Leuchter AF, Cook IA, Feifel D, Goethe JW, Husain M, Carpenter LL, Thase ME, Krystal AD, Philip NS, Bhati MT, Burke WJ, Howland RH, Sheline YI, Aaronson ST, Iosifescu DV, O'Reardon JP, Gilmer WS, Jain R, Burgoyne KS, Phillips B, Manberg PJ, Massaro J, Hunter AM, Lisanby SH, George MS. Efficacy and safety of low-field synchronized transcranial magnetic stimulation (sTMS) for treatment of major depression. Brain Stimul. 2015;8(4):787–94.
59. Larson J, Wong D, Lynch G. Patterned stimulation at the theta frequency is optimal for the induction of hippocampal long-term potentiation. Brain Res. 1986;368(2):347–50.
60. Huang YZ, Chen RS, Rothwell JC, Wen HY. The after-effect of human theta burst stimulation is NMDA receptor dependent. Clin Neurophysiol. 2007;118(5):1028–32.
61. Cole E, O'Sullivan SJ, Tik M, Williams NR. Accelerated theta burst stimulation: safety, efficacy, and future advancements. Biol Psychiatry. 2024;95(6):523–35.
62. Blumberger DM, Vila-Rodriguez F, Thorpe KE, Feffer K, Noda Y, Giacobbe P, Knyahnytska Y, Kennedy SH, Lam RW, Daskalakis ZJ, Downar J. Effectiveness of theta burst versus high-frequency repetitive transcranial magnetic stimulation in patients with depression (THREE-D): a randomised non-inferiority trial. Lancet. 2018;391(10131):1683–92.
63. Cole EJ, Stimpson KH, Bentzley BS, Gulser M, Cherian K, Tischler C, Nejad R, Pankow H, Choi E, Aaron H, Espil FM, Pannu J, Xiao X, Duvio D, Solvason HB, Hawkins J, Guerra A, Jo B, Raj KS, Phillips AL, Barmak F, Bishop JH, Coetzee JP, DeBattista C, Keller J, Schatzberg AF, Sudheimer KD, Williams NR. Stanford accelerated intelligent Neuromodulation therapy for treatment-resistant depression. Am J Psychiatry. 2020;177(8):716–26.
64. Cole EJ, Phillips AL, Bentzley BS, Stimpson KH, Nejad R, Barmak F, Veerapal C, Khan N, Cherian K, Felber E, Brown R, Choi E, King S, Pankow H, Bishop JH, Azeez A, Coetzee J, Rapier R, Odenwald N, Carreon D, Hawkins J, Chang M, Keller J, Raj K, DeBattista C, Jo B, Espil FM, Schatzberg AF, Sudheimer KD, Williams NR. Stanford Neuromodulation therapy (SNT): a double-blind randomized controlled trial. Am J Psychiatry. 2022b;179(2):132–41.
65. Williams NR, Sudheimer KD, Bentzley BS, Pannu J, Stimpson KH, Duvio D, Cherian K, Hawkins J, Scherrer KH, Vyssoki B, DeSouza D, Raj KS, Keller J, Schatzberg AF. High-dose spaced theta-burst TMS as a rapid-acting antidepressant in highly refractory depression. Brain. 2018;141(3):e18.
66. Zhao H, Jiang C, Zhao M, Ye Y, Yu L, Li Y, Luan H, Zhang S, Xu P, Chen X, Pan F, Shang D, Hu X, Jin K, Chen J, Mou T, Hu S, Fitzgibbon BM, Fitzgerald PB, Cash RFH, Che X, Huang M. Comparisons of accelerated continuous and intermittent theta burst stimulation for treatment-resistant depression and suicidal ideation. Biol Psychiatry. 2024;96(1):26–33.
67. Cai DB, Qin ZJ, Lan XJ, Liu QM, Qin XD, Wang JJ, Goya-Maldonado R, Huang XB, Ungvari GS, Ng CH, Zheng W, Xiang YT. Accelerated intermittent theta burst stimulation for major depressive disorder or bipolar depression: a systematic review and meta-analysis. Asian J Psychiatr. 2023;85:103618.
68. Wilkening J, Witteler F, Goya-Maldonado R. Suicidality and relief of depressive symptoms with intermittent theta burst stimulation in a sham-controlled randomized clinical trial. Acta Psychiatr Scand. 2022;146(6):540–56.
69. Cole EJ, Phillips AL, Bentzley BS. Stanford Neuromodulation therapy (SNT): a double-blind randomized controlled trial. Am J Psychiatry. 2022a;179(2):132–41.
70. Duprat R, Desmyter S, Rudi de R, van Heeringen K, Van den Abbeele D, Tandt H, Bakic J, Pourtois G, Dedoncker J, Vervaet M, Van Autreve S, Lemmens GM, Baeken C. Accelerated intermittent theta burst stimulation treatment in medication-resistant major depression: a fast road to remission? J Affect Disord. 2016;200:6–14.
71. Cai DB, Qin XD, Qin ZJ, Lan XJ, Wang JJ, Ng CH, Zheng W, Xiang YT. Adjunctive continuous theta burst stimulation for major depressive disorder or bipolar depression: a meta-analysis of randomized controlled studies. J Affect Disord. 2024;346:266–72.
72. Chistyakov AV, Kreinin B, Marmor S, Kaplan B, Khatib A, Darawsheh N, Koren D, Zaaroor M, Klein E. Preliminary assessment of the therapeutic efficacy of continuous theta-burst magnetic stimulation (cTBS) in major depression: a double-blind sham-controlled study. J Affect Disord. 2015;170:225–9.

73. Li CT, Chen MH, Juan CH, Huang HH, Chen LF, Hsieh JC, Tu PC, Bai YM, Tsai SJ, Lee YC, Su TP. Efficacy of prefrontal theta-burst stimulation in refractory depression: a randomized sham-controlled study. Brain. 2014;137(Pt 7):2088–98.
74. Chistyakov AV, Rubicsek O, Kaplan B, Zaaroor M, Klein E. Safety, tolerability and preliminary evidence for antidepressant efficacy of theta-burst transcranial magnetic stimulation in patients with major depression. Int J Neuropsychopharmacol. 2010;13(3):387–93.
75. Qin ZJ, Huang SQ, Lan XJ, Shi ZM, Huang XB, Ungvari GS, Jackson T, Zheng W, Xiang YT. Bilateral theta burst stimulation for patients with acute unipolar or bipolar depressive episodes: a systematic review of randomized controlled studies. J Affect Disord. 2023;340:575–82.
76. Prasser J, Schecklmann M, Poeppl TB, Frank E, Kreuzer PM, Hajak G, Rupprecht R, Landgrebe M, Langguth B. Bilateral prefrontal rTMS and theta burst TMS as an add-on treatment for depression: a randomized placebo controlled trial. World J Biol Psychiatry. 2015;16(1):57–65.
77. Tang VM, Blumberger DM, Throop A, McClintock SM, Voineskos D, Downar J, Knyahnytska Y, Mulsant BH, Fitzgerald PB, Daskalakis ZJ. Continuation magnetic seizure therapy for treatment-resistant unipolar or bipolar depression. J Clin Psychiatry. 2021;82(6)
78. El-Deeb FA, Gad EA, Kandeel AA, Haiba AA, Fathy SM, Peterson MM, Youssef NA. Comparative effectiveness clinical trial of magnetic seizure therapy and electroconvulsive therapy in major depressive disorder. Ann Clin Psychiatry. 2020;32(4):239–48.
79. Li J, Zhang X, Jiang J, Zhang B, Tang Y, Zhang T, Jia Y, Li Q, Xia M, Sheng J, Li C, Wang J. Comparison of electroconvulsive therapy and magnetic seizure therapy in schizophrenia: structural changes/neuroplasticity. Psychiatry Res. 2022;312:114523.
80. Cretaz E, Brunoni AR, Lafer B. Magnetic seizure therapy for unipolar and bipolar depression: a systematic review. Neural Plast. 2015;2015:521398.
81. Daskalakis ZJ, Dimitrova J, McClintock SM, Sun Y, Voineskos D, Rajji TK, Goldbloom DS, Wong AHC, Knyahnytska Y, Mulsant BH, Downar J, Fitzgerald PB, Blumberger DM. Magnetic seizure therapy (MST) for major depressive disorder. Neuropsychopharmacology. 2020;45(2):276–82.
82. Deng ZD, Luber B, McClintock SM, Weiner RD, Husain MM, Lisanby SH. Clinical outcomes of magnetic seizure therapy vs electroconvulsive therapy for major depressive episode: a randomized clinical trial. JAMA Psychiatry. 2024;81(3):240–9.
83. Wang J, Vila-Rodriguez F, Ge R, Gao S, Gregory E, Jiang W, Yang C, Wang G. Accelerated magnetic seizure therapy (aMST) for treatment of major depressive disorder: a pilot study. J Affect Disord. 2020;264:215–20.
84. Weissman CR, Blumberger DM, Dimitrova J, Throop A, Voineskos D, Downar J, Mulsant BH, Rajji TK, Fitzgerald PB, Daskalakis ZJ. Magnetic seizure therapy for suicidality in treatment-resistant depression. JAMA Netw Open. 2020;3(8):e207434.
85. Brunoni AR, Moffa AH, Fregni F, Palm U, Padberg F, Blumberger DM, Daskalakis ZJ, Bennabi D, Haffen E, Alonzo A, Loo CK. Transcranial direct current stimulation for acute major depressive episodes: meta-analysis of individual patient data. Br J Psychiatry. 2016;208(6):522–31.
86. Nikolin S, Moffa A, Razza L, Martin D, Brunoni A, Palm U, Padberg F, Bennabi D, Haffen E, Blumberger DM, Salehinejad MA, Loo CK. Time-course of the tDCS antidepressant effect: an individual participant data meta-analysis. Prog Neuro-Psychopharmacol Biol Psychiatry. 2023;125:110752.
87. Moffa AH, Brunoni AR, Fregni F, Palm U, Padberg F, Blumberger DM, Daskalakis ZJ, Bennabi D, Haffen E, Alonzo A, Loo CK. Safety and acceptability of transcranial direct current stimulation for the acute treatment of major depressive episodes: analysis of individual patient data. J Affect Disord. 2017;221:1–5.
88. Jaiswal A, Goyal N, Shreekantiah U. High-definition transcranial direct current stimulation-primed intermittent theta burst stimulation in treatment-resistant depression: a controlled study. J ECT. 2024;40(1):41–6.

89. Jog MS, Kim E, Anderson C, Kubicki A, Kayathi R, Jann K, Yan L, Leaver A, Hellemann G, Iacoboni M. In-vivo imaging of targeting and modulation of depression-relevant circuitry by transcranial direct current stimulation: a randomized clinical trial. Transl Psychiatry. 2021;11(1):138.
90. Zhou D, Li X, Wei S, Yu C, Wang D, Li Y, Li J, Liu J, Li S, Zhuang W, Li Y, Luo R, Liu Z, Liu J, Xu Y, Fan J, Zhu G, Xu W, Tang Y, Cho RY, Kosten TR, Zhang XY. Transcranial direct current stimulation combined with repetitive transcranial magnetic stimulation for depression: a randomized clinical trial. JAMA Netw Open. 2024;7(11):e2444306.
91. Li X, Liu J, Wei S, Yu C, Wang D, Li Y, Li J, Zhuang W, Luo RC, Li Y, Liu Z, Su Y, Liu J, Xu Y, Fan J, Zhu G, Xu W, Tang Y, Yan H, Cho RY, Kosten TR, Zhou D, Zhang X. Cognitive enhancing effect of rTMS combined with tDCS in patients with major depressive disorder: a double-blind, randomized, sham-controlled study. BMC Med. 2024;22(1):253.

Bipolar Disorder

4

Xinhu Yang, Xingbing Huang, Zhenjuan Qin, Yuping Ning, and Wei Zheng

Abstract

Bipolar disorder constitutes a substantial source of personal distress and economic burden, as evidenced by the Global Burden of Disease Study. Conventional therapeutic strategies, including pharmacotherapy and psychotherapy, are endorsed by clinical guidelines and expert consensus. However, a significant proportion of individuals with bipolar disorder exhibit inadequate treatment responses or develop treatment resistance, necessitating alternative or adjunctive interventions. Over the past two decades, various neuromodulatory techniques, such as transcranial magnetic stimulation (TMS) and transcranial direct current stimulation (tDCS), have been investigated to address the complexities of bipolar disorder, leading to increased interest in their therapeutic applications. This chapter provides an overview of advanced TMS and tDCS paradigms in the context of bipolar disorder.

Keywords

Bipolar disorder · Transcranial magnetic stimulation · Transcranial direct current stimulation · Efficacy · Safety

---

Xinhu Yang and Xingbing Huang have equally contributed to this chapter.

X. Yang · X. Huang · Y. Ning (✉) · W. Zheng (✉)
The Affiliated Brain Hospital, Guangzhou Medical University, Guangzhou, China

Key Laboratory of Neurogenetics and Channelopathies of Guangdong Province and the Ministry of Education of China, Guangzhou Medical University, Guangzhou, China

Z. Qin
The Brain Hospital of Guangxi Zhuang Autonomous Region, Liuzhou, China

W. Zheng, Y. Ning (eds.), *TMS and tDCS for Psychiatric Disorders*,
https://doi.org/10.1007/978-981-96-8504-2_4

Abbreviations

BDI-II	the Beck Depression Inventory-II
CANMAT	the Canadian Network for Mood and Anxiety Treatments
CANTAB	the Cambridge Neuropsychological Test Automated Battery
cTBS	continuous theta burst stimulation
DLPFC	dorsolateral prefrontal cortex
DSM-V	Statistical Manual of Mental Disorders, Fifth Edition
dTMS	deep transcranial magnetic stimulation
ECT	electroconvulsive therapy
EEG	electroencephalogram
FDA	The Food and Drug Administration
fMRI	functional magnetic resonance imaging
GDNF	glial cell-derived neurotrophic factor
HDRS	Hamilton Depression Rating Scale
HD-tDCS	high-definition tDCS
HF	high frequency
ISBD	the International Society for Bipolar Disorders
iTBS	intermittent theta burst stimulation
LF	low frequency
MADRS	Montgomery-Åsberg Depression Rating Scale
MDAUD	major depressive disorder with comorbid alcohol use disorder
MDD	major depressive disorder
MDE	major depressive episodes
MRI	magnetic resonance imaging
MST	magnetic seizure therapy
MT	the motor threshold
mPFC-dACC	the medial prefrontal cortex–dorsal anterior cingulate cortex
NIBS	noninvasive brain stimulation
OR	odds ratio
RCT	randomized controlled trial
RMT	the resting motor threshold
rTMS	repetitive transcranial magnetic stimulation
SAINT	The Stanford Accelerated Intelligent Neuromodulation Therapy
sgACC	subgenual anterior cingulate cortex
SMD	standardized mean difference
TBS	theta burst stimulation
tDCS	transcranial direct current stimulation
TMS	transcranial magnetic stimulation
TRD	treatment-resistant depression

4.1 Introduction

Bipolar disorder represents a complex group of chronic psychiatric conditions, with bipolar I disorder characterized by manic episodes and bipolar II disorder defined by hypomanic episodes accompanied by major depressive episodes (MDE), as specified in the Diagnostic and Statistical Manual of Mental Disorders, Fifth Edition (DSM-V) [1]. These conditions are further delineated by distinct mood episodes, including manic, depressive, and mixed states [2]. According to the Global Burden of Disease Study [3], bipolar disorder affects approximately 39.5 million individuals worldwide, with prevalence rates varying significantly across regions and demographic groups. The disorder results in profound alterations in mood, cognition, and behavior, often severely impairing social functioning and overall quality of life. For instance, individuals with bipolar depression frequently experience pervasive sadness, loss of interest in activities, and disruptions in sleep and appetite, which adversely impact their personal well-being and ability to participate in occupational and social activities. Moreover, bipolar disorder significantly increases the risk of suicidal behavior, posing a critical public health challenge [4].

The management of bipolar disorder involves diverse therapeutic strategies aimed at mitigating symptoms across manic and depressive episodes. Evidence-based consensus statements and treatment guidelines recommend the use of mood stabilizers, such as lithium and valproate, often in combination with atypical antipsychotics, as first-line pharmacological interventions [1]. Psychosocial interventions, such as cognitive-behavioral therapy, play an essential adjunctive role, with studies demonstrating their capacity to enhance health outcomes for patients with bipolar disorder [5]. Despite these interventions, 60.5–75.7% of patients fail to achieve satisfactory outcomes [6], underscoring the urgent need for innovative treatments, including noninvasive brain stimulation techniques to address these therapeutic gaps.

Among emerging approaches, noninvasive brain stimulation techniques such as transcranial magnetic stimulation (TMS) and transcranial direct current stimulation (tDCS) have garnered increasing attention for their potential in bipolar disorder treatment. A recent meta-analysis highlights growing evidence supporting the efficacy of TMS in bipolar depression [7], although its benefits for hypomania, mania, and mixed states remain limited [8]. Similarly, a meta-analysis of tDCS suggests moderate improvement in depressive symptoms (standardized mean difference (SMD) = 0.71) in bipolar depression [9]. However, these findings remain preliminary, necessitating further research to refine these techniques and develop more effective treatments. This chapter systematically explores the efficacy and safety of advanced repetitive TMS (rTMS) paradigms and tDCS protocols for bipolar disorder. A comprehensive search of English-language databases, including PubMed, Cochrane Library, EMBASE, and PsycINFO, was conducted from inception to November 24, 2024. The analysis encompasses various rTMS modalities—unilateral rTMS, bilateral rTMS, accelerated rTMS, deep TMS, priming TMS, synchronized TMS, theta burst stimulation (TBS) such as intermittent TBS (iTBS), accelerated iTBS, continuous TBS, and bilateral TBS—and magnetic seizure therapy (MST), as well as tDCS paradigms, including conventional tDCS and high-definition tDCS (HD-tDCS). This chapter provides an in-depth review of these technologies, assessing their therapeutic efficacy, safety, treatment regimens, and clinical applications, alongside recommendations for optimizing their use in managing this multifaceted disorder.

4.2 TMS

4.2.1 rTMS

4.2.1.1 Unilateral rTMS

Efficacy

Unilateral rTMS is a noninvasive brain stimulation modality designed to selectively target a single cerebral hemisphere, widely utilized in clinical settings. This technique employs electromagnetic induction to generate localized electrical currents within the cerebral cortex, thereby modulating neuronal membrane potentials and altering electrical activity and metabolic functions in targeted brain regions [10]. The United States Food and Drug Administration (FDA) has approved TMS as a therapeutic option for treatment-resistant depression (TRD), providing an alternative for patients unresponsive to conventional pharmacological interventions [11]. Meta-analyses have consistently validated the efficacy of unilateral rTMS in managing bipolar disorder [12–14]. For instance, a systematic review and meta-analysis of randomized controlled trials (RCTs) [14] demonstrated that high-frequency (HF) rTMS over the left dorsolateral prefrontal cortex (DLPFC) significantly improved response rates compared to sham stimulation (50.70% vs. 28.07%, $P = 0.02$) in patients with bipolar depression. Similarly, low-frequency (LF) rTMS over the right DLPFC showed a higher, albeit statistically marginal, response rate (92.11% vs. 69.44%, $P = 0.06$). However, due to insufficient data, the remission rate for bipolar mania remains indeterminate [13]. Furthermore, unilateral rTMS has been associated with reductions in the frequency and duration of depressive episodes, alongside improved mood stability in individuals with bipolar disorder [14]. These findings highlight the variability in treatment outcomes, necessitating individualized stimulation parameters and site selection to optimize therapeutic efficacy.

Safety

Unilateral rTMS is generally regarded as a safe intervention for bipolar disorder [15], especially when compared to other neuromodulation techniques such as electroconvulsive therapy (ECT), which is associated with a higher incidence of adverse effects. Commonly reported side effects of unilateral rTMS include headache, scalp discomfort, and mild muscle twitching [16]. A meta-analysis revealed that transient headaches were reported by 15.4–28.6% of patients undergoing this treatment [13]. Although rare, serious complications such as seizures and intracranial bleeding have been documented, emphasizing the importance of comprehensive patient assessments prior to initiation [17]. Given the potential variability in patient response and safety concerns, careful evaluation and monitoring remain critical for ensuring the safe and effective application of unilateral rTMS in bipolar disorder management.

Treatment Regimen

In clinical practice, the application of unilateral rTMS for bipolar I and II disorders primarily focuses on managing bipolar depression, with two predominant perspectives regarding therapeutic parameters. The first supports HF rTMS applied to the DLPFC as an effective strategy for reducing depressive symptoms in both bipolar I and II disorders. A systematic review has demonstrated the efficacy of HF rTMS in alleviating the severity of bipolar depression [14]. The second perspective, although supported by more limited evidence, advocates for LF rTMS, such as 1 Hz, targeting the right DLPFC for bipolar depression [14].

A critical consideration in the treatment of bipolar disorder is the potential risk of inducing mania. Existing studies suggest that rTMS does not significantly increase the risk of mania when compared to sham treatments [18]. However, further investigation into optimal rTMS parameters specific to bipolar disorder is necessary to enhance both safety and efficacy.

Clinical Recommendations

The Canadian Network for Mood and Anxiety Treatments (CANMAT) and the International Society for Bipolar Disorders (ISBD) Guidelines [5] offer comprehensive recommendations for managing bipolar disorder, categorizing therapeutic approaches into first-line, second-line, and third-line options, alongside those that are not recommended. Within this framework, rTMS is designated as a third-line treatment for both acute bipolar I depression and acute mania (Table 4.1). The definitions of evidence levels and treatment line ratings, as outlined by the CANMAT and ISBD Guidelines, are provided in Table 4.2 [5]. In contrast, the Chinese guidelines for the prevention and treatment of bipolar disorder do not include recommendations for the use of rTMS [19].

4.2.1.2 Bilateral rTMS

Efficacy

Bilateral rTMS has emerged as a promising therapeutic approach for bipolar disorder, although its efficacy remains mixed. Most research focuses on the depressive phase, with a network meta-analysis showing that bilateral rTMS had greater efficacy than sham therapy for acute treatment of major depressive disorder (MDD) or bipolar depression (summary odds ratio (OR) = 4.92). However, a systematic review [14] found no significant difference in response rates between bilateral rTMS and sham treatment (19.44% vs. 11.11%, $P = 0.32$). Additional systematic reviews [20] highlight the limited data on rTMS efficacy for other phases of bipolar disorder. These findings emphasize the need for further sham-controlled studies and

Table 4.1 TMS in bipolar disorder

TMS	Clinical Recommendations				Episode: Levels + References
	Target	Frequency	Length (min)	Duration (sessions)	
Unilateral rTMS	Unclear	Unclear	Unclear	Unclear	Depressive episode: Third-Line[a] [5]
	Unclear	Unclear	Unclear	Unclear	Mania episode: Third-Line[a] [5]
Bilateral rTMS	Unclear	Unclear	Unclear	Unclear	Depressive episode: Third-Line[a] [5]
	Unclear	Unclear	Unclear	Unclear	Mania episode: Third-Line[a] [5]
Accelerated rTMS	Unclear	Unclear	Unclear	Unclear	Both episode[b] [31]
Deep TMS	Unclear	Unclear	Unclear	Unclear	Both episode[b] [31]
Priming TMS	Unclear	Unclear	Unclear	Unclear	Both episode[b] [31]
Synchronized TMS	Unclear	Unclear	Unclear	Unclear	Both episode[b] [44]
iTBS	Unclear	Unclear	Unclear	Unclear	Both episode[b] [31]
Accelerated iTBS	Unclear	Unclear	Unclear	Unclear	Both episode[b] [31]
Continuous TBS	Unclear	Unclear	Unclear	Unclear	Both episode[b] [31]
Bilateral TBS	Unclear	Unclear	Unclear	Unclear	Both episode[b] [31]
MST	Unclear	Unclear	Unclear	Unclear	Both episode[b] [8]

Abbreviations: *iTBS* intermittent theta burst stimulation, *MST* magnetic seizure therapy, *rTMS* repetitive transcranial magnetic stimulation, *TBS* theta burst stimulation, *TMS* transcranial magnetic stimulation

[a]The CANMAT and ISBD Guidelines mention adjunctive rTMS as third-line treatment only for acute bipolar I depression and acute mania

[b]Both episodes including depressive episode and mania episode

large-scale clinical trials to determine the broader applicability of rTMS, including its potential to manage manic and mixed episodes, stabilize mood, and reduce suicidal behaviors.

Cognitive dysfunction is increasingly recognized as a critical target for rTMS interventions in bipolar disorder. Cognitive impairments are common among patients with bipolar depression. A study involving 64 patients with bipolar disorder assessed the impact of 20 sessions of bilateral rTMS (10 Hz on the left DLPFC and 1 Hz on the right DLPFC) on depressive symptoms and cognitive functions, utilizing the Beck Depression Inventory-II (BDI-II) and the Cambridge Neuropsychological Test Automated Battery (CANTAB) [21]. Results indicated that bilateral rTMS improved depressive symptoms and enhanced cognitive functions, including sustained attention, working memory, and executive functions. These effects are thought to stem from the modulation of cognitive control mechanisms (top-down processing) and the amelioration of negative emotional biases.

Safety

The adverse effects of bilateral rTMS include sensations such as scalp tingling, neck soreness, and a burning sensation, along with potential auditory disturbances [22]. These side effects are generally transient and tend to resolve upon discontinuation

Table 4.2 Definitions of level of evidence and line of treatment ratings according to the 2018 CANMAT/ISBD guidelines[a]

Level of evidence ratings	
Level	Evidence
1	Meta-analysis with narrow confidence interval or replicated double-blind (DB) randomized controlled trial (RCT) that includes a placebo or active control comparison ($N \geq 30$ in each active treatment arm)
2	Meta-analysis with wide confidence interval or one DB RCT with placebo or active control comparison condition ($N \geq 30$ in each active treatment arm)
3	At least one DB RCT with placebo or active control comparison condition ($N = 10–29$ in each active treatment arm) or health system administrative data
4	Uncontrolled trial, anecdotal reports, or expert opinion
Line of treatment ratings	
First	Level 1 or level 2 evidence for efficacy plus clinical support for safety/tolerability, and no risk of treatment-emergent switch[b]
Second	Level 3 or higher evidence for efficacy plus clinical support for safety and/or tolerability and low risk of treatment-emergent switch[b]
Third	Level 4 evidence or higher for efficacy plus clinical support for safety/tolerability
Not recommended	Level 1 evidence for lack of efficacy or level 2 evidence for lack of efficacy plus expert opinion

CANMAT Canadian Network for Mood and Anxiety Treatments, *ISBD* International Society for Bipolar Disorders

[a]This chapter is adapted from the study by Keramatian et al. [5]

[b]The text specifically notes when lack of clinical support for safety and/or tolerability, or risk of treatment-emergent switch, has affected recommendations

of the treatment. rTMS is contraindicated in patients with metallic implants or intracranial conductive devices, and caution is warranted when administering the therapy to individuals taking medications that lower the seizure threshold. Expert guidelines [23] estimate the incidence of rTMS-induced seizures at stimulation frequencies above 1 Hz to be less than 1%.

The risk of seizures associated with rTMS is influenced by the stimulation frequency and the total number of pulses delivered. Although rare cases of rTMS-induced status epilepticus have been reported, the treatment does not appear to increase the long-term risk of seizures in patients [24]. Careful assessment of the potential risks and benefits is essential when considering rTMS as a therapeutic option, ensuring patient safety while maximizing therapeutic efficacy.

Treatment Regimen

Bilateral rTMS is predominantly utilized in the treatment of bipolar depression rather than bipolar mania. This approach combines LF stimulation of the DLPFC with HF stimulation of the left DLPFC within a single treatment session [14]. Evidence from a recent study [25] indicates that bilateral stimulation demonstrates greater efficacy compared to unilateral protocols. In this regimen, the right DLPFC is stimulated at

1 Hz for 10 s with 2-s intervals, delivering 150 trains and 1500 pulses per session over 20 sessions, culminating in a total of 30,000 pulses. Following this, the left DLPFC is stimulated at 10 Hz for 5 s with 10-s intervals between pulses, resulting in 75 trains (3750 pulses per session). The stimulation intensity for the right DLPFC is set at 120% of the resting motor threshold (RMT), while the left DLPFC is stimulated at 100% RMT to enhance tolerability. Treatments are typically conducted once daily over 4–6 weeks [14]. Conversely, current evidence suggests that unilateral stimulation may be more suitable for patients experiencing mania [26].

Clinical Recommendations

The CANMAT and ISBD guidelines classify rTMS as a third-line treatment for acute bipolar I depression and acute mania [5]. However, the Chinese guidelines for the prevention and treatment of bipolar disorder do not include recommendations for rTMS application [19]. This divergence highlights regional differences in clinical practice and perspectives on the role of rTMS in bipolar disorder management (Table 4.1), underscoring the need for more globally unified treatment protocols informed by robust evidence.

4.2.1.3 Accelerated rTMS

Efficacy

Increasing the frequency of rTMS sessions to more than once daily, known as the accelerated rTMS protocol, seeks to enhance antidepressant efficacy while reducing the overall treatment duration, thereby alleviating the time burden on both patients and clinicians [27]. Although no targeted meta-analysis of accelerated rTMS for bipolar disorder currently exists, several studies suggest its potential effectiveness. In a real-world clinical setting, Massé-Leblanc et al. [28] evaluated 247 adults with treatment-resistant unipolar or bipolar depressive episodes (bipolar I or II disorder) diagnosed according to DSM-IV or DSM-V criteria. Posttreatment analysis revealed that 46.3% of patients met response criteria and 36.1% achieved remission based on the Montgomery-Åsberg Depression Rating Scale (MADRS). Similarly, George et al. [29] examined 41 patients experiencing depressive episodes and suicidal ideation, reporting a more rapid reduction in suicide ideation scores in the active stimulation group.

Safety

The accelerated rTMS protocol has demonstrated safety and tolerability in patients with depression [30], including older adults [31]. Common adverse effects observed in real-world clinical applications include headaches (37.1%), stimulation site pain (32.2%), and fatigue (22.4%) [28]. While generally well-tolerated, patient responses and potential side effects require careful monitoring to ensure safety and optimize treatment outcomes.

Treatment Regimen

In clinical applications, the protocol described by Massé-Leblanc et al. [28] utilized a figure-of-eight coil for TMS therapy. This included a unilateral 20 Hz stimulation protocol targeting the left DLPFC or a bilateral approach combining 20 Hz stimulation of the left DLPFC with continuous theta burst stimulation (cTBS) inhibiting the right DLPFC. In contrast, George et al. [29] implemented accelerated rTMS using a figure-eight solid core coil applied to the left prefrontal cortex at 120% of the motor threshold (MT). This protocol involved 10 Hz stimulation, a 5-s train duration, and a 10-s intertrain interval, delivering 6000 pulses per session over three sessions daily for 3 days, amounting to nine sessions and 54,000 stimulations in total.

Clinical Recommendations

Current research focuses primarily on the depressive phase of bipolar disorder, for which specific grade recommendations are lacking [31] (Table 4.1). There is insufficient evidence to extend recommendations to the treatment of manic or mixed states, highlighting the need for further clinical studies and validation. Clinicians are encouraged to adopt personalized treatment plans, tailoring rTMS protocols to individual patient needs and therapeutic goals.

4.2.2 Deep TMS

4.2.2.1 Efficacy

Deep TMS (dTMS), an advanced TMS modality, utilizes the H7 coil to target the medial prefrontal cortex–dorsal anterior cingulate cortex (mPFC-dACC) and other cortical regions with suprathreshold stimulation [32]. Compared to conventional TMS coils, the H-coil, integrated into a wearable helmet, offers broader and deeper neural stimulation [33]. In a clinical trial [34], patients with bipolar depression underwent 20 sessions of either active or sham dTMS targeting the left DLPFC with an H1 coil at 18 Hz, using 2-s trains at 120% of the MT. Active dTMS showed significantly greater improvements than sham treatment at the endpoint, although the benefit was not sustained at follow-up. The active group exhibited a higher response rate (48%) than the sham group (24%), with an OR of 2.92 ($P = 0.08$), but remission rates were comparable between the groups. Notably, no treatment-emergent mania switch episodes were reported.

In contrast, applying HF dTMS to the bilateral prefrontal cortex in patients with depressive episodes of bipolar disorder or MDD did not yield significant improvements in depressive symptoms [34]. However, another sham-controlled study [34] demonstrated short-term efficacy of dTMS targeting the left DLPFC in bipolar depressive episodes. Active dTMS significantly reduced Hamilton Depression Rating Scale (HDRS) scores compared to sham from baseline to week 4 ($P = 0.03$) and week 6 ($P = 0.02$), though no differences were observed at follow-up (week 8). Response rates at week 4 also showed a trend favoring the active treatment group (48%) over the sham group (24%, $P = 0.08$). These findings align with research by McGirr et al. [35], which demonstrated the superior efficacy of traditional figure-eight coil rTMS targeting the right DLPFC in acute bipolar depression compared to sham. Conversely, other studies have reported inconsistent results [36]. One study reported no significant clinical effects of figure-eight coil rTMS on the left DLPFC in patients with bipolar disorder [36, 37]. Similarly, Fitzgerald et al. [37] found no evidence to support the superior efficacy of figure-eight coil rTMS over sham treatment in patients with bipolar disorder. Given these inconsistencies and the limited number of dTMS trials for bipolar disorder, further research is necessary to establish its clinical efficacy. Specifically, studies should focus on whether H1-coil-induced DLPFC stimulation provides superior outcomes to sham.

4.2.2.2 Safety

In 2013, the United States FDA approved the dTMS protocol, utilizing H1 coils, for the treatment of MDD [38]. This regulatory milestone highlights the potential of Deep TMS as a well-tolerated adjunctive therapy for patients with treatment-resistant bipolar depression undergoing adequate pharmacotherapy [34]. Notably, the study by Tavares et al. reported a higher incidence of scalp pain in the active treatment groups (20%) compared to the sham groups (0%) ($P = 0.05$). Other adverse events, including headache, neck pain, burning sensations, hearing disturbances, and concentration difficulties, were observed at comparable rates between the two groups without significant differences. Additionally, research conducted by Rapinesi et al. [39] evaluated the safety of dTMS in a cohort of 82 patients experiencing various depressive episodes, encompassing MDE within MDD, bipolar I disorder, and MDD with comorbid alcohol use disorder (MDAUD). The absence of complications or adverse effects (0%) across all participants underscores the safety and broad applicability of this therapeutic modality in managing diverse depressive conditions.

4.2.2.3 Treatment Regimen

Rapinesi et al. [39] further demonstrated the efficacy of dTMS targeting the DLPFC at an intensity of 120% of the MT, delivered unilaterally or bilaterally (preferably on the left side) using an H1 coil. Stimulation was applied at a frequency of 20 Hz to maximize neural engagement, with each session comprising multiple 2-s pulse trains interspersed with 20-s intervals to facilitate neural recovery. This protocol was administered daily over a 4-week period, resulting in a comprehensive treatment regimen totaling 20 sessions.

4.2.2.4 Clinical Recommendations

Systematic reviews [38, 40] have suggested that dTMS may serve as an effective intervention for bipolar depression and, potentially, for TRD, positioning it as a promising alternative therapeutic option. However, the current body of clinical trial data remains insufficient [31] (Table 4.1), particularly in the contexts of bipolar depression and bipolar mania. This evidentiary gap underscores the necessity for more extensive research to evaluate the long-term durability of dTMS, its therapeutic efficacy, and potential predictors of treatment response.

4.2.3 Priming TMS

4.2.3.1 Efficacy

Priming TMS introduces an innovative approach within the domain of rTMS by utilizing subthreshold HF stimulation to enhance the efficacy of subsequent LF stimulation trains, thereby amplifying neural responsiveness to LF stimulation. Evidence from a network meta-analysis [40] supports the superior efficacy of priming TMS (effect size: 6.02) compared to sham therapy in the acute treatment of MDE in adults. In a clinical trial [41] evaluating sequential bilateral rTMS versus right-sided unilateral rTMS with a priming protocol, response rates were 57% and 55%, respectively ($P = 0.80$). Similarly, remission rates were nearly identical, with 40.2% in the bilateral group and 40.1% in the priming group ($P = 0.90$), demonstrating no significant differences between these modalities. Additional studies suggest that priming TMS achieves higher response rates compared to cTBS, further highlighting its therapeutic potential in managing depression [42].

4.2.3.2 Safety

Priming, a technique that enhances the neural response to rTMS through a preliminary phase of low-intensity HF stimulation, remains an underexplored area. However, initial studies suggest that priming TMS demonstrates a favorable safety profile and excellent tolerability, supporting its potential as a viable and safe therapeutic option [40, 42].

4.2.3.3 Treatment Regimen

Priming TMS remains an area of ongoing exploration within the therapeutic framework for bipolar disorder. A study by Fitzgerald et al. [41] involving 179 participants, including individuals with bipolar I disorder in a depressive episode ($n = 27$) and bipolar II disorder in a depressive episode ($n = 13$), has provided valuable insights into the potential efficacy of this approach. The treatment protocol began with a priming HF stimulation phase at 6 Hz, delivered at 90% of RMT intensity, and administered over 20 cycles, each lasting 5 s. This was followed by an LF phase, consisting of 1 Hz stimulation at 110% RMT intensity, continuously delivered for a total of 900 pulses. The treatment course extended up to 4 weeks, with five sessions per week, resulting in a total of 20 sessions. The stimulation site was precisely located 6 cm anteriorly to the optimal scalp location for eliciting maximum thumb abductor muscle response.

4.2.3.4 Clinical Recommendations

Although the clinical evidence supporting priming TMS is still evolving, it represents a promising strategy for enhancing the efficacy of rTMS. This approach holds significant potential to augment the clinical utility of rTMS, particularly in patients exhibiting insufficient responses to conventional therapies. However, the current data on its application for bipolar disorder, encompassing both manic and depressive episodes, are insufficient to offer definitive recommendations for priming TMS use [31] (Table 4.1). Consequently, future research should focus on optimizing rTMS parameters, including frequency, intensity, and session duration, to refine treatment efficacy. Additionally, exploring the synergistic effects of combining rTMS with other therapeutic modalities, such as pharmacological treatments, is essential to determine the most effective integrated approach.

4.2.4 Synchronized TMS

4.2.4.1 Efficacy

Synchronized TMS is a brain stimulation technique that utilizes rotating spherical neodymium magnets, synchronized with an individual's alpha wave frequency, to create a global magnetic field distribution along the midline of the scalp [40]. The underlying mechanism involves the modulation of oscillatory activity through the programmed frequency of synchronized stimulation, which resets thalamocortical oscillators and restores endogenous oscillatory rhythms [43]. A recent review found no compelling evidence to support the efficacy of synchronized TMS in the treatment of MDE, with a notable scarcity of studies specifically investigating its application in bipolar disorder [44]. Further investigation in this area is warranted.

4.2.4.2 Safety

Evidence suggests that synchronized TMS is associated with a reduced incidence of adverse effects compared to conventional rTMS, likely due to the absence of neural depolarization [40, 45]. Additionally, synchronized TMS consumes less energy than traditional rTMS, potentially leading to reduced treatment costs and enhanced safety profiles [46].

4.2.4.3 Treatment Regimen

Current research is predominantly focused on unipolar depression, with the field still in its early stages. The limited number of studies and small sample sizes underscore the need for further exploration. Notably, no studies have yet examined the application of synchronized TMS in bipolar disorder [45].

4.2.4.4 Clinical Recommendations

As of now, no clinical recommendations exist regarding the use of synchronized TMS for the treatment of bipolar disorder, including both depressive and manic episodes [44] (Table 4.1).

4.2.5 TBS

4.2.5.1 iTBS

Efficacy

A meta-analysis has demonstrated that both iTBS and HF-rTMS exhibit comparable efficacy, acceptability, and safety profiles for the treatment of treatment-resistant MDD. However, a randomized controlled trial by McGirr et al. [47] found that iTBS targeting the left DLPFC was ineffective for patients with acute bipolar depression. Preliminary evidence suggests that the mechanism of iTBS, when aimed at the frontal lobe, may be associated with physiological changes in the mPFC of patients with acute bipolar depression [48]. Despite these findings, no dedicated reviews or meta-analyses have specifically addressed the relationship between iTBS and bipolar disorder, leaving the effectiveness of iTBS in this context poorly defined due to limited research.

Safety

The safety profile of iTBS has become a central focus in neuromodulation research. Current clinical evidence supports that iTBS exhibits a safety profile comparable to conventional rTMS, with no reports of seizures or significant adverse events [47]. Extensive clinical trials and systematic reviews of other mental health conditions, such as schizophrenia, have affirmed the safety of iTBS as a noninvasive neuromodulation method, with minimal side effects and a very low risk of severe complications [49]. Nonetheless, ongoing research is monitoring the long-term safety of iTBS to ensure its safe and effective application in clinical settings.

Treatment Regimen

iTBS is a form of patterned rTMS characterized by bursts of three pulses at 50 Hz, occurring at a 5 Hz inter-burst frequency [50]. This technique is notable for its brief application time, sustained effects, and more accurate reflection of physiological neural activity patterns. The various TBS protocols affect neuronal excitability differently, with iTBS specifically increasing excitability through a 2-s stimulation train followed by an 8-s pause [51]. Optimal stimulation is applied over the left DLPFC at an intensity ranging from 80% to 120% of MT [51].

Clinical Recommendations

The current evidence base for iTBS in the treatment of bipolar depression or mania is insufficient to provide definitive clinical recommendations [31] (Table 4.1).

4.2.5.2 Accelerated iTBS

Efficacy

In a meta-analysis encompassing five RCTs involving 239 patients with bipolar disorder or MDD experiencing MDE, Cai et al. [52] demonstrated that the iTBS protocol achieved a superior therapeutic response rate compared to sham stimulation. Specifically, active accelerated iTBS showed a significantly higher response rate than sham stimulation (39.8% versus 19.0%), although the remission rates were comparable between the two groups (36.5% versus 16.2%) [52]. A systematic review of five double-blinded RCTs, involving a total of 239 individuals with MDD or bipolar disorder in depressive episodes, also revealed that active iTBS was more effective than sham stimulation in terms of therapeutic outcomes. In a recent RCT, Sheline et al. [53] examined the efficacy of accelerated iTBS for treatment-refractory bipolar disorder with moderate to severe depressive episodes. The primary outcome measure was the change in MADRS scores before and after the intervention. The findings indicated that accelerated iTBS significantly alleviated depressive symptoms compared to sham stimulation, with an estimated difference of −14.75, suggesting its potential as a promising treatment for individuals with treatment-resistant bipolar disorder [53].

Safety
A pilot RCT involving 26 individuals diagnosed with bipolar depression found that twice-daily iTBS was well-tolerated, with no adverse events reported, further supporting its safety and tolerability [54]. Moreover, a meta-analysis by Mutz et al. [40] showed that the discontinuation rates were similar between the active and sham iTBS groups, reinforcing the treatment's acceptability and tolerability in both conditions.

Treatment Regimen
In the context of TRD, Sheline et al. [53] conducted a study where participants received 10 daily sessions of image-guided active or sham accelerated iTBS over 5 days, with one session per hour. The stimulation was set at 90% of the RMT, accumulating to a total of 90,000 pulses.

In a study on bipolar depression, Bulteau et al. [54] applied iTBS to the left DLPFC at 80% of the RMT. The protocol utilized a 50 Hz frequency, with triplet pulses delivered at 200 ms intervals within bursts occurring every 10 s. Patients attended two daily sessions, spaced 3 h apart, amounting to either 30 sessions over 3 weeks or 10 sessions in cases of early remission. Each session involved 990 pulses organized into two burst sequences, each consisting of three pulses at 50 Hz, repeated every 200 ms, with 10-s intervals between bursts.

The Stanford Accelerated Intelligent Neuromodulation Therapy (SAINT) protocol, developed by Cole et al. at Stanford University [55, 56], is specifically designed for the treatment of TRD. It utilizes an accelerated iTBS protocol, grounded in neuroscience principles. Participants in the SAINT protocol underwent magnetic resonance imaging (MRI) and resting-state functional MRI (fMRI) scans, with neuronavigation used to identify the left DLPFC region most anticorrelated with the subgenual anterior cingulate cortex (sgACC). This region was then targeted for iTBS. Sessions were delivered at 90% of the MT, each consisting of 1800 pulses with a 50-min intersession interval. The pulses, delivered at 50 Hz in 2-s bursts, were spaced with an 8-s interval between bursts. A total of 10 daily sessions were conducted, delivering 18,000 pulses over 5 consecutive days, accumulating to a total of 90,000 pulses.

Clinical Recommendations

Due to the limited body of research, no clinical guidelines currently exist for the use of accelerated iTBS in the treatment of bipolar depression or mania [31] (Table 4.1).

4.2.5.3 Continuous TBS

Efficacy

The efficacy of cTBS remains uncertain. A systematic review [40] found no substantial evidence supporting the application of cTBS for MDE. Similarly, a meta-analysis evaluating the therapeutic effects of TBS in patients with MDD or bipolar disorder reported no significant clinical advantages [52]. Studies focusing on individuals with acute-phase bipolar depression have also yielded inconsistent results regarding the efficacy of intensive iTBS in this population [57]. While some analyses observed improvements in depressive symptoms within the total sample, active cTBS failed to outperform sham treatment in alleviating these symptoms.

Safety

From a safety perspective, cTBS is widely considered a safe and well-tolerated intervention, particularly in managing MDE among patients with bipolar disorder. Dellink et al. [59] identified headaches (38%), local pain (27%), and tingling (14%) as the most common adverse events. However, isolated case reports have indicated the potential for cTBS to induce generalized seizures, underscoring the need for vigilance [58]. Large-scale, methodologically robust studies are required to clarify the therapeutic potential and safety profile of cTBS in bipolar disorder.

Treatment Regimen

The standard target for continuous cTBS is the right DLPFC. Protocols vary significantly across studies. For instance, one approach involves delivering cTBS at 50 Hz in bursts of three pulses, repeated every 200 ms at a frequency of 5 Hz, with a total of 600 pulses per session administered three times daily (1800 pulses per day) over 5 days at 80% of the RMT [57]. In contrast, an alternative protocol employs cTBS at 110% of the RMT, using the same pulse pattern but with five sessions per day. Each session delivers 900 stimuli with 15-min intervals between sessions, resulting in 4500 stimuli daily and a cumulative total of 18,000 stimuli over 4 treatment days [59].

Clinical Recommendations

Due to limited sample sizes and short follow-up periods, current studies have failed to demonstrate significant differences in treatment outcomes, precluding specific recommendations for bipolar depression or mania at this time [31] (Table 4.1). Future investigations should prioritize larger sample sizes and extended follow-up durations to provide more definitive insights into the efficacy and long-term outcomes of cTBS in this patient population.

4.2.5.4 Bilateral TBS

Efficacy

The efficacy of bilateral TBS remains uncertain based on recent preliminary findings. A double-blind, 6-week, sham-controlled RCT [60] evaluated the impact of bilateral TBS targeting the right and left DLPFC in adults experiencing acute mixed episodes of bipolar disorder and MDD. Changes in MADRS scores at week 3 revealed no statistically significant differences between groups. Similarly, the response and remission rates at week 6, assessed using MADRS, demonstrated no significant distinction between active and sham groups (response rate: 35.7% vs. 43.7%; remission rate: 28.5% vs. 37.5%).

A systematic review [61] encompassing six RCTs with 285 participants—82% diagnosed with MDD and 18% with a depressive episode of bipolar disorder—compared active bilateral TBS with sham stimulation. The analysis indicated no significant superiority of active bilateral TBS over sham in response rates (48.3% vs. 29.5%, $P = 0.07$) or remission rates (32.9% vs. 26.7%, $P = 0.37$). However, the inclusion of mixed diagnoses underscores a critical gap in research exploring the exclusive application of bilateral TBS in bipolar disorder. This limitation highlights the need for dedicated studies to optimize protocols and evaluate potential benefits in this specific population.

Safety

The limited number of RCTs investigating rTMS for bipolar disorder remains a significant barrier to drawing definitive conclusions. Notably, the RCT conducted by Tavares et al. [60] represents a key investigation into the antidepressant efficacy and safety profile of bilateral TBS applied over the DLPFC in individuals with bipolar and unipolar mixed depression. The study reported no significant differences in the incidence of treatment-emergent manic switches between active and sham groups, suggesting a comparable safety profile for both interventions. These findings are essential for assessing rTMS as a viable therapeutic modality in bipolar disorder.

Treatment Regimen

In this study [60], the bilateral TBS sessions were administered in the following sequence: cTBS targeting the right DLPFC was followed by iTBS targeting the left DLPFC. The parameters for cTBS consisted of bursts of three pulses at a frequency of 50 Hz, with a 20-ms interval between stimuli, applied continuously for a duration of 120 s, resulting in a total of 1800 pulses delivered to the right DLPFC. For iTBS, bursts of three pulses at 50 Hz, with a 20-ms interval between stimuli, were administered for 2 s and repeated every 10 s over a total duration of 570 s, also culminating in 1800 pulses delivered to the left DLPFC. The stimulation intensity was set at 80% of the MT. The study spanned a total duration of 6 weeks. A total of 1800 pulses were administered per session, resulting in a cumulative dose of 75,600 pulses throughout the study.

Clinical Recommendations

At present, no clinical guidelines exist for the use of bilateral TBS in treating bipolar disorder, including both depressive and manic episodes [37] (Table 4.1).

4.2.6 MST

4.2.6.1 Efficacy

MST represents an innovative and promising neuromodulation therapy. The technique induces therapeutic seizures by applying alternating magnetic fields, which generate eddy currents within the cerebral cortex [62]. MST requires anesthetic administration and medical monitoring setups akin to those utilized for ECT. However, unlike ECT, which directly stimulates the brain via external electrodes, MST employs a modified TMS device to deliver high-frequency magnetic pulses, inducing seizures to achieve therapeutic effects [63].

MST has demonstrated potential efficacy in managing treatment-resistant bipolar disorder. A systematic review [64] evaluating MST for unipolar and bipolar depression consistently reported significant antidepressant effects, with remission rates ranging from 30% to 40%. In comparison, ECT has shown higher remission rates of 50–70% under similar conditions. Notably, MST was associated with minimal cognitive side effects, offering a superior cognitive safety profile compared to ECT.

An open-label clinical study [65] further validated MST's therapeutic potential for bipolar depression. Among 26 patients undergoing at least eight treatments, MST achieved a response rate of 38.5% and a remission rate of 23.1%, with minimal cognitive impairments reported. In a subsequent study [66], the same group observed a significant reduction in suicidal ideation among patients with bipolar depression treated with MST, underscoring its effectiveness in addressing this critical symptom.

A comparative study assessing the efficacy and cognitive side effects of MST versus ECT in bipolar mania treatment [67] provided additional insights. ECT demonstrated therapeutic effectiveness in 95.0% of patients (19 out of 20), while MST was effective in 86.4% (19 out of 22). Both interventions produced comparable reductions in Young Mania Rating Scale (YMRS) scores. However, cognitive outcomes favored MST; patients undergoing ECT experienced declines in verbal performance, whereas those treated with MST exhibited no significant changes in language function. Furthermore, attention-related outcomes did not differ significantly between the groups. These findings highlight MST's high efficacy and reduced cognitive side effects, positioning it as a promising alternative to ECT for managing bipolar mania.

4.2.6.2 Safety

Compared to ECT, MST offers comparable overall efficacy while resulting in significantly fewer cognitive side effects. The key advantage of MST lies in its capacity to induce seizures with greater specificity to targeted brain regions, minimizing cognitive impairments such as retrograde and anterograde amnesia and reducing other adverse effects. As highlighted by Singh et al. [68], this reduction in cognitive side effects represents a substantial advantage in clinical practice. These attributes enhance patient compliance and suggest that MST may shape the future trajectory of neuromodulation therapies. Nevertheless, further investigation is essential to validate MST's efficacy and safety for critical clinical presentations, such as acute phases of bipolar disorder. Future research should prioritize the exploration of novel stimulation targets and conduct large-scale RCTs to provide robust evidence.

4.2.6.3 Treatment Regimen

For depressive episodes, MST is typically administered at 100 Hz, using 100% of the device's maximum output, for a 10-s duration [69]. In contrast, for manic episodes, MST protocols involve treatment sessions two to three times per week over a 2-week period, with each session lasting 8–10 min [67]. The device operates at a frequency of 75 Hz at maximum output, with magnetic stimulation durations starting at 4 s and incrementally increasing by 4–8 s, not exceeding 20 s in total. Adjustments are made to achieve an optimal seizure duration, defined as 15 s or more, confirmed through electroencephalogram (EEG) monitoring.

4.2.6.4 Clinical Recommendations

Due to the limited body of research, no graded clinical recommendations currently exist for MST in the treatment of bipolar depression or mania [8] (Table 4.1).

4.3 tDCS

4.3.1 Conventional tDCS

4.3.1.1 Efficacy

Conventional tDCS has demonstrated efficacy in managing bipolar disorder, particularly bipolar depression. A systematic review and meta-analysis evaluating the effectiveness and tolerability of tDCS in bipolar depression reported significant reductions in depression scores, with a medium effect size observed post-acute treatment (SMD = 0.71) and a large effect size at the study endpoint (SMD = 1.27) [9]. This therapeutic effect is attributed to the modulation of neural synchronization within the emotional brain network at both cortical and subcortical levels [70]. Furthermore, emerging evidence indicates that tDCS can enhance cognitive function in patients with bipolar disorder [71, 72]. However, the limited data regarding its efficacy in hypomanic, manic, or mixed episodes underscores the need for further targeted investigations.

4.3.1.2 Safety

The safety profile of tDCS in bipolar disorder is generally favorable, with no serious adverse events reported [8]. In a meta-analysis of bipolar depression [9], six cases of affective switching were documented among 46 patients undergoing tDCS. These occurrences were typically mild and transient, arising during or shortly after stimulation. Moreover, functional magnetic resonance imaging studies have confirmed that tDCS does not cause brain tissue edema, disrupt the blood–brain barrier, or induce other structural alterations [73].

4.3.1.3 Treatment Regimen

The efficacy of tDCS is influenced by specific protocol parameters. Electrode configurations such as F3-F8 and F3-F4 are commonly used, as they generate robust electric fields in the DLPFC and have been associated with significant antidepressant effects in bipolar disorder [74]. Stimulation intensity typically ranges from 0.029 and 0.08 mA/cm^2, employing a constant direct current mode without pulsed or variable frequency [75]. Recommended treatment regimens include two 20-min sessions per day for 14 days [76]. However, higher intensities, such as 2.5 mA over 30-min sessions spanning 20 treatments, may exceed the optimal dose for many patients [77, 78]. A recent systematic review [74] further highlights the lack of consensus on ideal stimulation parameters, including intensity, duration, session frequency, and total current dose, for achieving optimal outcomes in bipolar disorder.

4.3.1.4 Clinical Recommendations

Given the current paucity of robust research, graded clinical recommendations for the use of conventional tDCS in treating bipolar disorder, including depressive and manic episodes, are not yet available [8, 79] (Table 4.3). Further studies are essential to delineate the role of tDCS in this context and establish evidence-based protocols for its application.

4.3.2 HD-tDCS

4.3.2.1 Efficacy

HD-tDCS represents a significant advancement in neuromodulation therapy, employing refined current flow models to improve the precision of stimulation delivery. This approach allows for personalized customization of electrode placement and current intensity, aligning with the unique brain anatomy and therapeutic needs of individual patients. Such precision not only enhances the efficacy of treatment but also reduces the likelihood of adverse effects. A recent network meta-analysis of randomized controlled trials [80] demonstrated that HD-tDCS targeting the F3 region significantly alleviates depressive symptoms compared to sham controls, with no notable increase in dropout rates or side effects. Additionally, emerging evidence suggests that glial cell-derived neurotrophic factor (GDNF) may serve as a predictive biomarker for manic episodes treated with HD-tDCS, offering a pathway for personalized therapeutic strategies [81].

4.3.2.2 Safety

Extensive evidence underscores the potential of HD-tDCS as an effective and safe neurostimulation technique for bipolar disorder. Studies have reported favorable outcomes across various brain targets in both short- and long-term follow-ups [70, 80]. For instance, in Zhang et al.'s study, no instances of epilepsy or seizures were observed. However, 3 out of 28 participants in the active group and 2 out of 27 in the sham group reported burning and itching sensations. Additionally, two active group participants experienced headaches, while one sham group participant reported tingling sensations. Transient hypomania was noted in two active group participants and one in the sham group. All adverse effects were deemed tolerable and resolved by the conclusion of the treatment period. Despite these promising findings, current research on HD-tDCS in bipolar disorder remains constrained by small sample sizes, single-center designs, and reliance on case studies. Future studies must address these limitations to validate the efficacy and safety of HD-tDCS in broader populations. Furthermore, identifying optimal brain targets is crucial, as therapeutic outcomes are likely to vary depending on the specific regions stimulated.

4.3.2.3 Treatment Regimen

Two studies [70, 81] investigating the application of tDCS for the treatment of depressive episodes in bipolar disorder reported differences in their protocols. Veenakumari et al. [81] implemented anodal stimulation of the right dorsolateral prefrontal cortex (DLPFC) using a 4 × 1 electrode configuration, with cathodes positioned at adjacent electrode sites (AF4, AF8, FC2, FC6) and the central electrode at F4. In contrast, Zhang et al. [70] placed the anode over F3, with return electrodes located at FP1, FZ, C3, and F7. The protocol involved 14 sessions, each lasting 20 min, with ramp-up and ramp-down periods of 30 s each. These differences in protocol design could influence therapeutic outcomes and patient adherence. Despite these discrepancies, both studies demonstrated the potential efficacy of tDCS as a noninvasive neuromodulation technique for bipolar disorder.

4.3.2.4 Clinical Recommendations

The lack of robust evidence-based studies has left the clinical application of HD-tDCS in bipolar disorder, including both depressive and manic episodes, an unresolved issue (Table 4.3). The scarcity of empirical data has hindered the

Table 4.3 tDCS in bipolar disorder

tDCS	Clinical recommendations					Episode: Levels + References
	Anode position	Cathode position	Intensity (mA)	Length (min)	Duration (sessions)	
Conventional tDCS	Unclear	Unclear	Unclear	Unclear	Unclear	Both episode[a] [8, 79]
HD-tDCS	Unclear	Unclear	Unclear	Unclear	Unclear	Both episode[a] [8, 79]

Abbreviations: *HD-tDCS* high-definition transcranial direct current stimulation, *tDCS* transcranial direct current stimulation

[a]Both episodes including depressive episode and mania episode

establishment of consensus on optimal clinical practices, leaving the field of HD-tDCS for bipolar disorder treatment open to ongoing debate [8, 79].

4.4 TMS vs. tDCS

4.4.1 Efficacy

TMS and tDCS are widely utilized noninvasive brain stimulation (NIBS) techniques for treating bipolar disorder, particularly bipolar depression. A recent network meta-analysis encompassing 113 studies demonstrated that both TMS and tDCS exhibit higher efficacy and acceptability compared to sham treatments for MDD and bipolar depression [40]. This analysis revealed superior response rates for several modalities of TMS and tDCS over sham therapy, ranked by OR as follows: priming TMS (OR = 6.02), bilateral rTMS (OR = 4.92), bilateral TBS (OR = 4.44), low-frequency right rTMS (OR = 3.65), intermittent TBS (OR = 3.20), high-frequency left rTMS (OR = 3.17), and tDCS (OR = 2.65). All treatments demonstrated acceptability comparable to sham therapy.

While both TMS and tDCS are pivotal in the management of bipolar disorder, they differ significantly in their characteristics. TMS is frequently employed for bipolar depression due to its rapid and robust efficacy, particularly in patients unresponsive to conventional antidepressants. In contrast, tDCS is an emerging modality with promising potential for bipolar disorder treatment, though further research is required to optimize its application and efficacy.

4.4.2 Safety

The safety profiles of TMS and tDCS have been rigorously evaluated, with findings indicating that both techniques are generally safe and well-tolerated [40]. No significant differences in dropout rates or severe adverse events were observed when comparing active treatments to sham controls. However, left-sided excitatory stimulation, such as high-frequency rTMS over F3 (HF-rTMS-F3) and HD-tDCS targeting F3, may carry a heightened risk of side effects, including the potential induction of hypomanic episodes, compared to right-sided inhibitory stimulation [40]. This underscores the importance of careful patient monitoring and the customization of stimulation parameters to ensure safety and efficacy.

Direct comparisons of safety profiles between TMS and tDCS remain challenging due to methodological discrepancies, variations in patient demographics, and differences in stimulation protocols across studies. To address these limitations, RCTs with larger sample sizes and standardized evaluation criteria are necessary. Such studies should systematically compare the safety and efficacy of TMS, tDCS, and specific TBS protocols to establish more definitive recommendations for clinical practice.

4.5 Conclusion

Both TMS and tDCS represent key components of a comprehensive treatment strategy for bipolar disorder, offering a viable alternative for patients who exhibit inadequate responses to traditional pharmacological interventions. It is crucial to emphasize that the application of either modality necessitates oversight by qualified medical professionals, in accordance with established clinical protocols, to safeguard patient well-being and maximize therapeutic outcomes.

As highlighted in this chapter, a variety of optimization strategies for TMS exist; however, the most effective approach for treating bipolar disorder has yet to be clearly defined. Moreover, while the combined use of TMS and tDCS has demonstrated efficacy in the treatment of MDD [82], its applicability to bipolar disorder remains inconclusive. Concurrently, China is working toward the establishment of rapid antidepressant centers, which will integrate advanced treatments such as MST and SAINT. However, the effectiveness of these therapies for bipolar disorder—particularly in cases of refractory bipolar depression or mania—remains uncertain. Furthermore, certain biomarkers, including stimulation parameters [83] and genetic markers [84], have shown potential in predicting the clinical effectiveness of TMS or tDCS in MDD. However, research on these indicators within the context of bipolar disorder remains limited [8]. In summary, the growing body of research and technological advancements indicate that TMS and tDCS will play an increasingly significant role in the management of bipolar disorder in the near future.

Acknowledgments None.

Disclosure/Conflicts of Interest The authors declare no conflicts of interest in conducting this study or preparing the manuscript.

Financial Support This study was funded by the Science and Technology Program of Guangzhou (20251A011047, 20252A011018, 2023A03J0839, and 2023A03J0436), Science and Technology Planning Project of Liwan District of Guangzhou (202201012), National Clinical Key specialty construction project [(2023) 33], the Natural Science Foundation Program of Guangdong (2023A1515011383, 2024A1515012578), the Science and Technology Program of Guangzhou (202206010077), Guangzhou Municipal Key Discipline in Medicine (2021–2023), Guangzhou Traditional Chinese Medicine and Integrated Traditional Chinese and Western Medicine Science and Technology Project (20232A010014), Guangzhou High-level Clinical Key Specialty, Department of Emergency Medicine of National clinical key specialty and Guangzhou Research-oriented Hospital. The funders had no role in study design, data collection and analysis, decision to publish, or preparation of the manuscript.

References

1. McIntyre RS, Berk M, Brietzke E, Goldstein BI, López-Jaramillo C, Kessing LV, Malhi GS, Nierenberg AA, Rosenblat JD, Majeed A, et al. Bipolar disorders. Lancet. 2020;396(10265):1841–56.
2. Tondo L, Vázquez GH, Baldessarini RJ. Depression and mania in bipolar disorder. Curr Neuropharmacol. 2017;15(3):353–8.

3. GBD 2019 Mental Disorders Collaborators. Global, regional, and national burden of 12 mental disorders in 204 countries and territories, 1990–2019: a systematic analysis for the global burden of disease study 2019. Lancet Psychiatry. 2022;9(2):137–50.
4. Jolin EM, Weller EB, Weller RA. The public health aspects of bipolar disorder in children and adolescents. Curr Psychiatry Rep. 2007;9(2):106–13.
5. Keramatian K, Chithra NK, Yatham LN. The CANMAT and ISBD guidelines for the treatment of bipolar disorder: summary and a 2023 update of evidence. Focus (Am Psychiatr Publ). 2023;21(4):344–53.
6. Tournier M, Neumann A, Pambrun E, Weill A, Chaffiol J-P, Alla F, Bégaud B, Maura G, Verdoux H. Conventional mood stabilizers and/or second-generation antipsychotic drugs in bipolar disorders: a population-based comparison of risk of treatment failure. J Affect Disord. 2019;257:412–20.
7. Kishi T, Ikuta T, Sakuma K, Hatano M, Matsuda Y, Kito S, Iwata N. Repetitive transcranial magnetic stimulation for bipolar depression: a systematic review and pairwise and network meta-analysis. Mol Psychiatry. 2024;29(1):39–42.
8. Mutz J. Brain stimulation treatment for bipolar disorder. Bipolar Disord. 2023;25(1):9–24.
9. Dondé C, Amad A, Nieto I, Brunoni AR, Neufeld NH, Bellivier F, Poulet E, Geoffroy P-A. Transcranial direct-current stimulation (tDCS) for bipolar depression: a systematic review and meta-analysis. Prog Neuro-Psychopharmacol Biol Psychiatry. 2017;78:123–31.
10. Jannati A, Oberman LM, Rotenberg A, Pascual-Leone A. Assessing the mechanisms of brain plasticity by transcranial magnetic stimulation. Neuropsychopharmacology. 2023;48(1):191–208.
11. Wen K-S, Zheng W. Optimization strategies of transcranial magnetic stimulation in major depressive disorder. Alpha Psychiatry. 2023;24(6):270–2.
12. Sehatzadeh S, Daskalakis ZJ, Yap B, Tu H-A, Palimaka S, Bowen JM, O'Reilly DJ. Unilateral and bilateral repetitive transcranial magnetic stimulation for treatment-resistant depression: a meta-analysis of randomized controlled trials over 2 decades. J Psychiatry Neurosci. 2019;44(3):151–63.
13. Tee MMK, Au CH. A systematic review and meta-analysis of randomized sham-controlled trials of repetitive transcranial magnetic stimulation for bipolar disorder. Psychiatry Q. 2020;91(4):1225–47.
14. Nguyen TD, Hieronymus F, Lorentzen R, McGirr A, Østergaard SD. The efficacy of repetitive transcranial magnetic stimulation (rTMS) for bipolar depression: a systematic review and meta-analysis. J Affect Disord. 2021;279:250–5.
15. Chen J-J, Liu Z, Zhu D, Li Q, Zhang H, Huang H, Wei Y, Mu J, Yang D, Xie P. Bilateral vs. unilateral repetitive transcranial magnetic stimulation in treating major depression: a meta-analysis of randomized controlled trials. Psychiatry Res. 2014;219(1):51–7.
16. Fitzgerald PB, Daskalakis ZJ. A practical guide to the use of repetitive transcranial magnetic stimulation in the treatment of depression. Brain Stimul. 2012;5(3):287–96.
17. O'Reardon JP, Solvason HB, Janicak PG, Sampson S, Isenberg KE, Nahas Z, McDonald WM, Avery D, Fitzgerald PB, Loo C, et al. Efficacy and safety of transcranial magnetic stimulation in the acute treatment of major depression: a multisite randomized controlled trial. Biol Psychiatry. 2007;62(11):1208–16.
18. Xia G, Gajwani P, Muzina DJ, Kemp DE, Gao K, Ganocy SJ, Calabrese JR. Treatment-emergent mania in unipolar and bipolar depression: focus on repetitive transcranial magnetic stimulation. Int J Neuropsychopharmacol. 2008;11(1):119–30.
19. Wang Z, Ma Y, Chen J. The Chinese guidelines for the prevention and treatment of bipolar disorders: evidence-based choices. Chin J Psychiatry. 2017;50(2):96–100.
20. Konstantinou G, Hui J, Ortiz A, Kaster TS, Downar J, Blumberger DM, Daskalakis ZJ. Repetitive transcranial magnetic stimulation (rTMS) in bipolar disorder: a systematic review. Bipolar Disord. 2022;24(1):10–26.
21. Rostami R, Kazemi R, Nasiri Z, Ataei S, Hadipour AL, Jaafari N. Cold cognition as predictor of treatment response to rTMS; a retrospective study on patients with unipolar and bipolar depression. Front Hum Neurosci. 2022;16:888472.

22. McClintock SM, Reti IM, Carpenter LL, McDonald WM, Dubin M, Taylor SF, Cook IA, O'Reardon J, Husain MM, Wall C, et al. Consensus recommendations for the clinical application of repetitive transcranial magnetic stimulation (rTMS) in the treatment of depression. J Clin Psychiatry. 2018;79(1):3651.
23. Rossi S, Antal A, Bestmann S, Bikson M, Brewer C, Brockmöller J, Carpenter LL, Cincotta M, Chen R, Daskalakis JD, et al. Safety and recommendations for TMS use in healthy subjects and patient populations, with updates on training, ethical and regulatory issues: expert guidelines. Clin Neurophysiol. 2021;132(1):269–306.
24. Rossi S, Hallett M, Rossini PM, Pascual-Leone A. Safety, ethical considerations, and application guidelines for the use of transcranial magnetic stimulation in clinical practice and research. Clin Neurophysiol. 2009;120(12):2008–39.
25. Kazemi R, Rostami R, Khomami S, Horacek J, Brunovsky M, Novak T, Fitzgerald PB. Electrophysiological correlates of bilateral and unilateral repetitive transcranial magnetic stimulation in patients with bipolar depression. Psychiatry Res. 2016;240:364–75.
26. Gold AK, Ornelas AC, Cirillo P, Caldieraro MA, Nardi AE, Nierenberg AA, Kinrys G. Clinical applications of transcranial magnetic stimulation in bipolar disorder. Brain Behav. 2019;9(10):e01419.
27. Sonmez AI, Camsari DD, Nandakumar AL, Voort JLV, Kung S, Lewis CP, Croarkin PE. Accelerated TMS for depression: a systematic review and meta-analysis. Psychiatry Res. 2019;273:770–81.
28. Massé-Leblanc C, Desbeaumes Jodoin V, Nguyen DK, Fournier-Gosselin M-P, Stip E, Lespérance P, Miron J-P. Evaluating real-world effectiveness of accelerated transcranial magnetic stimulation for treatment-resistant depression in a tertiary referral center based in Quebec, Canada. Psychiatry Res. 2024;332:115685.
29. George MS, Raman R, Benedek DM, Pelic CG, Grammer GG, Stokes KT, Schmidt M, Spiegel C, Dealmeida N, Beaver KL, et al. A two-site pilot randomized 3 day trial of high dose left prefrontal repetitive transcranial magnetic stimulation (rTMS) for suicidal inpatients. Brain Stimul. 2014;7(3):421–31.
30. Desbeaumes Jodoin V, Miron J-P, Lespérance P. Safety and efficacy of accelerated repetitive transcranial magnetic stimulation protocol in elderly depressed unipolar and bipolar patients. Am J Geriatr Psychiatry. 2019;27(5):548–58.
31. Lefaucheur J-P, Aleman A, Baeken C, Benninger DH, Brunelin J, Di Lazzaro V, Filipović SR, Grefkes C, Hasan A, Hummel FC, et al. Evidence-based guidelines on the therapeutic use of repetitive transcranial magnetic stimulation (rTMS): an update (2014-2018). Clin Neurophysiol. 2020;131(2):474–528.
32. Tzirini M, Roth Y, Harmelech T, Zibman S, Pell GS, Kimiskidis V, Tendler A, Zangen A, Samaras T. Electrical field measurements and simulations of the H7 and D-B80 coils: non-equivalence of the TMS coils for obsessive compulsive disorder. Brain Stimul. 2021;14(6):1525–7.
33. Tzirini M, Roth Y, Harmelech T, Zibman S, Pell GS, Kimiskidis VK, Tendler A, Zangen A, Samaras T. Detailed measurements and simulations of electric field distribution of two TMS coils cleared for obsessive compulsive disorder in the brain and in specific regions associated with OCD. PLoS One. 2022;17(8):e0263145.
34. Tavares DF, Myczkowski ML, Alberto RL, Valiengo L, Rios RM, Gordon P, de Sampaio-Junior B, Klein I, Mansur CG, Marcolin MA, et al. Treatment of bipolar depression with deep TMS: results from a double-blind, randomized, parallel group, sham-controlled clinical trial. Neuropsychopharmacology. 2017;42(13):2593–601.
35. McGirr A, Karmani S, Arsappa R, Berlim MT, Thirthalli J, Muralidharan K, Yatham LN. Clinical efficacy and safety of repetitive transcranial magnetic stimulation in acute bipolar depression. World Psychiatry. 2016;15(1):85–6.
36. Nahas Z, Kozel FA, Li X, Anderson B, George MS. Left prefrontal transcranial magnetic stimulation (TMS) treatment of depression in bipolar affective disorder: a pilot study of acute safety and efficacy. Bipolar Disord. 2003;5(1):40–7.
37. Fitzgerald PB, Hoy KE, Elliot D, McQueen S, Wambeek LE, Daskalakis ZJ. A negative double-blind controlled trial of sequential bilateral rTMS in the treatment of bipolar depression. J Affect Disord. 2016;198:158–62.

38. Di Passa A-M, Prokop-Millar S, Yaya H, Dabir M, McIntyre-Wood C, Fein A, MacKillop E, MacKillop J, Duarte D. Clinical efficacy of deep transcranial magnetic stimulation (dTMS) in psychiatric and cognitive disorders: a systematic review. J Psychiatr Res. 2024;175:287–315.
39. Rapinesi C, Kotzalidis GD, Ferracuti S, Girardi N, Zangen A, Sani G, Raccah RN, Girardi P, Pompili M, Del Casale A. Add-on high frequency deep transcranial magnetic stimulation (dTMS) to bilateral prefrontal cortex in depressive episodes of patients with major depressive disorder, bipolar disorder I, and major depressive with alcohol use disorders. Neurosci Lett. 2018;671:128–32.
40. Mutz J, Vipulananthan V, Carter B, Hurlemann R, Fu CHY, Young AH. Comparative efficacy and acceptability of non-surgical brain stimulation for the acute treatment of major depressive episodes in adults: systematic review and network meta-analysis. BMJ. 2019;364:l1079.
41. Fitzgerald PB, Hoy KE, Singh A, Gunewardene R, Slack C, Ibrahim S, Hall PJ, Daskalakis ZJ. Equivalent beneficial effects of unilateral and bilateral prefrontal cortex transcranial magnetic stimulation in a large randomized trial in treatment-resistant major depression. Int J Neuropsychopharmacol. 2013;16(9):1975–84.
42. Li H, Cui L, Li J, Liu Y, Chen Y. Comparative efficacy and acceptability of neuromodulation procedures in the treatment of treatment-resistant depression: a network meta-analysis of randomized controlled trials. J Affect Disord. 2021;287:115–24.
43. Leuchter AF, Cook IA, Feifel D, Goethe JW, Husain M, Carpenter LL, Thase ME, Krystal AD, Philip NS, Bhati MT, et al. Efficacy and safety of low-field synchronized transcranial magnetic stimulation (sTMS) for treatment of major depression. Brain Stimul. 2015;8(4):787–94.
44. Cheng C-M, Li C-T, Tsai S-J. Current updates on newer forms of transcranial magnetic stimulation in major depression. Adv Exp Med Biol. 2021;1305:333–49.
45. Jin Y, Phillips B. A pilot study of the use of EEG-based synchronized transcranial magnetic stimulation (sTMS) for treatment of major depression. BMC Psychiatry. 2014;14:13.
46. Philip NS, Leuchter AF, Cook IA, Massaro J, Goethe JW, Carpenter LL. Predictors of response to synchronized transcranial magnetic stimulation for major depressive disorder. Depress Anxiety. 2019;36(3):278–85.
47. McGirr A, Vila-Rodriguez F, Cole J, Torres IJ, Arumugham SS, Keramatian K, Saraf G, Lam RW, Chakrabarty T, Yatham LN. Efficacy of active vs sham intermittent theta burst transcranial magnetic stimulation for patients with bipolar depression: a randomized clinical trial. JAMA Netw Open. 2021;4(3):e210963.
48. Diederichs C, DeMayo MM, Cole J, Yatham LN, Harris AD, McGirr A. Intermittent theta-burst stimulation transcranial magnetic stimulation increases GABA in the medial prefrontal cortex: a preliminary sham-controlled magnetic resonance spectroscopy study in acute bipolar depression. Front Psych. 2021;12:665402.
49. Zhang X, Yang X, Shi Z, Xu R, Tan J, Yang J, Huang X, Huang X, Zheng W. A systematic review of intermittent theta burst stimulation for neurocognitive dysfunction in older adults with schizophrenia. J Pers Med. 2023;13(3):485.
50. Diaz AP, Fernandes BS, Quevedo J, Sanches M, Soares JC. Treatment-resistant bipolar depression: concepts and challenges for novel interventions. Braz J Psychiatry. 2022;44(2):178–86.
51. Lan X-J, Yang X-H, Qin Z-J, Cai D-B, Liu Q-M, Mai J-X, Deng C-J, Huang X-B, Zheng W. Efficacy and safety of intermittent theta burst stimulation versus high-frequency repetitive transcranial magnetic stimulation for patients with treatment-resistant depression: a systematic review. Front Psych. 2023;14:1244289.
52. Cai D-B, Qin Z-J, Lan X-J, Liu Q-M, Qin X-D, Wang J-J, Goya-Maldonado R, Huang X-B, Ungvari GS, Ng CH, et al. Accelerated intermittent theta burst stimulation for major depressive disorder or bipolar depression: a systematic review and meta-analysis. Asian J Psychiatr. 2023;85:103618.
53. Sheline YI, Makhoul W, Batzdorf AS, Nitchie FJ, Lynch KG, Cash R, Balderston NL. Accelerated intermittent theta-burst stimulation and treatment-refractory bipolar depression: a randomized clinical trial. JAMA Psychiatry. 2024;81(9):936–41.
54. Bulteau S, Beynel L, Marendaz C, Dall'Igna G, Peré M, Harquel S, Chauvin A, Guyader N, Sauvaget A, Vanelle J-M, et al. Twice-daily neuronavigated intermittent theta burst stimu-

lation for bipolar depression: a randomized sham-controlled pilot study. Neurophysiol Clin. 2019;49(5):371–5.
55. Cole EJ, Stimpson KH, Bentzley BS, Gulser M, Cherian K, Tischler C, Nejad R, Pankow H, Choi E, Aaron H, et al. Stanford accelerated intelligent neuromodulation therapy for treatment-resistant depression. Am J Psychiatry. 2020;177(8):716–26.
56. Cole EJ, Phillips AL, Bentzley BS, Stimpson KH, Nejad R, Barmak F, Veerapal C, Khan N, Cherian K, Felber E, et al. Stanford neuromodulation therapy (SNT): a double-blind randomized controlled trial. Am J Psychiatry. 2022;179(2):132–41.
57. Mallik G, Mishra P, Garg S, Dhyani M, Tikka SK, Tyagi P. Safety and efficacy of continuous theta burst "intensive" stimulation in acute-phase bipolar depression: a pilot, exploratory study. J ECT. 2023;39(1):28–33.
58. Kelkar RS, Sreeraj VS, Mehta UM, Patwardhan AA, Thirthalli J, Blumberger DM. Generalized seizure induced by continuous theta-burst stimulation (cTBS) in a patient with bipolar depression: a case report. Brain Stimul. 2023;16(3):693–4.
59. Dellink A, Hebbrecht K, Zeeuws D, Baeken C, De Fré G, Bervoets C, De Witte S, Sabbe B, Morrens M, Coppens V. Continuous theta burst stimulation for bipolar depression: a multicenter, double-blind randomized controlled study exploring treatment efficacy and predictive potential of kynurenine metabolites. J Affect Disord. 2024;361:693–701.
60. Tavares DF, Suen P, Rodrigues Dos Santos CG, Moreno DH, et al. Treatment of mixed depression with theta-burst stimulation (TBS): results from a double-blind, randomized, sham-controlled clinical trial. Neuropsychopharmacology. 2021;46(13):2257–65.
61. Qin Z-J, Huang S-Q, Lan X-J, Shi Z-M, Huang X-B, Ungvari GS, Jackson T, Zheng W, Xiang Y-T. Bilateral theta burst stimulation for patients with acute unipolar or bipolar depressive episodes: a systematic review of randomized controlled studies. J Affect Disord. 2023;340:575–82.
62. Allan CL, Ebmeier KP. The use of ECT and MST in treating depression. Int Rev Psychiatry. 2011;23(5):400–12.
63. Wu H, Jiang J, Cao X, Wang J, Li C. Magnetic seizure therapy for people with schizophrenia. Cochrane Database Syst Rev. 2023;6(6):CD012697.
64. Cretaz E, Brunoni AR, Lafer B. Magnetic seizure therapy for unipolar and bipolar depression: a systematic review. Neural Plast. 2015;2015:521398.
65. Tang VM, Blumberger DM, Dimitrova J, Throop A, McClintock SM, Voineskos D, Downar J, Knyahnytska Y, Mulsant BH, Fitzgerald PB, et al. Magnetic seizure therapy is efficacious and well tolerated for treatment-resistant bipolar depression: an open-label clinical trial. J Psychiatry Neurosci. 2020;45(5):313–21.
66. Tang VM, Blumberger DM, Hill AT, Weissman CR, Voineskos D, Rajji TK, Downar J, Knyahnytska Y, Mulsant BH, Fitzgerald PB, et al. Magnetic seizure therapy for the treatment of suicidality in bipolar depression. Biol Psychiatry. 2021;90(10):e51–3.
67. Chen S, Sheng J, Yang F, Qiao Y, Wang W, Wen H, Yang Q, Chen X, Tang Y. Magnetic seizure therapy vs modified electroconvulsive therapy in patients with bipolar mania: a randomized clinical trial. JAMA Netw Open. 2024;7(4):e247919.
68. Singh R, Sharma R, Prakash J, Chatterjee K. Magnetic seizure therapy. Ind Psychiatry J. 2021;30(Suppl 1):S320–1.
69. Fitzgerald PB, Hoy KE, Elliot D, McQueen S, Wambeek LE, Chen L, Clinton AM, Downey G, Daskalakis ZJ. A pilot study of the comparative efficacy of 100 Hz magnetic seizure therapy and electroconvulsive therapy in persistent depression. Depress Anxiety. 2018;35(5):393–401.
70. Zhang L, Li Q, Du Y, Gao Y, Bai T, Ji G-J, Tian Y, Wang K. Effect of high-definition transcranial direct current stimulation on improving depression and modulating functional activity in emotion-related cortical-subcortical regions in bipolar depression. J Affect Disord. 2023;323:570–80.
71. Bersani FS, Minichino A, Bernabei L, Spagnoli F, Corrado A, Vergnani L, Mannarelli D, Pauletti C, Fattapposta F, Biondi M, et al. Prefronto-cerebellar tDCS enhances neurocognition in euthymic bipolar patients. Findings from a placebo-controlled neuropsychological and psychophysiological investigation. J Affect Disord. 2017;209:262–9.

72. Martin DM, Chan H-N, Alonzo A, Green MJ, Mitchell PB, Loo CK. Transcranial direct current stimulation to enhance cognition in euthymic bipolar disorder. Bipolar Disord. 2015;17(8):849–58.
73. Mayur P, Howari R, Byth K, Vannitamby R. Concomitant transcranial direct current stimulation with ultrabrief electroconvulsive therapy: a 2-week double-blind randomized sham-controlled trial. J ECT. 2018;34(4):291–5.
74. D'Urso G, Toscano E, Barone A, Palermo M, Dell'Osso B, Di Lorenzo G, Mantovani A, Martinotti G, Fornaro M, Iasevoli F, et al. Transcranial direct current stimulation for bipolar depression: systematic reviews of clinical evidence and biological underpinnings. Prog Neuro-Psychopharmacol Biol Psychiatry. 2023;121:110672.
75. Nitsche MA, Cohen LG, Wassermann EM, Priori A, Lang N, Antal A, Paulus W, Hummel F, Boggio PS, Fregni F, et al. Transcranial direct current stimulation: state of the art 2008. Brain Stimul. 2008;1(3):206–23.
76. Tang R, Song H. Chinese experts′ consensus on clinical application of transcranial direct current stimulation in the treatment of common neurological diseases and mental disorders. Chin J Psychiatry. 2022;05:327–82.
77. Loo CK, Alonzo A, Martin D, Mitchell PB, Galvez V, Sachdev P. Transcranial direct current stimulation for depression: 3-week, randomised, sham-controlled trial. Br J Psychiatry. 2012;200(1):52–9.
78. Loo CK, Husain MM, McDonald WM, Aaronson S, O'Reardon JP, Alonzo A, Weickert CS, Martin DM, McClintock SM, Mohan A, et al. International randomized-controlled trial of transcranial direct current stimulation in depression. Brain Stimul. 2018;11(1):125–33.
79. Cho H, Razza LB, Borrione L, Bikson M, Charvet L, Dennis-Tiwary TA, Brunoni AR, Sudbrack-Oliveira P. Transcranial electrical stimulation for psychiatric disorders in adults: a primer. Focus (Am Psychiatr Publ). 2022;20(1):19–31.
80. Hsu C-W, Chou P-H, Brunoni AR, Hung K-C, Tseng P-T, Liang C-S, Carvalho AF, Vieta E, Tu Y-K, Lin P-Y, et al. Comparing different non-invasive brain stimulation interventions for bipolar depression treatment: a network meta-analysis of randomized controlled trials. Neurosci Biobehav Rev. 2024;156:105483.
81. Veenakumari M, Goyal N, Kumar M, Kshitiz KK, Kumar P. Serum glial cell derived neurotrophic factor (GDNF) as a predictor of response to HD-tDCS in bipolar affective disorder. Asian J Psychiatr. 2022;68:102965.
82. Zhou D, Li X, Wei S, Yu C, Wang D, Li Y, Li J, Liu J, Li S, Zhuang W, et al. Transcranial direct current stimulation combined with repetitive transcranial magnetic stimulation for depression: a randomized clinical trial. JAMA Netw Open. 2024;7(11):e2444306.
83. Woodham RD, Selvaraj S, Lajmi N, Hobday H, Sheehan G, Ghazi-Noori A-R, Lagerberg PJ, Rizvi M, Kwon SS, Orhii P, et al. Home-based transcranial direct current stimulation treatment for major depressive disorder: a fully remote phase 2 randomized sham-controlled trial. Nat Med. 2024;31(1):87–95.
84. Elder GJ, Taylor J-P. Transcranial magnetic stimulation and transcranial direct current stimulation: treatments for cognitive and neuropsychiatric symptoms in the neurodegenerative dementias? Alzheimer's Res Ther. 2014;6(9):74.

Schizophrenia

5

Hu Deng, Xinyue Yang, Shuting Li, Shujuan Pan, and Chundi Wang

Abstract

Schizophrenia is a chronic and debilitating psychiatric disorder characterized by heterogeneous clinical manifestations and a complex etiology involving genetic predisposition, environmental triggers, and neurodevelopmental abnormalities. This makes it difficult to treat schizophrenia in a clinical setting. Although some patients with schizophrenia may benefit from clinical drug therapy, many individuals still suffer from treatment resistance and poor disease prognosis. Transcranial magnetic stimulation (TMS), utilizing pulsed magnetic fields to induce focal neuronal depolarization, and transcranial direct current stimulation (tDCS), employing weak direct currents to modulate resting membrane potentials, constitute cutting-edge neuromodulation modalities with growing evidence for efficacy in refractory neuropsychiatric conditions. Over the past two decades, many clinical trials have been conducted to find a better and more effective therapy for schizophrenia. Unfortunately, the results are mixed and no consensus exists. Hence, additional studies are needed to determine the optimal stimulus parameters and to identify key variables that might influence individual responses to TMS or tDCS. An overview of TMS and tDCS for schizophrenia is provided in this chapter.

Keywords

Noninvasive neurostimulation · Schizophrenia · TMS · tDCS

H. Deng · X. Yang · S. Li · S. Pan
Beijing Huilongguan Hospital, Peking University Huilongguan Clinical Medical School, Beijing, China

C. Wang (✉)
Department of Psychology, School of Humanities and Social Sciences, Beihang University, Beijing, China
e-mail: wangchundi@buaa.edu.cn

W. Zheng, Y. Ning (eds.), *TMS and tDCS for Psychiatric Disorders*,
https://doi.org/10.1007/978-981-96-8504-2_5

Abbreviations

ADRs	adverse drug reactions
AF3	Left Anterior Frontal 3
AH	auditory hallucinations
AHRS	the auditory hallucination rating scale
AMT	active motor threshold
AtDCS	Anodal transcranial direct current stimulation
AVH	auditory verbal hallucinations
BDNF	brain-derived neurotrophic factor
cTBS	continuous theta burst stimulation
DLPFC	dorsolateral prefrontal cortex
dTMS	deep transcranial magnetic stimulation
ECT	electroconvulsive therapy
EEG	electroencephalogram
F7	Left Frontal 7
FC1	Left Frontal-Central 1
FC3	Left Frontal-Central 3
FC5	Left Frontal-Central 5
FT7	Left Frontal-Temporal 7
GDNF	glial cell line-derived neurotrophic factor
HD-tDCS	high-definition transcranial direct current stimulation
HF-rTMS	high-frequency repetitive transcranial magnetic stimulation
Hz	hertz
iCiTBS	intensive cerebellar intermittent theta burst stimulation
iTBS	intermittent theta burst stimulation
kg	kilogram
LF-rTMS	low-frequency repetitive transcranial magnetic stimulation
M1 area	primary motor cortex
mA	milliampere
MCCB	the MATRICS consensus cognitive battery
MCF-B65	a figure-eight coil
MDD	major depressive disorder
MEP	motor-evoked potential
mg	milligram
min	minute
mm	millimeter
MST	magnetic seizure therapy
MT	motor threshold
NIBS	noninvasive brain stimulation
P1	Left Parietal 1
PANSS	Positive and Negative Syndrome Scale
PO7	Left Parieto-Occipital 7
PSYRATS-AH	Psychotic Symptom Rating Scale—auditory hallucinations
PTSD	posttraumatic stress disorder

RCTs	randomized controlled trials
RMT	resting motor threshold
rTMS	repetitive transcranial magnetic stimulation
SANS	Scale for the Assessment of Negative Symptoms
SNT	Stanford Neuromodulation Therapy
sTMS	synchronized transcranial magnetic stimulation
TBS	theta burst stimulation
tDCS	transcranial direct current stimulation
TMS	transcranial magnetic stimulation
TMS-EMG	transcranial magnetic stimulation-electromyography
TP3	An area between the Left Temporal 3 electrode and Left Parietal 3 electrode
TPC	temporal–parietal cortex
TPJ	temporoparietal junction

5.1 Schizophrenia

Schizophrenia is a highly debilitating psychiatric condition with a global lifetime prevalence of around 1% [1, 2]. The disorder typically appears in early adulthood and is characterized by five primary dimensions: positive symptoms, negative symptoms, cognitive deficits, anxiety and depression, and aggressive behavior [3]. Cognitive impairment, which includes difficulties with executive function, working memory and cognitive processing, has a significant impact on patients' social functioning and overall disease prognosis in schizophrenia [4]. A substantial body of evidence has emerged to support the efficacy of transcranial magnetic stimulation and transcranial direct current stimulation in the treatment of schizophrenia. However, the optimal stimulation parameters for each disorder are unknown.

Antipsychotic medications are the prevailing treatment for schizophrenia. While these medications have been shown to be effective in mitigating the positive symptoms of schizophrenia, such as delusions and hallucinations, their efficacy is limited in addressing the negative symptoms and cognitive impairments associated with the illness. Notably, a considerable number of patients exhibit treatment-resistant schizophrenia and do not adequately respond to antipsychotic drugs. In addition, the adverse effects of antipsychotics may lead to reduced adherence in some patients with schizophrenia.

In such cases, noninvasive brain stimulation, such as transcranial magnetic stimulation (TMS) and transcranial direct current stimulation (tDCS), can be promising treatment options. TMS is a technique that utilizes a magnetic field to generate an electric current in a specific region of the brain. This process is initiated by the application of a pulsed magnetic field over the scalp. The resulting electric field modulates neuronal activity in the targeted area. tDCS entails the application of a weak electrical current through two or more electrodes positioned on the scalp, with the purpose of stimulating underlying brain tissue. A substantial body of evidence has emerged to support the efficacy of TMS or tDCS in the treatment of schizophrenia. The biological mechanisms underlying the effects of TMS and tDCS on neuropsychiatric disorders are intricate and not fully understood. However, the optimal stimulation parameters for schizophrenia remain to be elucidated. Therefore, this chapter aims to provide a comprehensive review of the extant literature on the efficacy and safety of TMS and tDCS paradigms for patients with schizophrenia by systematically identifying and reviewing relevant studies in the field.

5.2 TMS

5.2.1 rTMS

5.2.1.1 Unilateral rTMS

Efficacy

Repetitive transcranial magnetic stimulation (rTMS) is a noninvasive neuromodulation technique that has been used in psychiatry for its therapeutic potential. Introduced in the mid-1980s, rTMS has gained recognition as a treatment for major depression [5] and has also been investigated as a potential intervention for patients with schizophrenia [6, 7]. The first randomized controlled trials investigating the efficacy of unilateral rTMS in schizophrenia patients for both positive and negative symptoms were conducted in the late 1990s [8]. However, the application of 1 Hz rTMS to right dorsolateral prefrontal cortex (DLPFC) (1200 total pulses) did not produce any therapeutic benefit. Subsequent research has modified the target site and stimulation frequency for unilateral rTMS in the treatment of schizophrenia. A recent meta-analysis suggests that targeting the left DLPFC with stimulation frequencies above 1 Hz may produce superior therapeutic outcomes [9]. To be noted, a meta-analysis also found that working memory in schizophrenia could be significantly improved by high-frequency rTMS (HF-rTMS) over the left DLPFC [10]. Unilateral rTMS shows promise as a viable treatment option for schizophrenia patients with both positive and negative symptoms, although the optimal parameters for rTMS therapy remain to be established.

Safety

In addition to efficacy, the safety of rTMS in schizophrenia has been evaluated in numerous clinical trials and meta-analyses [11–13]. In terms of safety, unilateral rTMS has been shown to be a noninvasive and safe method of brain stimulation for schizophrenia, with minimal serious adverse effects. Commonly reported side effects include mild headache and localized scalp discomfort, which have been documented in the majority of studies [14, 15]. Mild headache can usually be relieved with oral analgesics such as paracetamol. Currently, HF-rTMS, rather than low-frequency rTMS (LF-rTMS), may be a viable treatment option for schizophrenia. The safety and tolerability of HF-TMS is receiving increasing attention from researchers. HF-rTMS over the DLPFC has been shown to be safe and tolerable in some recent clinical trials and systematic reviews on the safety of rTMS [16, 17].

Apart from mild headache and discomfort, seizures are a major reason for discontinuation of rTMS treatment in patients with schizophrenia. According to a meta-analysis of safety evaluations of rTMS treatments in neuropsychiatric disorders, the odds of having a seizure were 4.7 times higher with TMS compared to sham stimulation [18]. This suggests that TMS should be considered with caution, with close monitoring for possible adverse effects and further research into the safety of TMS.

Treatment Regimen

Most studies fall into two stimulation camps regarding the efficacy of unilateral rTMS in schizophrenia. The first camp is stimulation of the temporal–parietal cortex (TPC) for the improvement of auditory hallucinations, which are a cardinal symptom of schizophrenia. Hoffman et al. found that persistent auditory hallucinations were significantly improved with a 1 Hz rTMS protocol [12], and this phenomenon was reproducible [19, 20]. However, unilateral rTMS targeting the TPC did not show a significant effect in the active rTMS group compared to the sham group in certain clinical trials [21, 22]. The variability of findings regarding the efficacy of unilateral rTMS has necessitated further investigation of alternative parameters for this treatment modality.

The second camp targets negative symptoms or cognitive dysfunction by stimulating the DLPFC. Cohen et al. found that negative symptom scores and cognitive performance could be significantly reduced by 20 Hz rTMS targeting the left DLPFC [23]. A similar beneficial effect of unilateral rTMS, and a specificity of stimulation frequency, has been observed in subsequent studies [24–26]. However, similar to the inconsistencies in TPC, some studies have shown inconsistent replication in DLPFC [27, 28]. Therefore, the challenge of interpreting the inconsistent findings persists, suggesting that advancements in technology may offer a potential solution.

Clinical Recommendations

Table 5.1 shows the clinical recommendation levels for this regimen. In favor of a possible benefit of low-frequency unilateral rTMS of the TPC on negative symptoms and auditory hallucinations, it seems reasonable to give a level C recommendation.

5.2.1.2 Bilateral rTMS

Efficacy

Bilateral rTMS has been investigated as a potential treatment for schizophrenia in addressing both positive and negative symptoms. Unlike unilateral rTMS, which targets a single hemisphere, bilateral rTMS stimulates both hemispheres, often focusing on the DLPFC. Studies suggest that bilateral rTMS may offer enhanced therapeutic outcomes compared to unilateral rTMS, particularly for

Table 5.1 Clinical recommendations of TMS in schizophrenia

TMS	Clinical recommendations					Levels + references
		Target	Frequency	Length (min)	Duration (sessions)	
Unilateral rTMS	Negative symptoms	Left DLPFC	10–20 Hz	20	10–20	Third-line [23–26, 29–34]
	Auditory-verbal hallucinations	Left TPC	1 Hz	20	10–12	Third-line [12, 19, 20, 35]
Bilateral rTMS	Negative symptoms	Bilateral DLPFC	HF (10 or 20 Hz)	20	20, 30	Third-line [24, 36–38]
	Auditory-verbal hallucinations	Wernicke's and right homologous regions	LF (1 Hz)	16	15	Third-line [39]
Accelerated rTMS	Auditory-verbal hallucinations	Left or bilateral TPJ; the temporoparietal area	LF (1 Hz) or HF (20 Hz)	10 or 20	10 (twice daily)	Third-line [40–45]
Deep TMS	Negative symptoms	Left DLPFC; bilateral PFC	HF (20 Hz)	15	20	Third-line [46–48]
Priming TMS	Auditory-verbal hallucinations	Left temporoparietal region	6 Hz (priming) and 1 Hz (treatment)	20	10, 20	Third-line [49, 50]
Synchronized TMS	Unclear	Unclear	Unclear	Unclear	Unclear	N/A
TBS	Unclear	Unclear	Unclear	Unclear	Unclear	N/A
iTBS	Unclear	Unclear	Unclear	Unclear	Unclear	N/A
Accelerated iTBS	Unclear	Unclear	Unclear	Unclear	Unclear	N/A
Continuation TBS	Unclear	Unclear	Unclear	Unclear	Unclear	N/A
Bilateral TBS	Unclear	Unclear	Unclear	Unclear	Unclear	N/A
MST	Unclear	Unclear	Unclear	Unclear	Unclear	N/A

Abbreviations: *N/A* not applicable, *rTMS* repetitive transcranial magnetic stimulation, *TMS* transcranial magnetic stimulation reducing negative symptoms, which are often resistant to other treatments. Some studies found that bilateral stimulation of the prefrontal cortex significantly improved working memory performance and reduced negative symptoms as measured by the Scale for the Assessment of Negative Symptoms (SANS) [24, 36, 51]. Additionally, another study also demonstrated that bilateral prefrontal rTMS may improve negative symptoms in that it can normalize an increased brain response to ambiguous emotional stimuli [37]. However, not all studies report consistent results, with some showing no significant differences between bilateral and unilateral or sham treatments [38, 52], especially for positive symptoms such as auditory hallucinations [40, 41, 53].

Safety

Most adverse effects associated with bilateral rTMS are transient and mild. For example, a study on HF-rTMS reported no severe adverse reactions, with only a small percentage of patients discontinuing due to discomfort such as persistent headaches [38, 52]. However, like any therapeutic intervention, it is essential to understand the potential risks and side effects associated with its use. The most frequently reported side effects of bilateral rTMS are twitching of the facial muscles during rTMS stimulation and transient, mild headache after rTMS stimulation [24]. These sensations are typically benign and diminish shortly after the session concludes [40]. Additionally, although the overall risk of seizures with rTMS is low, it remains one of the most serious potential side effects. Bilateral rTMS has been associated with a slightly increased risk of seizures compared to sham stimulation, but this risk is generally very low, especially when patients are carefully screened for seizure risk factors prior to treatment [13].

Treatment Regimen

Most studies on bilateral rTMS for schizophrenia focus on two primary areas of stimulation: the DLPFC and, less commonly, the TPC. These studies aim to address both positive symptoms, such as auditory hallucinations, and negative symptoms, such as cognitive deficits and emotional withdrawal. The DLPFC is commonly targeted for alleviating negative symptoms such as emotional withdrawal, lack of motivation, and cognitive impairments. Studies have employed high-frequency stimulation (usually 10–20 Hz) applied bilaterally to the DLPFC. For example, some studies have shown a significant reduction in negative symptom severity after patients received 3 weeks of 10 Hz bilateral rTMS treatment, compared to sham treatment [24, 36]. However, the results of another study suggest that HF-rTMS (20 Hz) targeting the bilateral DLPFC does not alleviate negative symptoms in patients with schizophrenia [38]. Bilateral rTMS targeting the TPC has also been explored for treating persistent auditory hallucinations, a prominent positive symptom of schizophrenia. However, a study showed that rTMS of the left temporoparietal region appeared more efficient in reducing auditory hallucinations when compared to bilateral or sham stimulation [54].

Taken together, despite some promising findings, the efficacy of bilateral rTMS for schizophrenia remains variable across studies. Some trials have reported significant improvements, while others have shown limited or no effect compared to sham treatments. This variability highlights the need for further research to optimize stimulation parameters, such as frequency, intensity, and session duration, to achieve more consistent outcomes [38].

Clinical Recommendations

As the DLPFC is implicated in the negative symptoms of schizophrenia, the use of bilateral rTMS with high-frequency stimulation (10–20 Hz) to this region may be effective in reducing symptoms such as cognitive impairments and emotional withdrawal. Table 5.1 shows the clinical recommendation levels for this regimen.

5.2.1.3 Accelerated rTMS

Efficacy

Accelerated rTMS is an emerging technique aimed at enhancing the therapeutic effects of standard rTMS by increasing the frequency and intensity of treatment sessions. They can provide the same number of sessions and pulses as a traditional 6-week treatment course in just 1 week because of the increased number of daily stimulation sessions, which may accelerate treatment response and enhance treatment outcomes [55]. The efficacy of accelerated rTMS in treating schizophrenia, particularly for alleviating negative symptoms and cognitive impairments, has been explored in several studies. Some studies showed that LF-rTMS (2 sessions/day) to the left temporoparietal cortex significantly improved hallucinatory symptoms in patients with schizophrenia [42, 56]. Studies with HF-rTMS treatment targeting the temporoparietal junction (TPJ) also showed significant improvement in hallucinatory symptoms [43], while some studies showed no significant improvement [44, 45]. Another study exploring accelerated rTMS for the treatment of negative symptoms showed significant improvement in the majority of patients who received 3–5 sessions of treatment per day [57]. This suggests that accelerated rTMS holds promise as a faster-acting intervention compared to conventional rTMS schedules, especially for patients with treatment-resistant symptoms.

However, there are also findings that auditory verbal hallucinations (AVH) in patients with schizophrenia did not significantly improve with LF-rTMS treatment targeting the TPJ [40]. A study using intensive cerebellar intermittent theta burst stimulation (iCiTBS) in a treatment-resistant schizophrenia patient has found similar results, with no significant difference in schizophrenia psychopathology, cognitive functions, and global improvement between active and sham treatment [58]. Therefore, there remain gaps in understanding the long-term efficacy and optimal protocols for accelerated rTMS in schizophrenia.

Safety

Studies demonstrate that accelerated rTMS, which involves multiple daily sessions, is generally well-tolerated. Common side effects include mild headaches and scalp discomfort, with no significant reports of serious adverse effects like seizures [56, 59]. In a comprehensive review, Caulfield et al. highlighted that increasing session frequency does not significantly elevate the risk of adverse outcomes, making accelerated rTMS comparable in safety to standard protocols [55]. Additionally, an open-label study by Sverak et al. found that accelerated rTMS was well-received, with most patients showing no serious side effects [57]. The consensus across studies is that accelerated rTMS can be safely implemented in schizophrenia treatment, offering faster symptom alleviation without compromising patients' safety [60]. These findings demonstrate the potential of accelerated rTMS as an efficient therapeutic tool in psychiatric treatment, particularly for patients needing more rapid interventions.

Treatment Regimen

In an accelerated protocol, sessions are often scheduled twice or thrice daily for up to 5 days per week, resulting in a total of 20–30 sessions over 2–3 weeks [24, 36, 40, 42, 43, 45, 55, 56]. Similar to other rTMS treatment regimens, the main targets of accelerated rTMS are the DLPFC for negative symptoms of schizophrenia [24, 36, 57] and the TPJ for hallucinatory symptoms [40, 42, 43, 45, 56].

Clinical Recommendations

Accelerated rTMS shows promise as a faster alternative to traditional rTMS protocols, offering a condensed treatment course with 2–5 sessions per day over a span of 1–2 weeks. This approach is particularly beneficial for patients with treatment-resistant symptoms, such as negative symptoms and cognitive impairments, as studies have demonstrated significant improvements in these domains. More importantly, given its potential to accelerate symptom alleviation without increasing adverse outcomes, it is recommended for patients requiring more rapid intervention. Table 5.1 shows the clinical recommendation levels for this regimen.

5.2.2 Deep TMS

5.2.2.1 Efficacy

Compared with traditional TMS, deep transcranial magnetic stimulation (dTMS) can target deeper and larger brain regions, making it a very promising treatment option [61–63]. The first study to apply dTMS to the treatment of schizophrenia showed significant improvement in negative symptoms and cognitive function after patients received 4 weeks of dTMS treatment [46]. Some subsequent studies have also demonstrated improvements of negative symptoms and cognitive deficits in schizophrenia with dTMS treatment [47, 48]. However, in the study by Rabany et al., there was no significant between-group difference in the improvement of negative symptoms [48]. In other studies, no significant reduction in negative symptoms was observed in either active or sham treatment groups [64, 65].

Another key area of research is the use of dTMS for positive symptoms, such as hallucinations. Rosenberg et al. conducted an open-label study using dTMS over the left temporoparietal cortex, which exhibited no significant reductions in auditory hallucinations after treatment [65]. Conversely, the study by Rabany et al.

demonstrated a positive shift in positive symptoms following dTMS treatment [48]. However, no significant between-group differences were observed in either study.

5.2.2.2 Safety

The safety of dTMS in treating schizophrenia has been the subject of extensive research. The consensus across numerous studies is that dTMS is generally well-tolerated, with few severe adverse events. The most commonly reported adverse effects are typically mild, including headaches, scalp discomfort and transient fatigue and are consistent with those reported in other noninvasive brain stimulation techniques such as traditional rTMS [46, 66]. Importantly, dTMS has not been associated with an increased risk of seizures or cognitive impairments, even in patients receiving daily treatments for extended periods [61, 65]. Studies focusing on schizophrenia have confirmed the safety of dTMS, particularly when used as an add-on therapy for improving negative symptoms and auditory hallucinations. Patients receiving dTMS reported minor and reversible side effects, while serious complications remained rare across trials [48, 67]. Overall, the safety profile of dTMS is favorable, with most studies supporting its continued use as a noninvasive and effective option for treating schizophrenia symptoms, especially when traditional pharmacological treatments have proven insufficient.

5.2.2.3 Treatment Regimen

The treatment regimen of dTMS in schizophrenia is typically characterized by a standardized protocol that is designed to target negative symptoms and cognitive impairments. Most studies utilize high-frequency (typically 10–20 Hz) dTMS applied to the DLPFC, an area implicated in the negative and cognitive symptoms of schizophrenia [46–48]. The treatment is typically administered in the form of daily sessions, with each session lasting approximately 20 min, over a period of 2–5 weeks. This results in a total of 15–30 sessions. Additionally, several studies have explored the use of low-frequency dTMS over regions like the TPJ to reduce auditory hallucinations, another key symptom of schizophrenia [68, 69]. Patients undergoing this protocol receive 1 Hz stimulation for about 20 min per session [65]. This approach has shown efficacy, particularly in treatment-resistant patients who did not respond to antipsychotic medications. The efficacy of dTMS is contingent on various factors, including the frequency of stimulation, the location of the stimulation site, and the duration of the treatment regimen. However, a substantial body of research has reached a consensus that the most substantial effects in terms of the reduction of schizophrenia symptoms are observed in cases where the treatment is administered over an extended period of time [61, 66, 67].

5.2.2.4 Clinical Recommendations

The capacity of dTMS to access deeper brain regions in comparison to conventional TMS makes it a suitable option for patients exhibiting treatment-resistant symptoms. The clinical recommendation levels for this regimen are shown in Table 5.1. However, further exploration and validation in studies with larger samples are required to fully explore its specific treatment protocols and effects.

5.2.3 Priming TMS

5.2.3.1 Efficacy

Priming TMS is defined as the application of high-frequency stimulation (typically 6–10 Hz) prior to LF-rTMS. The purpose of priming is to enhance cortical excitability and thereby augment the therapeutic effects of rTMS [70]. This approach is designed to modulate cortical excitability and increase the therapeutic effects of subsequent LF-rTMS, which is known to reduce the cortical hyperactivity associated with schizophrenia symptoms like auditory hallucinations [48, 52].

Studies indicate that priming before LF-rTMS enhances the depressant effect on cortical excitability, making it more effective in reducing both positive and negative symptoms in schizophrenia patients [67, 70]. The enhanced cortical plasticity induced by the priming protocol is believed to potentiate the effects of LF-rTMS, leading to more sustained symptom relief [49]. However, other studies have found that priming TMS did not improve treatment response in patients with treatment-resistant schizophrenia, particularly in reducing auditory hallucinations and cognitive dysfunction [50].

5.2.3.2 Safety

The safety profile of priming TMS in treating schizophrenia has been widely studied, with the majority of findings indicating that it is a safe, well-tolerated intervention. Studies consistently report that this approach does not increase the risk of serious adverse effects such as seizures, which are a major concern with some neurostimulation techniques [70, 71]. Common side effects associated with priming TMS include mild headaches, scalp discomfort, and temporary fatigue, all of which typically resolve shortly after treatment and are consistent with traditional rTMS protocols [50, 72]. Importantly, long-term studies indicate that repeated sessions of priming TMS do not lead to cognitive impairments or neurological complications, even when administered over multiple weeks or at higher intensities [73]. Given the relatively mild and transient nature of the side effects, priming TMS continues to be considered a safe alternative or adjunct treatment for patients with schizophrenia.

5.2.3.3 Treatment Regimen

The treatment regimen for priming TMS in schizophrenia typically involves a two-phase protocol aimed at enhancing the efficacy of LF-rTMS [70]. The priming TMS protocol consists of an initial "priming" stimulation phase of HF-rTMS (6–10 Hz), followed by low-frequency (1 Hz) stimulation applied to specific cortical regions, such as the DLPFC or TPJ. The treatment regimen generally comprises daily sessions over a period of 2–4 weeks, with each session lasting approximately

20–30 min. The high-frequency priming phase is scheduled to last for approximately 10 min, followed by 20 min of low-frequency stimulation [73].

5.2.3.4 Clinical Recommendations

The administration of priming TMS has been demonstrated to enhance cortical excitability and promote symptom alleviation in individuals diagnosed with schizophrenia. The clinical recommendation levels for this regimen are shown in Table 5.1. This approach is generally safe, with mild side effects such as headaches and scalp discomfort. While effective for many, responses may vary, particularly in treatment-resistant cases.

5.2.4 Synchronized TMS

5.2.4.1 Efficacy

Synchronized transcranial magnetic stimulation (sTMS) is an innovative neuromodulation technique that involves delivering magnetic pulses synchronized to the brain's intrinsic oscillatory rhythms. This method aims to improve the precision of brain stimulation by aligning with the individual's natural cortical activity, making it a promising intervention for schizophrenia [74]. Patients diagnosed with schizophrenia have been shown to exhibit impaired cortical inhibition of gamma oscillations in the DLPFC when compared to both healthy subjects and patients diagnosed with bipolar disorder [75–77]. Studies have shown that sTMS significantly enhances gamma oscillations in the frontal cortex, leading to improvements in working memory and auditory processing, both of which are compromised in schizophrenia patients [78, 79]. However, further research is required to optimize long-term outcomes and to determine the most effective stimulation parameters.

5.2.4.2 Safety

sTMS, being synchronized to the patient's brain oscillations, is designed to optimize efficacy while minimizing overstimulation, thereby reducing the risk of adverse effects [80, 81]. The most common side effects, such as mild headaches or scalp discomfort, tend to be transient and resolve shortly after the treatment session [75]. The risk of more serious side effects, such as seizures, remains low with sTMS, which aligns with the safety profile seen in traditional TMS protocols. A substantial body of research, encompassing trials on treatment-resistant schizophrenia, reports an absence of major neurological complications, even with extended or frequent sessions [82, 83]. Overall, sTMS provides a favorable safety profile with minimal side effects, supporting its ongoing use in the treatment of schizophrenia symptoms.

5.2.4.3 Treatment Regimen

The treatment regimen for sTMS typically involves the delivery of pulses synchronized to the brain's natural oscillatory frequencies, which has been demonstrated to enhance the effect of TMS. By synchronizing the magnetic pulses with the patient's cortical activity, sTMS provides real-time feedback on cortical excitability and synchrony, enabling the customization of stimulation parameters to the individual's specific neurophysiological requirements. This combination has shown promise in enhancing cortical plasticity and improving neural circuit function, particularly in early-course schizophrenia [81, 84]. The primary focus of sTMS is the DLPFC, a region implicated in executive function and cognitive processing [82]. A substantial body of research supports the utilization of sTMS for the enhancement of cognitive performance and the alleviation of symptoms in cases of treatment-resistant schizophrenia. The underlying mechanisms of action of these protocols are thought to involve the targeting of gamma oscillations and alpha synchrony, which have been observed to be frequently disrupted in patients, resulting in improvements in cognitive domains and a reduction in hallucinations [78, 79].

5.2.4.4 Clinical Recommendations

sTMS has been shown to align magnetic pulses with the brain's natural oscillatory rhythms, thereby enhancing the precision of brain stimulation. This technique has shown promise in enhancing gamma oscillations in the DLPFC, leading to enhanced working memory and auditory processing in patients with schizophrenia. Despite the effectiveness of this technique, further research is necessary to optimize long-term outcomes and stimulation parameters (Table 5.1).

5.2.5 TBS

Noninvasive brain stimulation (NIBS) has been demonstrated to modulate cortical activity in humans [85]. Notably, a modified form of rTMS known as theta burst stimulation (TBS) has garnered significant attention due to its effectiveness in eliciting responses with brief stimulation durations at low intensities [86]. Recently, the TBS protocol has been proposed as a more efficient method for neuroplasticity across various clinical rehabilitation contexts. The rTMS waveform is transmitted to the brain via a magnetic field generated by a coil positioned over the scalp. This magnetic field traverses the skull without obstruction and is ultimately converted into electrical current, stimulating the underlying brain tissue to facilitate neuroplasticity. Thus, the TBS rTMS waveform promotes neuroplasticity through the conversion of magnetic to electrical current [86].

In summary, TBS comprises bursts of three pulses delivered at a frequency of 50 Hz, with an interburst interval of 5 Hz [87]. There are two primary paradigms of TBS: intermittent theta burst stimulation (iTBS) and continuous theta burst stimulation (cTBS). Variations in stimulation parameters, such as pulse delivery frequency and dosage (number of pulses), can significantly influence the strength and duration of the after-effects of TBS [88–90]. Research examining plasticity changes induced by TBS indicates that it is one of the most potent NIBS techniques available for neuromodulation [85].

rTMS has been explored in patients with schizophrenia to address refractory auditory hallucinations, cognitive deficits, and negative symptoms [91–94]. The iTBS paradigm has been shown to enhance cortical excitability (inducing long-term potentiation), whereas the cTBS paradigm has been associated with a reduction in cortical excitability (inducing long-term depression) [87].

From a therapeutic perspective, TBS sessions are generally shorter than standard rTMS protocols. The optimal stimulation parameters remain undetermined, although tolerability appears comparable to standard rTMS, with increased discomfort reported at higher stimulation intensities. Researchers have expressed interest in utilizing TBS to treat symptoms of schizophrenia [95]. TBS is a well-tolerated therapeutic intervention and neurophysiological tool, with potential applications for targeting specific symptom domains of schizophrenia. Notably, iTBS has shown promise as an intervention for alleviating negative symptoms of the disorder.

To date, several TBS protocols, including iTBS, have been proposed; however, previous systematic reviews have highlighted inconsistent efficacy findings across individual TBS studies in the context of schizophrenia [96].

5.2.5.1 iTBS

Efficacy

Currently, rTMS has gained significant traction in psychiatry as a NIBS technique, with iTBS emerging as a more novel form of rTMS [97]. The iTBS is characterized by the delivery of 600 pulses within a brief duration of 3 min [98]. This method produces a rapid increase in motor cortex excitability, and its excitatory effects may surpass those of traditional rTMS [99, 100].

Recently, as TBS paradigms have expanded in the treatment of major depressive disorder (MDD), researchers have also shown interest in utilizing iTBS to address symptoms of schizophrenia. The most commonly used theta burst paradigm, iTBS, delivers triplet burst pulses at a frequency of 50 Hz, with bursts occurring every 5 Hz. This paradigm has been found to enhance cortical excitability, inducing long-term potentiation [95]. Evidence suggests that the TBS protocol mimics the coupling of theta and gamma rhythms, resulting in alterations in cortical excitability [87]. One randomized controlled trials (RCTs) study demonstrated a significant advantage of active iTBS over sham iTBS in improving neurocognitive performance in older adults with schizophrenia [101].

Additionally, research has indicated that iTBS applied to the left DLPFC is associated with reductions in negative symptom scores, overall symptom scores, Positive and Negative Syndrome Scale (PANSS) general subscale scores, depressive symptom scores, and anxiety symptom scores, as well as improvements in overall cognitive impairment scores compared to sham treatment [96]. Lisman and Idiart reported that theta-frequency oscillations modulate high-frequency gamma oscillations, which are linked to cognitive processing in human recognition memory [102]. Furthermore, by influencing the intrinsic circuitry of the motor cortex, iTBS enhances the amplitude of later I-waves and induces long-term potentiation like changes at synaptic connections within the motor cortex [103, 104].

iTBS paradigms can effectively modulate corticospinal excitability in healthy individuals, resulting in increased excitability with iTBS. A previous study on TBS [105] found that the potentiating effect of iTBS was greater than the inhibitory effect of cTBS. The most pronounced effect sizes for iTBS are typically observed at mid-time points, specifically 20–30 min poststimulation. Notably, delivering 1200 pulses of iTBS at this mid-time point yielded a more substantial effect compared to 600 pulses, whereas the differences associated with cTBS were relatively marginal [85].

However, research has revealed an unexpected outcome with 1200 pulses of iTBS at later time points, showing a nonsignificant trend toward motor-evoked potential (MEP) suppression. This finding suggests that the optimal interval for iTBS to elicit longer-lasting effects has yet to be determined, and the underlying mechanisms of action for iTBS remain incompletely understood.

Preliminary evidence indicates that adjunctive iTBS may confer benefits in the treatment of neurocognitive function in older patients with schizophrenia. Future RCTs with larger sample sizes focusing on the neurocognitive effects of adjunctive iTBS in this population are warranted to verify these findings [101].

In summary, the results suggest that iTBS is associated with therapeutic benefits for schizophrenia, particularly in alleviating negative symptoms, depressive symptoms, anxiety symptoms, and cognitive impairment. Larger-scale, long-term studies on TBS protocols are necessary to evaluate the durability of their effects (e.g., through continuation studies).

Safety

To ensure safety, the stimulator was powered by a 5 V direct current source, incorporating a 5-s ramp up and ramp down period at the beginning and end of each session to minimize discomfort for participants [106]. Common adverse events associated with transcranial magnetic stimulation (TMS) included headache (6.6%), arm/hand/other pain (2.6%), numbness or tingling (2.6%), other sensations (2.6%), weakness (1.3%), and miscellaneous events (1.3%) [107].

A previous study found that changes in positive symptom scores, all-cause discontinuation rates, discontinuation rates due to adverse events, incidence of headaches, and incidence of dizziness did not significantly differ between any TBS protocols and sham treatments [96].

It is important to note that while the adverse effects reported in the majority of case reports and clinical trials were limited to mild headaches, fatigue, and minor neck pain, two case reports documented serious adverse events. The first involved a 15-year-old who experienced a seizure during an iTBS session [108]. Although seizures are considered a very rare adverse effect of rTMS, their occurrence should not be underestimated [109]. Adhering to specific safety protocols can minimize, but not entirely eliminate, this risk. Specifically, accurate intensity settings based on motor threshold (MT) and careful screening of participants for risk factors, such as a history of epilepsy, should be taken into account [110, 111].

Other researchers found that the side effect profile was generally mild, with the exception of one incident of hypertension resulting from iTBS [112].

Future studies should systematically collect data on adverse events using standardized rating scales and incorporate long-term follow-up assessments (6–12 months) to evaluate adverse events, hearing, and neurocognitive outcomes [113].

Treatment Regimen

The treatment protocol consisted of 20 sessions of iTBS at 80% resting motor threshold (RMT), delivered over 10 workdays (990 pulses per session, repeated twice daily) targeting the left DLPFC. Thirty-three trains of iTBS were administered as 10 bursts of stimulation for 2 s, followed by an 8-s pause. Each burst comprised three pulses at 50 Hz. Previous study reported a significant reduction in the Scale for the SANS score in the active group compared with the sham group at the 6-month follow-up [114].

Another treatment protocol involved 10 sessions of iTBS at 80% RMT, also delivered over 10 workdays (a total of 600 pulses per session). Twenty trains of iTBS were delivered as 10 bursts of stimulation for 2 s, followed by an 8-s pause. Each burst consisted of three pulses at 50 Hz. Researchers observed significant improvement in the negative symptoms of patients in the study group compared with the sham group [115].

In a different protocol, 42 sessions of iTBS were administered at 80% RMT over 14 consecutive days (totaling 600 pulses per session, with 3 sessions per day, and 15 min between each session, resulting in 1800 total pulses per day). Twenty trains of iTBS were delivered as 10 bursts of stimulation for 2 s, followed by an 8-s pause. Each burst comprised three pulses at 50 Hz. Both overall and negative symptom scores improved in the active arm, but not in the sham group [116].

In another study, all patients were scheduled to attend once daily for 4 weeks, totaling 20 sessions. For 10 and 20 Hz stimulation, the initial stimulation intensity was set at 80% RMT and gradually increased to 110%. A total of 1500 pulses were delivered in each session, consisting of 5 s of stimulation trains (at either 10 or 20 Hz) followed by 30 s of rest. The TBS protocol included prolonged iTBS sessions (2400 pulses) delivered at 80% MT. The trains of iTBS were administered as 10 bursts of stimulation for 2 s, with each burst comprising three pulses at 50 Hz. In all three active arms of the study, there were significant decreases in the PANSS and SANS scores, but not in the sham arm [29].

In a different protocol, the treatment protocol consisted of 10 sessions of iTBS at 100% active motor threshold (AMT), delivered over 5 workdays (a total of 600 pulses per session, with one session in the morning and another in the afternoon). Twenty trains of iTBS were administered as 10 bursts of stimulation for 2 s, followed by an 8-s pause. Each burst comprised three pulses at 50 Hz. There was a significant improvement in negative symptom scores when baseline measurements were compared with posttreatment and follow-up assessments (PANSS sub-scores). Furthermore, improvements in mood and cognition were significant when comparing baseline with posttreatment and follow-up evaluations [112].

In another study, participants received 3 daily sessions of iTBS treatment, separated by 15 min, over a duration of 4 weeks. The stimulation parameters included 600 pulses delivered over 189 s. The pulses were administered at 50 Hz and repeated at 5 Hz. The target area was determined using an MRI-neuro-navigation system. The study reported a significant difference in measures of social cognition between participants in the active and sham groups at both the 2-week and 4-week follow-ups. Additionally, there were significant differences in negative symptom scores between the active and sham groups, while no significant differences were observed in positive symptom scores between the two groups.

In another study, the treatment protocol involved 3 daily sessions of iTBS, separated by 15 min, with a total of 1800 pulses delivered per day. The trains of iTBS were administered as 10 bursts of stimulation for 2 s, followed by an 8-s pause, with each burst consisting of three pulses at 50 Hz. This study reported a significant effect of active iTBS compared with sham on *n*-back task performance, particularly on the 3-back task [117].

Clinical Recommendations

The iTBS-like protocol emits 10 bursts and repeats every 10 s. In DC mode, direct current is delivered with an adjustable intensity ranging from 0 to 3 mA, generated by an adjustable current source and controlled by eight switches along with specific resistors. The duration of electrical stimulation can be set from 0 to 60 min, with increments of 1 min [86].

The most commonly used theta burst paradigm, iTBS, delivers triplet burst pulses at a frequency of 50 Hz, with bursts occurring every 5 Hz. This paradigm has been found to enhance cortical excitability, inducing long-term potentiation [95]. Further research is needed to investigate optimal TBS stimulation parameters, particularly for iTBS, to ensure these paradigms can be effectively translated into clinical settings [85]. The clinical recommendation levels for iTBS still requires further exploration and consensus (Table 5.1).

5.2.5.2 Accelerated iTBS

To improve response rates, accelerated stimulation protocols are currently being investigated. Unlike traditional daily sessions, which typically extend over 2–4 weeks, these accelerated protocols deliver the same number of stimulation sessions within a shorter timeframe. Recently, accelerated iTBS, a form of rTMS that utilizes bursts of high-frequency stimulation at repeated intervals, has been successfully implemented [118]. This method has been found to elicit rapid clinical responses after just 4 days of stimulation and significantly shorten the overall duration of the treatment protocol, providing advantages for both patients and caregivers [98, 119, 120].

However, randomized controlled studies on the effectiveness of a iTBS for improving social cognition and negative symptoms in individuals with schizophrenia are very limited [100].

In a case series report, accelerated iTBS was delivered to the TPJ over 5 days, with 5 sessions conducted each day, aimed at controlling AVH in three patients [121]. Two of the patients demonstrated significant improvement in AVH [95]. Another study found that accelerated iTBS can effectively ameliorate social cognition and negative symptoms in individuals with schizophrenia.

These findings suggest that accelerated iTBS may be a safe and effective neuromodulation technique for enhancing the overall functional recovery of individuals with schizophrenia, indicating a promising clinical application [100].

Efficacy

Research has demonstrated that accelerated iTBS administration can benefit individuals with schizophrenia. These accelerated protocols not only have the potential to treat a larger number of patients within the same timeframe but also to reduce the frequency of patient visits to rTMS clinics. As a result, this enhances overall accessibility and promotes better treatment adherence [122, 123].

However, no significant changes were observed in positive symptoms or general psychopathology [100]. In another study, accelerated iTBS was applied over the left TPJ for 5 days, with 5 sessions per day, aimed at controlling AVH in three patients [121]. Notably, two of these patients exhibited significant improvements in their AVH.

Future studies should explore the long-term effects of accelerated iTBS, considering both the duration of intervention and the intensity of stimulation, on social cognition and negative symptoms in individuals with chronic schizophrenia.

Safety

The accelerated iTBS protocol demonstrated good tolerability and a favorable safety profile. Adverse side effects were minimal; notably, no instances of seizures, fainting, speech difficulties, comprehension issues, or cognitive impairments were observed. The most commonly reported side effect was the occasional mild headache, which typically resolved either spontaneously or after the administration of acetaminophen [124].

These findings suggest that accelerated iTBS may improve the overall functional recovery of individuals with schizophrenia, and that it is a safe and effective neuromodulation technique with good clinical application prospects [100].

Treatment Regimen

Accelerated iTBS was conducted using a Brain Ultimate M-100 transcranial magnetic stimulator (Brain Ultimate, Atlanta, USA) equipped with a 70-mm air-cooled butterfly coil. The parameters for accelerated iTBS included: three pulses per burst, 50 Hz bursts, with a 2-s on and 8-s off cycle; totaling 600 pulses per session, resulting in a total duration of 3 min and 9 s, at an RMT intensity of 120% [99]. Given that high-dose TBS may induce discomfort in patients [125], the protocol involved three iTBS rounds per treatment day, separated by 15 min (totaling 1800 pulses per day).

Simultaneously, the TMS Navigation System (Brain Ultimate, Atlanta, USA) was employed to precisely locate the coil, which was positioned using the Visor 2.0, tangential to the skull with the handle oriented in a posterolateral direction. The stimulation target was identified in the left DLPFC, which may serve as an effective target for individuals with schizophrenia [126]. Treatment sessions were conducted in a sitting or semirecumbent position, occurring once daily, five times per week, for 4 weeks, culminating in a total of 20 sessions. All equipment and parameters were standardized and calibrated by the manufacturer. During treatment, subjects were required to wear noise-canceling earplugs and remove all objects sensitive to magnetic fields, including watches, cell phones, jewelry, and metal dentures.

In a case series report, accelerated intermittent iTBS was administered over the left TPJ for 5 days, with 5 sessions each day, to control AVH in three patients [121]. Each session comprised 1800 pulses delivered over 2 min, with bursts of stimulation at 50 Hz repeated every 200 ms. The target location was determined using neuronavigation. Notably, two patients, who had experienced only one antipsychotic treatment failure, demonstrated significant improvement in their AVH [121].

Previous study also found that accelerated iTBS can effectively improve social cognition and alleviate negative symptoms in individuals with schizophrenia [100].

Clinical Recommendations

To date, the most rapid iTBS (left DLPFC) treatment protocol is the Stanford Neuromodulation Therapy (SNT) [127]. This protocol involves 10 sessions of iTBS (left DLPFC) delivered daily, culminating in a total of 18,000 pulses per day (i.e., equivalent to the total number of pulses in the conventional 6-week iTBS protocol) over 5 consecutive days [123, 127]. The levels of clinical recommendation for accelerated iTBS require further investigation and consensus (Table 5.1).

5.2.5.3 Continuation TBS

Efficacy

TBS protocols, developed patterns of rTMS, exhibit an excitatory effect when the stimulation is intermittent (iTBS) and a suppressive effect when the stimulation is continuous (cTBS). cTBS, a refined paradigm of rTMS, has been shown to offer greater clinical benefits compared to traditional rTMS methods, demonstrating more profound and lasting effects on cortical excitability. This enhanced efficacy is attributed to its lower intensity, fewer pulses, and shorter stimulation duration [128, 129]. The suppressive effect of cTBS on cortical excitability is achieved by delivering three pulses at 50 Hz, repeated every 200 ms for a total duration of 40 s, at an intensity of 80% of the AMT [130].

In a previous study investigating the safety and tolerability of TBS, the safety, tolerability, and antidepressant efficacy of cTBS targeting the right DLPFC were validated. This was accomplished using various stimulation protocols ranging from 600 to 900 stimuli per train, resulting in a total of 1200–3600 stimuli per session [131]. Additionally, it has been observed that repeating the cTBS sequence after 15 min can significantly extend the duration of its effects [132, 133].

One study demonstrates that both cTBS and rTMS can reduce symptoms of schizophrenia-related AVH; however, cTBS generally provides greater therapeutic benefits. cTBS was associated with more significant improvements in clinical symptoms, as assessed by the PANSS and the Auditory Hallucination Rating Scale (AHRS), as well as in cognitive functions evaluated by the MATRICS Consensus Cognitive Battery (MCCB). Additionally, cTBS treatments resulted in increased serum levels of brain-derived neurotrophic factor (BDNF) and glial cell line-derived neurotrophic factor (GDNF), with a more notable increase observed following cTBS [132]. The cTBS paradigm has also been found to decrease cortical excitability, inducing long-term depression [87].

Safety

cTBS was demonstrated to be well tolerated, and notably, no adverse events or serious adverse events were reported [113].

Moreover, cTBS treatment did not result in an elevation in the occurrence of adverse events, nor did it cause an increase in treatment discontinuations due to any reason. This clearly suggests that cTBS is not only safe but also well tolerated when it comes to the treatment of auditory hallucinations [134].

Treatment Regimen

The previous study employed a cTBS intervention consisting of 25 sessions delivered over 5 consecutive days to the left primary motor cortex (M1 area), defined as the point of maximal stimulation of the abductor pollicis brevis. Each cTBS session involved a total of 600 pulses delivered over 200 bursts, utilizing a protocol of 50 Hz bursts of three pulses every 200 ms for a duration of 40 s, with an intensity set at 80% of the individual RMT. The RMT intensity was measured as the minimum stimulus required to induce contraction of the right thumb at least five out of ten times. The study demonstrated the feasibility and potential of cTBS intervention as an effective preventive measure against olanzapine-induced weight gain by enhancing cognitive restraint in eating behavior among drug-naive first-episode individuals with schizophrenia [135].

Researchers also found that when the number of stimulation sessions exceeded 10 and the total number of pulses surpassed 6000, cTBS treatment could significantly improve auditory hallucination symptoms in patients. This finding underscores the importance of appropriate parameters in determining the clinical efficacy of cTBS for the treatment of auditory hallucinations.

In a case involving a patient with very late-onset schizophrenia-like psychosis, cTBS was applied over the TPJ, specifically at TP3 (an area between the T3 and P3 electrodes). A MagPro X100 stimulator was used to deliver the stimulation with a figure-eight coil (MCF-B65) for 10 working days. During the first 3 days, the cTBS group received two double trains of stimulation (with a 15-min interval), while from days 4 to 10, they received one double train. Each train consisted of 801 pulses delivered in 267 bursts, with each burst comprising three pulses at 30 Hz, repeated every 100 ms, and delivered at 90% of the RMT. The researchers reported that the stimulation facilitated the effects of medication and reduced AVH. They speculated that cTBS may serve as a beneficial adjunct to pharmacological treatment to enhance its efficacy [136].

In another case report, the effects of 1 Hz rTMS and cTBS on AVH were described [137]. A 40-year-old female patient suffering from AVH initially received 1 Hz rTMS twice daily for 10 days, but no improvement was observed in her symptoms. After 1 month, cTBS was applied over electrode TP3 (approximately located at the temporoparietal cortex), following a similar regimen as rTMS, for 10 days at 2 sessions per day. A MagPro R30 stimulator delivered the stimulations via a figure-eight coil (MCF-B65) over 10 working days. She received double trains of stimulation with a 30-min interval between them. Each train consisted of 600 pulses delivered in 200 bursts, each comprising three pulses at 30 Hz, repeated every

200 ms. The cTBS sessions were applied at 80% of the RMT. After the 16th treatment, the patient reported complete suppression of AVH for several hours, and by the ninth day of treatment, she noted a significant reduction in her AVH symptoms. Following the treatment, she received maintenance sessions and reported very low AVH scores (below 5) [137].

In another study, the stimulations were delivered at a 48-h interval, and the stimulation site was determined based on the 10–20 EEG system. The iTBS treatment and sham protocols consisted of a single session at 80% RMT, totaling 600 pulses. The 20 trains of iTBS were delivered as 10 bursts of stimulation over 2 s, followed by 8 s off. Each burst comprised three pulses at 50 Hz and was repeated every 200 ms. The cTBS group received one train of stimulation at 100% RMT, consisting of 801 pulses delivered in 267 bursts, with each burst comprising three pulses at 30 Hz, repeated every 100 ms. The authors reported that cTBS resulted in a meaningful and significant difference compared to sham stimulation, along with improvements in manual dexterity [95]. To explore motor responses in adolescent patients with schizophrenia, transcranial magnetic stimulation-electromyography (TMS-EMG) was used to measure the MEP responses in 20 patients and 20 healthy control individuals before and after receiving cTBS and iTBS [138]. While both iTBS and cTBS produced excitatory and inhibitory plastic effects, the inhibitory effect of cTBS in patients was reported to be significantly larger compared to the healthy control group, whereas the excitatory effect of iTBS was significantly lower in the patient group compared to healthy control.

Auditory hallucinations are the most commonly occurring psychotic symptoms in schizophrenia. cTBS has been employed as an adjunctive treatment for these hallucinations.

Subgroup analyses revealed that patients receiving adjuvant cTBS for more than 10 stimulation sessions, with a total number of pulses exceeding 6000, experienced statistically significant improvements in their hallucination symptoms. Each cTBS session was conducted at an intensity of 80% of the RMT, at a frequency of 5 Hz, lasting for 40 s. Each stimulation consisted of three bursts at 50 Hz, amounting to a total of 600 pulses, which resulted in a session duration of 15 min. Participants underwent a 15-min rTMS session five times per week for 4 weeks.

Overall, cTBS provides greater therapeutic benefits compared to rTMS. It was associated with more significant improvements in clinical symptoms, as assessed by the PANSS and the AHRS, as well as in cognitive functions evaluated by the MCCB. The cTBS group generally exhibited a higher therapeutic effect than the rTMS group.

Clinical Recommendations

cTBS is a form of continuous transcranial stimulation that emits 10 bursts and repeats every 2 s. cTBS involves uninterrupted TBS trains lasting 20 s (300 pulses) or 40 s (600 pulses) and has been shown to decrease cortical excitability for up to 60 min [87]. Researchers also found that when the number of stimulation sessions exceeded 10 and the total number of pulses surpassed 6000, cTBS treatment could significantly improve auditory hallucination symptoms in patients. The cTBS utilizes a protocol of 50 Hz bursts or 30 Hz bursts. The levels of clinical recommendation for cTBS require further investigation and consensus (Table 5.1).

5.2.5.4 Bilateral TBS

Efficacy

To enhance the efficacy and efficiency of rTMS for auditory hallucinations, researchers provided initial evidence for the effectiveness of continuous bilateral theta-burst stimulation, a patterned form of rTMS that exerts its inhibitory effects after a significantly shorter stimulation period (40 s) compared to the commonly applied 1 Hz rTMS (10–30 min) [139]. Additional case studies [137, 140, 141], case series [142], comparisons with 1-Hz rTMS [143], and a sham-controlled pilot study [144] indicate an even larger effect size.

In this first controlled pilot study, some of the time saved by the shorter cTBS stimulation trains was utilized for a sequential bilateral treatment approach. Stimulation of the right temporoparietal area through sequential bilateral stimulation was included to address the hypothesis that right temporoparietal hyperactivity may also contribute to the genesis of auditory hallucinations (AH) [39]. After 3 weeks (15 sessions) of blinded treatment involving 16 patients, an analysis of covariance with baseline scores on the Psychotic Symptom Rating Scale—Auditory hallucinations (PSYRATS-AH) as a covariate revealed a statistical trend toward a reduction in AH with bilateral cTBS compared to sham stimulation. This finding suggests the efficacy of bilateral cTBS in treating AH. However, a recent trial did not find significant differences between left unilateral cTBS and sham treatment after a 5-day regimen (two treatments with a 30-min break) in a group of 71 patients with AH [145]. Therefore, the 15 sessions of once-daily bilateral cTBS are likely to be more effective [144]. Nevertheless, the equal or additional benefits of right temporoparietal [146, 147] or bilateral [39] TMS have not yet been convincingly demonstrated [148].

The brief stimulation trains of cTBS [145] allow for bilateral application, which may enhance the treatment effects, particularly in light of findings from imaging studies [39]. The 3-week treatment period is based on the results of a pilot study [144], which demonstrated that extending the treatment duration did not significantly improve the efficacy of cTBS. Another study found that reducing stimulation time to less than 2 min for bilateral stimulation could significantly enhance its practicality.

This evidence supports the notion that the bilateral application of cTBS represents an innovative, briefer, and potentially more effective stimulation paradigm compared to placebo stimulation.

Safety

Studies have demonstrated that the short-term application of bilateral TBS-DBS is not only safe but also effective [149]. In a recent controlled clinical trial using bilateral cTBS, researchers have demonstrated that theta burst stimulations of the temporal and temporoparietal cortex are safe with respect to hearing and speech perception [148].

Treatment Regimen

For unilateral applications, the amplitude could remain constant for both cTBS and TBS-DBS; however, slight adjustments are necessary for bilateral TBS-DBS to align with observed effects and side effects. Safety and tolerability of TBS-DBS are clinically evaluated by clinicians during a prior monopolar review, which involved assessing side effect thresholds (including dysarthria, paresthesia, tetanic contractions, or malaise) and gathering feedback from subjects regarding any side effects experienced during the 30-min DBS session [149].

Previous studies have utilized treatment regimens involving bilateral TBS. Patients will be randomly assigned to two treatment arms in a 1:1 ratio. Randomized patients will receive a total of 15 sessions of either bilateral TBS or sham stimulation (placebo) during a treatment period of 3 consecutive weeks (1 session per weekday). In each session, stimulation (or sham stimulation) will be applied successively to the left and right temporoparietal cortex, or vice versa. To control for order effects, the stimulation sequence (left–right/right–left) will be randomized in the first session and reversed in each subsequent session [148]. Results indicated a reduction in the mean PSYRATS-AH score following bilateral TBS compared to sham treatment.

Another study reported the use of bilateral cTBS over the temporoparietal cortex in a 52-year-old man, administered daily for 9 weeks, which resulted in the suppression of AVH. The patient had a longstanding history of paranoid schizophrenia. The stimulation was delivered at 80% of the AMT, targeting the TPJ between the P3/T3 and P4/T4 electrodes based on the 10–20 EEG system. Each cTBS session consisted of a continuous 40-s train, with each burst comprising three pulses at 50 Hz. While a reduction in AVH was observed during the early sessions, complete attenuation of AVH was noted by the sixth week of treatment [95].

Clinical Recommendations

Each bilateral TBS session consisted of a continuous 40-s train, with each burst comprising three pulses at 50 Hz. Further research and consensus are needed on the levels of clinical recommendation for bilateral TBS (Table 5.1).

5.2.6 MST

Magnetic seizure therapy (MST) presents a promising alternative to electroconvulsive therapy (ECT) [150]. To address the inherent limitations of electrical stimuli, the concept of using magnetic pulses as substitutes for ECT was proposed in 1994 [150].

MST is a noninvasive physical treatment developed as an enhancement over conventional convulsive therapies [151]. It integrates the characteristics of both ECT and rTMS. MST generates electromagnetic stimulation through rapidly alternating magnetic fields, similar to rTMS, but with greater intensity [151] This stimulation induces a seizure to alleviate symptoms [152]. One notable advantage of magnetic stimuli is their ability to penetrate the scalp and skull without resistance, facilitating targeted application to superficial cortical regions, which may help minimize cognitive side effects [150].

MST is administered under general anesthesia, requiring short-acting anesthetics and muscle relaxants [151, 153]. A twin coil positioned at the vertex (or, in some cases, a figure-eight coil over the right prefrontal cortex) generates magnetic fields that create an indirect electric current in the brain, inducing a seizure when it surpasses the individual seizure threshold [154–157].

The magnetic seizure threshold is typically titrated during the first treatment session by progressively increasing the duration of the train [154]. EEGs are monitored and assessed to ensure the occurrence of the required seizure [156], and precautions are taken to prevent side effects. Earplugs are provided for both patients and staff to prevent tinnitus or potential hearing damage [154], and patients are fitted with a bite block to protect their teeth [158]. Session arrangements generally mirror those of an ECT treatment plan, occurring twice a week for a duration of 5–6 weeks [157, 158].

MST was first employed in a study aimed at treating a patient with major depression in 2001 [159]. Since then, a growing number of studies have investigated the antidepressant effects of MST for depression, yielding promising results [154], although these findings remain somewhat unclear [160]. However, evidence regarding its efficacy for schizophrenia is limited. It remains uncertain whether MST is more or less effective than ECT or nonconvulsive TMS for schizophrenia, as well as whether MST offers additional benefits for individuals with this condition.

Schizophrenia is one of the most prevalent and debilitating mental disorders. Approximately 20% of individuals with schizophrenia do not respond to antipsychotics, which currently serve as the primary treatment for this condition, necessitating the exploration of alternative treatment options. MST is one of the novel NIBS techniques that have been investigated in recent years [152].

Pilot studies have indicated that MST is safe for individuals with schizophrenia [161]. Additionally, because it may result in less cognitive impairment, MST could be particularly valuable for older populations [162].

The research also found that MST, when compared to ECT, may lead to fewer delayed memory deficits and less cognitive deterioration (as measured by the number of individuals with delayed memory deficits and the mean change in global cognitive function). Furthermore, MST may enhance cognitive function to a greater extent (as indicated by the number of individuals showing any cognitive improvement) [152]. Researchers have also discovered that MST and ECT may be comparable in improving the overall condition of individuals with schizophrenia, as well as in alleviating both positive and negative symptoms of the disorder [152].

5.2.6.1 Efficacy

MST is an innovative neurotherapeutic intervention currently under development for treating major affective disorders. It has been designed as a more targeted form of convulsive therapy, specifically inducing seizures in the prefrontal cortex while sparing medial temporal structures such as the hippocampus [156, 159, 163–165].

This novel approach seeks to combine the effectiveness of ECT with the minimal cognitive side effects associated with rTMS [166].

MST functions by inducing therapeutic seizures using HF-rTMS [167, 168]. Clinical evidence suggests that it offers significant benefits, particularly in reducing neurocognitive adverse effects, making it a promising alternative to ECT [169]. The treatment has demonstrated response rates ranging from 40% to 70%, with remission rates between 15% and 46% [170]. Preliminary evidence also points to potential antipsychotic effects of MST in schizophrenia, with negligible neurocognitive adverse effects.

MST and ECT may be comparable in enhancing the overall condition of adults with schizophrenia, as well as in alleviating both positive and negative symptoms. Notably, MST may result in fewer delayed memory deficits and less cognitive deterioration while also improving cognitive function compared to ECT.

A systematic review synthesized two single-arm self-controlled studies and found low certainty of evidence, suggesting that MST improved psychotic symptoms, albeit with a high discontinuation rate. The review also reported inconsistent evidence regarding the cognitive adverse effects of MST. This aligns with our findings, highlighting the current lack of high-quality evidence investigating the effects of MST in individuals with schizophrenia [171].

As a novel NIBS technique, MST was developed as an alternative to existing treatment options for schizophrenia. However, no recommendations for this new therapy can be made until a thorough assessment is conducted. The available data are too scarce to draw meaningful conclusions for individuals with schizophrenia, clinicians, and policymakers. Those considering MST should remain aware of the unknown risks involved, and caution should be exercised before its application [152].

Further investigation is essential to address these questions with certainty, with particular emphasis on evaluating cognitive function in relation to the side effects profile.

Despite the limited data, current evidence does not rule out the potential for MST to serve as a promising treatment alternative, potentially offering desirable efficacy and a benign side effect profile for individuals with schizophrenia. Well-designed RCTs are needed to inform clinical practice regarding the efficacy and tolerability of MST in this population. The optimal information size should be achieved to ensure good precision, and efforts should be made to implement blinding to avoid bias.

5.2.6.2 Safety

Research on the neurocognitive effects of magnetic seizure therapy consistently indicates that it is a cognitively safe neurotherapeutic intervention. Animal studies have demonstrated that MST does not lead to cognitive dysfunction [172, 173] or induce pathological changes in the brain [174, 175]. Furthermore, both preclinical and clinical evidence support the conclusion that MST does not adversely affect neurocognitive functions [176]. Additionally, pilot studies have shown that MST is safe for individuals with schizophrenia [161].

A convergence of findings from virtual computer simulations, preclinical studies, and clinical investigations reveals that MST has a negligible impact on cortical and subcortical structures, thereby helping to preserve cognitive abilities. Data from virtual models show that MST stimulation influences only 21% of total brain volume, with effects confined to gray matter. Preclinical studies further reinforce this notion, indicating that even at suprathreshold levels, MST has minimal effects on cortical areas and cognitive functions. These results are echoed in clinical trials, providing a robust foundation for the safety of MST across various research methodologies [166].

In terms of adverse events, no significant complications were identified. The most frequently reported adverse drug reactions (ADRs) included dizziness and subjective memory loss [176].

5.2.6.3 Treatment Regimen

MST employs a TMS device along with specialized coils to deliver a series of magnetic pulses at intensities sufficient to induce a seizure while the patient is under general anesthesia. Key parameters that can be adjusted in MST include the configuration and placement of the magnetic coil, as well as specific characteristics of the magnetic pulse, such as frequency, pulse width, amplitude, duration, and dosage (threshold or suprathreshold). Similar to ECT, these parameters are essential for optimizing the balance between therapeutic efficacy and adverse effects. For example, ECT administered with right unilateral electrode placement, ultra-brief pulse widths, and titrated doses has demonstrated high efficacy with minimal cognitive side effects compared to other configurations [177]. A recent review by Hoy and Fitzgerald (2010) highlighted that MST parameters significantly influence treatment outcomes, noting that seizure induction is more reliable at 100 Hz than at 50 Hz [155].

The MST setting is designed to closely mirror ECT clinical practices in China [178]. Participants in the study were scheduled for 10 sessions of MST over a 4-week period, consisting of 3 sessions per week during the first 2 weeks and 2 sessions per week in the following weeks. MST was administered under general anesthesia using intravenous etomidate (0.21–0.3 mg/kg) and propofol (1.82–2.44 mg/kg) for sedation. Muscle relaxation was achieved with intravenous succinylcholine (1 mg/kg), and intravenous atropine (0.5 mg) was administered to reduce airway secretions. The procedure utilized a MagPro X100 (MagVenture A/S, Denmark) set to 25 Hz and 100% output, with a pulse width of 370 us and a peak magnetic field intensity of 4.2 Tesla.

To account for the likelihood that seizure thresholds may increase with ongoing treatment [179], a titration method was employed to determine the duration of magnetic stimulation. The initial duration began at 4 s and was increased by 4 s in each subsequent session, up to a maximum of 20 s (corresponding to 100–500 pulses per session). If the quality of the seizure was deemed inadequate (defined as a duration of less than 15 s) during a session, the duration increment for the next session would be adjusted to 8 s. In instances where no seizure was induced, an additional stimulation lasting 20 s was administered immediately. The magnetic stimulation was delivered via a twin coil positioned with its midline on the vertex, allowing for the induction of a stronger and deeper electric field compared to other MST coil configurations [150].

In the other study, participants were placed under general anesthesia using intravenous etomidate and propofol. Intravenous succinylcholine was administered as a muscle relaxant, while intravenous atropine was used to reduce airway secretions. MST was delivered at a frequency of 50 Hz and 100% output, with a pulse width of 370 us and a peak magnetic field intensity of 4.2 Tesla. The duration of stimulation was titrated for each participant based on the quality of the induced seizure, with a maximum duration of 20 s. If no seizures were generated, an additional stimulation lasting 20 s was administered immediately. Magnetic stimulation was delivered via a twin coil positioned with its midline on the verte [152]. The findings suggest that MST and ECT may be comparable in enhancing the overall condition of adults with schizophrenia, as well as in alleviating both positive and negative symptoms. Additionally, MST may result in fewer delayed memory deficits and less cognitive deterioration while improving cognitive function compared to ECT. There appeared to be no significant difference between the two groups regarding early study termination due to any reason, adverse effects, or perceived inefficacy.

5.2.6.4 Clinical Recommendations

The findings indicate that improving seizure quality and treatment response may be possible by employing higher magnetic pulse dosages and frequencies. Nevertheless, further research is needed to explore the impact of frequency, treatment frequency, and other parameters on cognitive functioning [150]. The levels of clinical recommendation for MST require further investigation and consensus (Table 5.1).

5.3 tDCS

Nearly 15 years ago, researchers revitalized a field of study examining the effects of applying a weak electric current between two electrodes placed on the scalps of healthy individuals [180]. This technique, termed transcranial direct current stimulation, demonstrated a significant, time-dependent, and polarity-specific modulation of neuronal populations beneath the electrodes. Since this initial neurophysiological discovery, an increasing number of researchers and clinicians have been investigating the effects of tDCS across various cognitive and behavioral domains [181].

Modern tDCS devices typically consist of an adjustable direct current stimulator and two stimulating electrodes (an anode and a cathode). These electrodes are usually attached to two distinct locations on the scalp—either directly or via larger sponge electrodes and a weak current (0.5–2.0 mA) is passed between them. As these current flows between the electrodes, it is believed that a small portion of it penetrates the brain. This current flow is thought to modulate neural activity beneath the electrode and, to a lesser extent, influence diffuse areas of the brain [181–183].

5.3.1 Conventional tDCS

5.3.1.1 Efficacy

As a novel neuromodulation technique, tDCS has been used extensively in the treatment of schizophrenia. Active conventional tDCS significantly reduced symptom severity compared to baseline and compared to sham conventional tDCS in patients with schizophrenia [184]. Although accumulating evidence indicates that tDCS has not demonstrated effectiveness in diminishing overall positive symptoms in patients with schizophrenia, a meta-analysis revealed that active tDCS with $\geq$10 stimulation sessions could significantly improve auditory hallucinations compared to sham [185]. However, conventional tDCS protocols have exhibited a notable enhancement in negative symptoms, yielding a medium effect size. Furthermore, there is substantial evidence supporting the use of conventional tDCS in enhancing attention and working memory in patients with schizophrenia [5, 186]. Additionally, emerging studies suggest that conventional tDCS may be efficacious in improving attention and vigilance [187].

5.3.1.2 Safety

The general impression is that conventional tDCS is a safe technique, with adverse effects that are mild and transient [188]. A meta-analysis by Brunoni and colleagues [189] showed that conventional tDCS has a benign side-effect profile when administered in 1–2 sessions to healthy volunteers. However, only 56% of all reviewed studies reported the presence or absence of adverse effects, indicating an underreporting of adverse effects in conventional tDCS clinical research. According to this meta-analysis, the most common adverse effects are observed in the active group, including itching, tingling, headache, burning sensation, and discomfort. While conventional tDCS has been well studied in adults, there is currently no specific guidance on conventional tDCS dosage for children. Limited studies of conventional tDCS in the pediatric population have shown that adverse effects are similar to those in adults, primarily limited to itching or tingling at the site of stimulation, with no reports of serious side effects [190]. One study showed that 10 sessions of conventional tDCS were well tolerated, with the most common side effect being irritability, followed by acute mood changes, tingling and itching [191]. However, due to anatomical and neurophysiological differences in the developing brain, dosage parameters considered safe and effective in adults should be adapted to achieve comparable results in children [192]. The researchers underscored the importance of exercising caution when employing current intensities of 2 mA or higher in pediatric populations, as the average current dose over the cortical surface following conventional tDCS stimulation may be higher in children than in adults.

5.3.1.3 Treatment Regimen

The most common cortical targets in schizophrenia are the left DLPFC and the left TPJ. Dysfunctional activation of the prefrontal cortex has been associated with negative symptoms and cognitive deficits [193], while auditory hallucinations are thought to stem from spontaneous hyperactivity in temporal regions that is not

Table 5.2 Optimization strategies of tDCS in schizophrenia

tDCS	Clinical recommendations					Levels + references
	Anode position	Cathode position	Intensity (mA)	Length (min)	Duration (sessions)	
Conventional tDCS	The midpoint between the anode-F3 and FP1	The midpoint between the cathode-T7 and P3	≥0.515	≥20	10–40	Second-line [7, 195, 198–212]
HD-tDCS	F3	Unclear	2 mA	≥20	≥5	N/A [195, 211, 213–216]

Abbreviations: *HD-tDCS* high-definition transcranial direct current stimulation, *N/A* not applicable, *tDCS* transcranial direct current stimulation

sufficiently regulated by reduced prefrontal activity [186, 194]. Typically, conventional tDCS protocols position the anode to increase neural activity in the left DLPFC and the cathode to decrease neural activity in the left TPJ, enabling the simultaneous targeting of negative symptoms and hallucinations [186, 195–198]. Brain stimulation with conventional tDCS involves the application of weak electric fields at currents of 1–2 mA for 20–30 min. The recommended treatment regimen typically involves between 10 and 40 sessions. Ultimately, it appears that conventional tDCS may offer potential benefits for alleviating negative symptoms in patients with schizophrenia.

5.3.1.4 Clinical Recommendations

Conventional tDCS for schizophrenia recommends stimulating the midpoint between the anode-F3 and FP1 and the midpoint between the cathode-T7 and P3. The current density used was 0.515 A/m^2 for 20 min for 10–40 treatments, with great potential to improve hallucinatory symptoms (level B evidence). Table 5.2 summarizes the optimization strategies for this treatment regimen.

5.3.2 HD-tDCS

High-definition transcranial direct current stimulation (HD-tDCS), an optimized form of tDCS, has the potential for more focused neuromodulation [217].

A notable challenge of conventional tDCS is its use of large electrodes positioned far apart, which leads to the stimulation of both the targeted region and adjacent structures [218]. To overcome this limitation, a more focused technique has been developed called HD-tDCS. This approach employs a 4*1 ring electrode configuration that effectively narrows the area of stimulation, enhancing the precision and effectiveness of the treatment [219].

HD-tDCS is grounded in the technical methodology of conventional tDCS but offers a more precise, focused, and longer-lasting effect [220].

5.3.2.1 Efficacy

HD-tDCS represents an advancement over conventional tDCS, offering the potential for precise and focused neuromodulation. HD-tDCS employs arrays of small electrodes (12 mm), and utilizing these electrodes in a 4*1 ring configuration is predicted to concentrate stimulation based on finite element model analyses derived from high-resolution magnetic resonance imaging. Studies have demonstrated that conventional tDCS montages induce diffuse distributions of current flow across widespread brain areas, where the highest current density may not occur directly beneath the electrodes. In contrast, HD-tDCS ensures elevated current densities primarily in the target area, resulting in effects that appear stronger, last longer, and are more focused than those of conventional tDCS [221, 222].

HD-tDCS represents a cutting-edge NIBS technique based on the principle that weak electric currents, when precisely directed at specific areas of the scalp, can stimulate the underlying cortical regions. While HD-tDCS shares a fundamental technical framework with conventional tDCS—utilizing a montage that consists of one anode and one cathode—it incorporates several modifications that enhance its focus and prolong the neuromodulation effects [223].

Similar to conventional tDCS, which establishes a unidirectional circuit through the placement of electrodes on the scalp, HD-tDCS also utilizes an anode and a cathode. However, its key advantage lies in the targeted delivery of current to the desired site. Among various montage configurations, the 4*1 ring montage is the most frequently employed, allowing HD-tDCS to achieve specific therapeutic effects while minimizing the influence of cortical stimulation from opposing polarities [219, 223]. HD-tDCS is designed to provide more precise cortical stimulation or inhibition, along with improved cortical penetration due to its unique montage design, which confers notable advantages.

The clinical utility of HD-tDCS has been validated in various settings [224]. Three open-label studies involving larger sample sizes have investigated its efficacy across psychiatric disorders, including schizophrenia [225], depression [226], and post-traumatic stress disorder (PTSD). Together, these studies provide independent evidence supporting the clinical effectiveness of HD-tDCS [227].

Currently, while in psychiatric contexts, it has primarily focused on treating auditory hallucinations [227]. Recent studies have indicated the potential utility of HD-tDCS in treating tinnitus [228] and auditory hallucinations in schizophrenia [225]. However, we did not find any published research on the application of HD-tDCS for the treatment of negative symptoms in schizophrenia [217].

Compared to healthy controls, individuals with schizophrenia exhibit reduced integrity of the white matter tracts in the corpus callosum and corona radiata. HD-tDCS has been shown to enhance integrity in the corpus callosum as well as the anterior and superior corona radiata, which correlates with improvements in cognitive performance. Thus, HD-tDCS presents a promising approach to ameliorate cognitive deficits in schizophrenia through its modulatory effects on white matter tracts. Given the lack of approved treatments for cognitive deficits, these findings hold significant clinical relevance [220].

5.3.2.2 Safety

Research indicates that HD-tDCS is both safe and well-tolerated at current intensities of up to 2 mA [229]. Adverse effects were measured using the checklist developed by Eryilmaz et al. [230]. Itching at the stimulation site was reported by two patients in a total of 3 sessions. Burning sensation at the stimulation site was reported by one patient during a session. All these adverse effects were reported by patients receiving active stimulation. However, both itching and burning sensation were short lasting, and none of those adverse effects warranted discontinuation of HD-tDCS sessions. No other adverse effects were noted [217].

5.3.2.3 Treatment Regimen

HD-tDCS was administered using the Starstim 32 device (Neuroelectrics BE, Spain). Starstim 32 is a wearable device that wirelessly transmits 32-channel data via Wi-Fi. The study utilized a 4*1 ring montage with NG Pistim electrodes, which are sintered Ag/AgCl ring electrodes with a 1 cm radius. Electrodes were arranged on the scalp with the assistance of an electroencephalogram (EEG) cap and a plastic casing to secure their positions. The central electrode, designated as the anode, was placed at F3 (left DLPFC) according to the 10/10 montage of the EEG electrode placement system. This electrode was surrounded by four return electrodes (FC1, F7, FC5, and AF3), defining the modulation area. All sessions were monitored for adverse effects, and at the end of each session, the HD-tDCS adverse effect checklist was applied [230]. In line with previous studies, the stimulation level for the active group was set at 2 mA, with 2 sessions administered each day for 5 days, featuring a 3-s ramp-up and ramp-down. The total stimulation duration was 20 min [195, 231]. Two consecutive sessions were separated by a minimum interval of 5 h. For the sham stimulation, electrodes were positioned identically to those in the active stimulation, but a 1 mA direct current was applied with a 3 s ramp-up and ramp-down for 30 s [232], providing a sensation of stimulation while minimizing actual effects. Allocation to the active or sham group was conducted through envelope randomization [217].

In a case of persistent AVH in schizophrenia with comorbid alopecia universalis, central cathodal stimulation was applied to CP5 (temporoparietal junction, TPJ) using the 10/10 EEG system, with four return electrodes at FT7, FC3, PO7, and P1. The stimulation was set at 2 mA for a 20-min session, with a 3-h intersession interval over 5 days, totaling 10 sessions. Remarkably, there was a complete alleviation of auditory hallucinations following 5 days of HD-tDCS treatment [227]. No electrode-based side effects were noted, despite the absence of scalp hair due to comorbid alopecia universalis [227].

In another open-label study involving 19 participants with persistent AVH in schizophrenia, central cathodal stimulation was again directed to CP5 (TPJ) using the 10/10 EEG system and four return electrodes at FT7, FC3, PO7, and P1. This study reported a significant reduction in auditory hallucinations after 5 days of HD-tDCS treatment [225].

Persistent AVH in schizophrenia case report central cathodal stimulation to CP5 (TPJ) employing the 10/10 EEG system, with four return electrodes positioned at

FT7, C5, CP3, and P5. The intervention led to a significant reduction in auditory hallucinations following 5 days of HD-tDCS treatment, utilizing a 2 mA current for 20-min sessions, with a 3-h intersession interval across 10 sessions [227].

In a previous study, the intervention employed a 4*1 HD-tDCS device (Soterix Medical Inc., USA). Dedicated Soterix software (HD-Targets and HD-Explore™) was utilized to configure electrode montages and map electrical fields, allowing for the maximization of stimulation in the target area [233]. Specifically, a 4*1 ring configuration was implemented, featuring one central anodal electrode surrounded by four return cathodal electrodes. The anode was positioned over the left DLPFC at F3 (10/20 EEG system), with the four cathodes placed at AF3, F5, F1, and FC3, forming a circular arrangement. This HD-tDCS montage is widely employed to target the left DLPFC in neuropsychiatric disorders [227]. Conductive gel was carefully applied to each individual electrode. Subsequently, the electrodes were firmly affixed to the scalp by means of an electroencephalogram (EEG) cap. Before the initiation of stimulation and throughout every session, impedance monitoring of the interface between the electrodes and the scalp was meticulously carried out.

In the active group, a consistent 2 mA current was administered for a duration of 20 min. In contrast, in the sham group, the same current was delivered for merely 30 s. It is worth noting that for both groups, there were 20-s "ramp up" and "ramp down" periods incorporated. Participants were explicitly instructed to maintain a state of relaxation and to refrain from engaging in any form of activity during the entire stimulation period [220].

In other study, participant received 10 sessions of HD-tDCS at 2 mA for 20 min, administered twice daily over 5 days. The improvement in negative symptoms within the active group was statistically significant, with a *p*-value of 0.05 when compared to the sham group. These results suggest that HD-tDCS may contribute to the improvement of negative symptoms in schizophrenia, indicating that its use as an adjunct to pharmacological treatment for these symptoms may be worth considering [217].

5.3.2.4 Clinical Recommendations

Table 5.2 summarizes the optimization strategies for this treatment regimen. In a comprehensive review of seven studies, all but one [234] consistently utilized a current intensity of 2 mA at the stimulation site. In the exception, which compared the effects of tDCS and HD-tDCS, a lower intensity of 0.5 mA was employed for HD-tDCS. The authors posited that this reduced intensity could account for the lack of significant effects observed in their findings.

Moreover, another study that varied both current intensity (1 mA vs. 2 mA) and stimulation duration (10 min vs. 20 min) demonstrated that the combination of 2 mA for 20 min yielded significantly greater efficacy [235].

5.4 TMS vs. tDCS

5.4.1 Efficacy

NIBS has been proposed as an intervention strategy for mental disorders. NIBS includes TMS and tDCS [186]. The stimulation frequency of TMS can either increase or decrease cortical excitability and ranges from 1 to 50 Hz. On the other hand, tDCS is another noninvasive neurostimulation technique that uses direct electrical currents to stimulate specific cortical regions. Anodal tDCS (AtDCS) is generally thought to enhance the function of the underlying cortical areas, while cathodal tDCS has a suppressive effect. The sustained effects of tDCS are thought to share a similar neurobiological basis with TMS, involving the induction of long-term potentiation like plasticity [186, 236, 237].

In patients with schizophrenia, tDCS and TMS show different treatment outcomes in reducing positive symptoms. For auditory hallucinations, tDCS produced a small but highly heterogeneous positive effect, and TMS produced a negligible but homogeneous positive effect. However, both TMS and tDCS protocols significantly improved negative symptoms with a medium effect size. It should be noted that there was a high degree of heterogeneity in these results. Some caution is needed in interpreting the results of NIBS for negative symptoms in schizophrenia. In terms of the efficacy on cognitive impairments in schizophrenia, active TMS was not superior to sham for cognitive improvement. In contrast, tDCS proved effective in enhancing attention and working memory in schizophrenia patients [186]. Furthermore, research indicates that tDCS may be more effective than TMS in enhancing attention/vigilance [187]. In conclusion, it seems that neurostimulation could be used in patients with schizophrenia, in particular for the negative symptoms and cognitive impairments of the disorder.

5.4.2 Safety

Both TMS and tDCS have been widely reported as safe techniques with minimal adverse effects [209, 238]. In terms of safety, unilateral rTMS has been shown to be a safe, NIBS technique with no serious adverse effects other than the well-known side effects of mild headache and localized scalp discomfort observed in the majority of studies. The general impression is that tDCS is a safe technique, with adverse effects that are mild and transient [188]. The most common adverse effects are observed in the active group and include itching, tingling, headache, burning, and discomfort.

5.5 Conclusion

The diverse and complex nature of schizophrenia has hindered the establishment of consensus regarding the efficacy of novel noninvasive neurostimulation techniques, specifically TMS and tDCS. Despite mixed findings on their effectiveness, both TMS and tDCS show promise as treatment options for patients with schizophrenia. However, additional research is necessary to determine the optimal stimulation parameters and identify key variables that may impact individual responses to either TMS or tDCS. In the design of future clinical trials, a priority should be the examination of the efficacy of higher doses of tDCS in schizophrenia. These trials should entail the administration of stimulation sessions twice, and potentially even three times, per day or using larger current intensity. Moreover, conducting follow-up assessments 1, 2, and 3 months following the conclusion of treatment is imperative. Furthermore, there is a need for further exploration into the potential of neuronavigation in conjunction with TMS, as this modification of TMS has shown promise in the treatment of schizophrenia. It is anticipated that these methods will contribute significantly to enhancing our understanding of the underlying pathological mechanisms and treatment of schizophrenia, ultimately leading to the achievement of sustained and effective outcomes.

Acknowledgments None.

Disclosure/Conflicts of Interest The authors declare no conflicts of interest in conducting this study or preparing the manuscript.

Financial Support This work was supported by the National Natural Science Foundation of China, 82301693 and 31900751; the Capital's Funds for Health Improvement and Research, CFH2024-4-2132; Beijing Municipal Science & Technology Commission, Z221100007422047. The funders had no role in study design, data collection and analysis, decision to publish, or preparation of the manuscript.

References

1. Huang Y, Wang Y, Wang H, et al. Prevalence of mental disorders in China: a cross-sectional epidemiological study. Lancet Psychiatry. 2019;6:211–24.
2. McCutcheon RA, Reis Marques T, Howes OD. Schizophrenia-an overview. JAMA Psychiatry. 2020;77:201–10.
3. Salokangas RKR. Symptom dimensions and outcome in schizophrenia. World Psychiatry. 2003;2:172–8.
4. Bowie CR, Harvey PD. Cognitive deficits and functional outcome in schizophrenia. Neuropsychiatr Dis Treat. 2006;2:531–6.
5. Sun C-H, Mai J-X, Shi Z-M, Zheng W, Jiang W-L, Li Z-Z, Huang X-B, Yang X-H, Zheng W. Adjunctive repetitive transcranial magnetic stimulation for adolescents with first-episode major depressive disorder: a meta-analysis. Front Psych. 2023;14:1200738.
6. Aleman A, Enriquez-Geppert S, Knegtering H, Dlabac-de Lange JJ. Moderate effects of noninvasive brain stimulation of the frontal cortex for improving negative symptoms in schizophrenia: meta-analysis of controlled trials. Neurosci Biobehav Rev. 2018;89:111–8.

7. Osoegawa C, Gomes JS, Grigolon RB, et al. Non-invasive brain stimulation for negative symptoms in schizophrenia: an updated systematic review and meta-analysis. Schizophr Res. 2018;197:34–44.
8. Klein E, Kolsky Y, Puyerovsky M, Koren D, Chistyakov A, Feinsod M. Right prefrontal slow repetitive transcranial magnetic stimulation in schizophrenia: a double-blind sham-controlled pilot study. Biol Psychiatry. 1999;46:1451–4.
9. Lorentzen R, Nguyen TD, McGirr A, Hieronymus F, Østergaard SD. The efficacy of transcranial magnetic stimulation (TMS) for negative symptoms in schizophrenia: a systematic review and meta-analysis. Schizophrenia (Heidelb Ger). 2022;8:35.
10. Jiang Y, Guo Z, Xing G, He L, Peng H, Du F, McClure MA, Mu Q. Effects of high-frequency transcranial magnetic stimulation for cognitive deficit in schizophrenia: a meta-analysis. Front Psych. 2019;10:135.
11. Gan J, Duan H, Chen Z, Shi Z, Gao C, Zhu X, Liang X. Effectiveness and safety of high dose transcranial magnetic stimulation in schizophrenia with refractory negative symptoms: a randomized controlled study. Zhonghua Yi Xue Za Zhi. 2015;95:3808–12.
12. Hoffman RE, Gueorguieva R, Hawkins KA, Varanko M, Boutros NN, Wu Y, Carroll K, Krystal JH. Temporoparietal transcranial magnetic stimulation for auditory hallucinations: safety, efficacy and moderators in a fifty patient sample. Biol Psychiatry. 2005;58:97–104.
13. Wang J, Zhou Y, Gan H, Pang J, Li H, Wang J, Li C. Efficacy towards negative symptoms and safety of repetitive transcranial magnetic stimulation treatment for patients with schizophrenia: a systematic review. Shanghai Arch Psychiatry. 2017;29:61–76.
14. Cole J, Bright K, Gagnon L, McGirr A. A systematic review of the safety and effectiveness of repetitive transcranial magnetic stimulation in the treatment of peripartum depression. J Psychiatr Res. 2019;115:142–50.
15. Loo CK, McFarquhar TF, Mitchell PB. A review of the safety of repetitive transcranial magnetic stimulation as a clinical treatment for depression. Int J Neuropsychopharmacol. 2008;11:131–47.
16. Arumugham SS, Thirthalli J, Andrade C. Efficacy and safety of combining clozapine with electrical or magnetic brain stimulation in treatment-refractory schizophrenia. Expert Rev Clin Pharmacol. 2016;9:1245–52.
17. Singh S, Kumar N, Verma R, Nehra A. The safety and efficacy of adjunctive 20-Hz repetitive transcranial magnetic stimulation for treatment of negative symptoms in patients with schizophrenia: a double-blinded, randomized, sham-controlled study. Indian J Psychiatry. 2020;62:21–9.
18. Tikka SK, Godi SM, Siddiqui MA, Garg S. Evidence from Indian studies on safety and efficacy of therapeutic transcranial magnetic stimulation across neuropsychiatric disorders – a systematic review and meta-analysis. Indian J Psychiatry. 2023;65:18–35.
19. Chibbaro G, Daniele M, Alagona G, Di Pasquale C, Cannavò M, Rapisarda V, Bella R, Pennisi G. Repetitive transcranial magnetic stimulation in schizophrenic patients reporting auditory hallucinations. Neurosci Lett. 2005;383:54–7.
20. Poulet E, Brunelin J, Bediou B, Bation R, Forgeard L, Dalery J, d'Amato T, Saoud M. Slow transcranial magnetic stimulation can rapidly reduce resistant auditory hallucinations in schizophrenia. Biol Psychiatry. 2005;57:188–91.
21. Rosa MO, Gattaz WF, Rosa MA, et al. Effects of repetitive transcranial magnetic stimulation on auditory hallucinations refractory to clozapine. J Clin Psychiatry. 2007;68:1528–32.
22. Saba G, Verdon CM, Kalalou K, Rocamora JF, Dumortier G, Benadhira R, Stamatiadis L, Vicaut E, Lipski H, Januel D. Transcranial magnetic stimulation in the treatment of schizophrenic symptoms: a double blind sham controlled study. J Psychiatr Res. 2006;40:147–52.
23. Cohen E, Bernardo M, Masana J, et al. Repetitive transcranial magnetic stimulation in the treatment of chronic negative schizophrenia: a pilot study. J Neurol Neurosurg Psychiatry. 1999;67:129–30.
24. Dlabac-de Lange JJ, Bais L, van Es FD, Visser BGJ, Reinink E, Bakker B, van den Heuvel ER, Aleman A, Knegtering H. Efficacy of bilateral repetitive transcranial magnetic stimula-

tion for negative symptoms of schizophrenia: results of a multicenter double-blind randomized controlled trial. Psychol Med. 2015;45:1263–75.
25. Prikryl R, Ustohal L, Prikrylova Kucerova H, Kasparek T, Venclikova S, Vrzalova M, Ceskova E. A detailed analysis of the effect of repetitive transcranial magnetic stimulation on negative symptoms of schizophrenia: a double-blind trial. Schizophr Res. 2013;149:167–73.
26. Schneider AL, Schneider TL, Stark H. Repetitive transcranial magnetic stimulation (rTMS) as an augmentation treatment for the negative symptoms of schizophrenia: a 4-week randomized placebo controlled study. Brain Stimul. 2008;1:106–11.
27. Holi MM, Eronen M, Toivonen K, Toivonen P, Marttunen M, Naukkarinen H. Left prefrontal repetitive transcranial magnetic stimulation in schizophrenia. Schizophr Bull. 2004;30:429–34.
28. Mittrach M, Thünker J, Winterer G, et al. The tolerability of rTMS treatment in schizophrenia with respect to cognitive function. Pharmacopsychiatry. 2010;43:110–7.
29. Zhao S, Kong J, Li S, Tong Z, Yang C, Zhong H. Randomized controlled trial of four protocols of repetitive transcranial magnetic stimulation for treating the negative symptoms of schizophrenia. Shanghai Arch Psychiatry. 2014;26:15–21.
30. Li Z, Yin M, Lyu X-L, Zhang L-L, Du X-D, Hung GC-L. Delayed effect of repetitive transcranial magnetic stimulation (rTMS) on negative symptoms of schizophrenia: findings from a randomized controlled trial. Psychiatry Res. 2016;240:333–5.
31. Tseng P-T, Zeng B-S, Hung C-M, et al. Assessment of noninvasive brain stimulation interventions for negative symptoms of schizophrenia: a systematic review and network meta-analysis. JAMA Psychiatry. 2022;79:770–9.
32. Wobrock T, Guse B, Cordes J, et al. Left prefrontal high-frequency repetitive transcranial magnetic stimulation for the treatment of schizophrenia with predominant negative symptoms: a sham-controlled, randomized multicenter trial. Biol Psychiatry. 2015;77:979–88.
33. Wölwer W, Lowe A, Brinkmeyer J, Streit M, Habakuck M, Agelink MW, Mobascher A, Gaebel W, Cordes J. Repetitive transcranial magnetic stimulation (rTMS) improves facial affect recognition in schizophrenia. Brain Stimul. 2014;7:559–63.
34. Yi S, Wang Q, Wang W, Hong C, Ren Z. Efficacy of repetitive transcranial magnetic stimulation (rTMS) on negative symptoms and cognitive functioning in schizophrenia: an umbrella review of systematic reviews and meta-analyses. Psychiatry Res. 2024;333:115728.
35. Paillère-Martinot M-L, Galinowski A, Plaze M, et al. Active and placebo transcranial magnetic stimulation effects on external and internal auditory hallucinations of schizophrenia. Acta Psychiatr Scand. 2017;135:228–38.
36. Dlabac-de Lange JJ, Liemburg EJ, Bais L, Van De Poel-Mustafayeva AT, De Lange-de Klerk ESM, Knegtering H, Aleman A. Effect of bilateral prefrontal rTMS on left prefrontal NAA and Glx levels in schizophrenia patients with predominant negative symptoms: an exploratory study. Brain Stimul. 2017;10:59–64.
37. Liemburg EJ, Dlabac-De Lange JJ, Bais L, Knegtering H, Aleman A. Effects of bilateral prefrontal rTMS on brain activation during social-emotional evaluation in schizophrenia: a double-blind, randomized, exploratory study. Schizophr Res. 2018;202:210–1.
38. Barr MS, Farzan F, Tran LC, Fitzgerald PB, Daskalakis ZJ. A randomized controlled trial of sequentially bilateral prefrontal cortex repetitive transcranial magnetic stimulation in the treatment of negative symptoms in schizophrenia. Brain Stimul. 2012;5:337–46.
39. Hoffman RE, Wu K, Pittman B, Cahill JD, Hawkins KA, Fernandez T, Hannestad J. Transcranial magnetic stimulation of Wernicke's and right homologous sites to curtail "voices": a randomized trial. Biol Psychiatry. 2013;73:1008–14.
40. Bais L, Vercammen A, Stewart R, van Es F, Visser B, Aleman A, Knegtering H. Short and long term effects of left and bilateral repetitive transcranial magnetic stimulation in schizophrenia patients with auditory verbal hallucinations: a randomized controlled trial. PLoS One. 2014;9:e108828.
41. Kim E-J, Yeo S, Hwang I, Park J-I, Cui Y, Jin H-M, Kim HT, Hwang T-Y, Chung Y-C. Bilateral repetitive transcranial magnetic stimulation for auditory hallucinations in patients with

schizophrenia: a randomized controlled, cross-over study. Clin Psychopharmacol Neurosci. 2014;12:222–8.
42. Brunelin J, Galvao F, Mondino M. Twice daily low frequency rTMS for treatment-resistant auditory hallucinations. Int J Clin Health Psychol. 2023;23:100344.
43. Montagne-Larmurier A, Etard O, Razafimandimby A, Morello R, Dollfus S. Two-day treatment of auditory hallucinations by high frequency rTMS guided by cerebral imaging: a 6 month follow-up pilot study. Schizophr Res. 2009;113:77–83.
44. Dollfus S, Jaafari N, Guillin O, et al. High-frequency neuronavigated rTMS in auditory verbal hallucinations: a pilot double-blind controlled study in patients with schizophrenia. Schizophr Bull. 2018;44:505–14.
45. Kimura H, Kanahara N, Takase M, Yoshida T, Watanabe H, Iyo M. A randomized, sham-controlled study of high frequency rTMS for auditory hallucination in schizophrenia. Psychiatry Res. 2016;241:190–4.
46. Levkovitz Y, Rabany L, Harel EV, Zangen A. Deep transcranial magnetic stimulation add-on for treatment of negative symptoms and cognitive deficits of schizophrenia: a feasibility study. Int J Neuropsychopharmacol. 2011;14:991–6.
47. Linsambarth S, Jeria A, Avirame K, Todder D, Riquelme R, Stehberg J. Deep transcranial magnetic stimulation for the treatment of negative symptoms in schizophrenia: beyond an antidepressant effect. J ECT. 2019;35:e46–54.
48. Rabany L, Deutsch L, Levkovitz Y. Double-blind, randomized sham controlled study of deep-TMS add-on treatment for negative symptoms and cognitive deficits in schizophrenia. J Psychopharmacol (Oxf Engl). 2014;28:686–90.
49. Ray P, Sinha VK, Tikka SK. Adjuvant low-frequency rTMS in treating auditory hallucinations in recent-onset schizophrenia: a randomized controlled study investigating the effect of high-frequency priming stimulation. Ann General Psychiatry. 2015;14:8.
50. Blumberger DM, Christensen BK, Zipursky RB, Moller B, Chen R, Fitzgerald PB, Daskalakis ZJ. MRI-targeted repetitive transcranial magnetic stimulation of Heschl's gyrus for refractory auditory hallucinations. Brain Stimul. 2012;5:577–85.
51. Barr MS, Farzan F, Rajji TK, Voineskos AN, Blumberger DM, Arenovich T, Fitzgerald PB, Daskalakis ZJ. Can repetitive magnetic stimulation improve cognition in schizophrenia? Pilot data from a randomized controlled trial. Biol Psychiatry. 2013;73:510–7.
52. Fitzgerald PB, Herring S, Hoy K, McQueen S, Segrave R, Kulkarni J, Daskalakis ZJ. A study of the effectiveness of bilateral transcranial magnetic stimulation in the treatment of the negative symptoms of schizophrenia. Brain Stimul. 2008;1:27–32.
53. Kozak K, Sharif-Razi M, Morozova M, Gaudette EV, Barr MS, Daskalakis ZJ, Blumberger DM, George TP. Effects of short-term, high-frequency repetitive transcranial magnetic stimulation to bilateral dorsolateral prefrontal cortex on smoking behavior and cognition in patients with schizophrenia and non-psychiatric controls. Schizophr Res. 2018;197:441–3.
54. Vercammen A, Knegtering H, Bruggeman R, Westenbroek HM, Jenner JA, Slooff CJ, Wunderink L, Aleman A. Effects of bilateral repetitive transcranial magnetic stimulation on treatment resistant auditory-verbal hallucinations in schizophrenia: a randomized controlled trial. Schizophr Res. 2009;114:172–9.
55. Caulfield KA, Fleischmann HH, George MS, McTeague LM. A transdiagnostic review of safety, efficacy, and parameter space in accelerated transcranial magnetic stimulation. J Psychiatr Res. 2022;152:384–96.
56. Brunelin J, Poulet E, Bediou B, Kallel L, Dalery J, D'amato T, Saoud M. Low frequency repetitive transcranial magnetic stimulation improves source monitoring deficit in hallucinating patients with schizophrenia. Schizophr Res. 2006;81:41–5.
57. Sverak T, Mayerova M, Obdrzalkova M, Ustohal L. Accelerated repetitive transcranial magnetic stimulation in the treatment of negative symptoms of schizophrenia: an open-label study. J ECT. 2022;38:e24–5.

58. Chauhan P, Garg S, Tikka SK, Khattri S. Efficacy of intensive cerebellar intermittent theta burst stimulation (iCiTBS) in treatment-resistant schizophrenia: a randomized placebo-controlled study. Cerebellum (Lond Engl). 2021;20:116–23.
59. Rachid F. Accelerated transcranial magnetic stimulation for the treatment of patients with depression: a review. Asian J Psychiatry. 2019;40:71–5.
60. Desbeaumes Jodoin V, Miron J-P, Lespérance P. Safety and efficacy of accelerated repetitive transcranial magnetic stimulation protocol in elderly depressed unipolar and bipolar patients. Am J Geriatr Psychiatry. 2019;27:548–58.
61. Bersani FS, Minichino A, Enticott PG, et al. Deep transcranial magnetic stimulation as a treatment for psychiatric disorders: a comprehensive review. Eur Psychiatry. 2013;28:30–9.
62. Di Passa A-M, Prokop-Millar S, Yaya H, Dabir M, McIntyre-Wood C, Fein A, MacKillop E, MacKillop J, Duarte D. Clinical efficacy of deep transcranial magnetic stimulation (dTMS) in psychiatric and cognitive disorders: a systematic review. J Psychiatr Res. 2024;175:287–315.
63. Lu M, Ueno S. Comparison of the induced fields using different coil configurations during deep transcranial magnetic stimulation. PLoS One. 2017;12:e0178422.
64. Moeller SJ, Gil R, Weinstein JJ, et al. Deep rTMS of the insula and prefrontal cortex in smokers with schizophrenia: proof-of-concept study. Schizophrenia. 2022;8:6.
65. Rosenberg O, Gersner R, Klein LD, Kotler M, Zangen A, Dannon P. Deep transcranial magnetic stimulation add-on for the treatment of auditory hallucinations: a double-blind study. Ann General Psychiatry. 2012;11:13.
66. Cole JC, Green Bernacki C, Helmer A, Pinninti N, O'Reardon JP. Efficacy of transcranial magnetic stimulation (TMS) in the treatment of schizophrenia: a review of the literature to date. Innov Clin Neurosci. 2015;12:12–9.
67. Soltani E, Bateni H, Shad MU. Different modalities of transcranial magnetic stimulation to manage schizophrenia. Prim Care Companion CNS Disord. 2022;24:21r03144.
68. Birdi A, Kumar PK, Nebhinani N, Gupta T, Tomo S, Purohit P, Banerjee M, Mitra P, Sharma P, Yadav D. Association of circulatory Klotho levels and its expression with miRNA-339 in patients with schizophrenia. Behav Brain Res. 2023;445:114359.
69. Rosenberg O, Roth Y, Kotler M, Zangen A, Dannon P. Deep transcranial magnetic stimulation for the treatment of auditory hallucinations: a preliminary open-label study. Ann General Psychiatry. 2011;10:3.
70. Iyer MB, Schleper N, Wassermann EM. Priming stimulation enhances the depressant effect of low-frequency repetitive transcranial magnetic stimulation. J Neurosci. 2003;23:10867–72.
71. Slotema CW, Blom JD, Hoek HW, Sommer IEC. Should we expand the toolbox of psychiatric treatment methods to include repetitive transcranial magnetic stimulation (rTMS)? A meta-analysis of the efficacy of rTMS in psychiatric disorders. J Clin Psychiatry. 2010;71:873–84.
72. Hasan A, Strube W, Palm U, Wobrock T. Repetitive noninvasive brain stimulation to modulate cognitive functions in schizophrenia: a systematic review of primary and secondary outcomes. Schizophr Bull. 2016;42:S95–109.
73. Marzouk T, Winkelbeiner S, Azizi H, Malhotra AK, Homan P. Transcranial magnetic stimulation for positive symptoms in schizophrenia: a systematic review. Neuropsychobiology. 2020;79:384–96.
74. Di Hou M, Santoro V, Biondi A, Shergill SS, Premoli I. A systematic review of TMS and neurophysiological biometrics in patients with schizophrenia. J Psychiatry Neurosci. 2021;46:E675–701.
75. Barr MS, Farzan F, Arenovich T, Chen R, Fitzgerald PB, Daskalakis ZJ. The effect of repetitive transcranial magnetic stimulation on gamma oscillatory activity in schizophrenia. PLoS One. 2011;6:e22627.
76. Farzan F, Barr MS, Levinson AJ, Chen R, Wong W, Fitzgerald PB, Daskalakis ZJ. Evidence for gamma inhibition deficits in the dorsolateral prefrontal cortex of patients with schizophrenia. Brain J Neurol. 2010;133:1505–14.

77. Ferrarelli F, Massimini M, Peterson MJ, et al. Reduced evoked gamma oscillations in the frontal cortex in schizophrenia patients: a TMS/EEG study. Am J Psychiatry. 2008;165:996–1005.
78. Barr MS, Farzan F, Rusjan PM, Chen R, Fitzgerald PB, Daskalakis ZJ. Potentiation of gamma oscillatory activity through repetitive transcranial magnetic stimulation of the dorsolateral prefrontal cortex. Neuropsychopharmacology. 2009;34:2359–67.
79. Farzan F, Barr MS, Sun Y, Fitzgerald PB, Daskalakis ZJ. Transcranial magnetic stimulation on the modulation of gamma oscillations in schizophrenia. Ann N Y Acad Sci. 2012;1265:25–35.
80. Levit-Binnun N, Handzy NZ, Moses E, Modai I, Peled A. Transcranial magnetic stimulation at M1 disrupts cognitive networks in schizophrenia. Schizophr Res. 2007;93:334–44.
81. Mehta UM, Naik SS, Thanki MV, Thirthalli J. Investigational and therapeutic applications of transcranial magnetic stimulation in schizophrenia. Curr Psychiatry Rep. 2019;21:89.
82. Ferrarelli F, Phillips ML. Examining and modulating neural circuits in psychiatric disorders with transcranial magnetic stimulation and electroencephalography: present practices and future developments. Am J Psychiatry. 2021;178:400–13.
83. Guo Q, Li C, Wang J. Updated review on the clinical use of repetitive transcranial magnetic stimulation in psychiatric disorders. Neurosci Bull. 2017;33:747–56.
84. Santoro V, Hou MD, Premoli I, et al. Investigating cortical excitability and inhibition in patients with schizophrenia: a TMS-EEG study. Brain Res Bull. 2024;212:110972.
85. Chung SW, Hill AT, Rogasch NC, Hoy KE, Fitzgerald PB. Use of theta-burst stimulation in changing excitability of motor cortex: a systematic review and meta-analysis. Neurosci Biobehav Rev. 2016;63:43–64.
86. Li Y-T, Chen S-C, Yang L-Y, Hsieh T-H, Peng C-W. Designing and implementing a novel transcranial electrostimulation system for neuroplastic applications: a preliminary study. IEEE Trans Neural Syst Rehabil Eng. 2019;27:805–13.
87. Huang Y-Z, Edwards MJ, Rounis E, Bhatia KP, Rothwell JC. Theta burst stimulation of the human motor cortex. Neuron. 2005;45:201–6.
88. Gamboa OL, Antal A, Moliadze V, Paulus W. Simply longer is not better: reversal of theta burst after-effect with prolonged stimulation. Exp Brain Res. 2010;204:181–7.
89. Goldsworthy MR, Pitcher JB, Ridding MC. The application of spaced theta burst protocols induces long-lasting neuroplastic changes in the human motor cortex. Eur J Neurosci. 2012;35:125–34.
90. Goldsworthy MR, Pitcher JB, Ridding MC. A comparison of two different continuous theta burst stimulation paradigms applied to the human primary motor cortex. Clin Neurophysiol. 2012;123:2256–63.
91. Daskalakis ZJ, Fitzgerald PB, Christensen BK. The role of cortical inhibition in the pathophysiology and treatment of schizophrenia. Brain Res Rev. 2007;56:427–42.
92. Poulet E, Haesebaert F, Saoud M, Suaud-Chagny MF, Brunelin J. Treatment of shizophrenic patients and rTMS. Psychiatr Danub. 2010;22(Suppl 1):S143–6.
93. Rossini PM, Burke D, Chen R, et al. Non-invasive electrical and magnetic stimulation of the brain, spinal cord, roots and peripheral nerves: basic principles and procedures for routine clinical and research application. An updated report from an I.F.C.N. Committee. Clin Neurophysiol. 2015;126:1071–107.
94. Voineskos D, Daskalakis ZJ. A primer on the treatment of schizophrenia through repetitive transcranial magnetic stimulation. Expert Rev Neurother. 2013;13:1079–82.
95. Poorganji M, Goeke K, Zomorrodi R, Voineskos D, Rajji TK, Daskalakis ZJ, Blumberger DM. The use of theta burst stimulation in patients with schizophrenia—a systematic review. Schizophr Res. 2023;261:245–55.
96. Kishi T, Ikuta T, Sakuma K, Hamanaka S, Nishii Y, Hatano M, Kito S, Iwata N. Theta burst stimulation protocols for schizophrenia. JAMA Netw Open. 2024;7:e2441159.
97. Chu C-S, Li C-T, Brunoni AR, et al. Cognitive effects and acceptability of non-invasive brain stimulation on Alzheimer's disease and mild cognitive impairment: a component network meta-analysis. J Neurol Neurosurg Psychiatry. 2021;92:195–203.

98. Di Lazzaro V, Dileone M, Pilato F, et al. Modulation of motor cortex neuronal networks by rTMS: comparison of local and remote effects of six different protocols of stimulation. J Neurophysiol. 2011;105:2150–6. https://doi.org/10.1152/jn.00781.2010.
99. Blumberger DM, Vila-Rodriguez F, Thorpe KE, et al. Effectiveness of theta burst versus high-frequency repetitive transcranial magnetic stimulation in patients with depression (THREE-D): a randomised non-inferiority trial. Lancet. 2018;391:1683–92.
100. Jin Y, Tong J, Huang Y, Shi D, Zhu N, Zhu M, Liu M, Liu H, Sun X. Effectiveness of accelerated intermittent theta burst stimulation for social cognition and negative symptoms among individuals with schizophrenia: a randomized controlled trial. Psychiatry Res. 2023;320:115033.
101. Zhang X, Yang X, Shi Z, Xu R, Tan J, Yang J, Huang X, Huang X, Zheng W. A systematic review of intermittent theta burst stimulation for neurocognitive dysfunction in older adults with schizophrenia. J Pers Med. 2023;13:485.
102. Lisman JE, Idiart MA. Storage of 7 +/− 2 short-term memories in oscillatory subcycles. Science. 1995;267:1512–5.
103. Di Lazzaro V, Pilato F, Saturno E, et al. Theta-burst repetitive transcranial magnetic stimulation suppresses specific excitatory circuits in the human motor cortex. J Physiol. 2005;565:945–50.
104. Di Lazzaro V, Pilato F, Dileone M, et al. The physiological basis of the effects of intermittent theta burst stimulation of the human motor cortex. J Physiol. 2008;586:3871–9.
105. Wischnewski M, Schutter DJLG. Efficacy and time course of theta burst stimulation in healthy humans. Brain Stimul. 2015;8:685–92.
106. Nguyen V-T, Wu C-W, Chen C-A, Lo C-C, Chen F-Y, Wu C-I, Sung P-S, Lin C-C, Chen J-J. Modulation of interhemispheric synchronization and cortical activity in healthy subjects by high-definition theta-burst electrical stimulation. Neural Plast. 2022;2022:3593262.
107. Hong YH, Wu SW, Pedapati EV, Horn PS, Huddleston DA, Laue CS, Gilbert DL. Safety and tolerability of theta burst stimulation vs. single and paired pulse transcranial magnetic stimulation: a comparative study of 165 pediatric subjects. Front Hum Neurosci. 2015; https://doi.org/10.3389/fnhum.2015.00029.
108. Purushotham A, Sinha VK, Goyal N, Tikka SK. Intermittent theta burst stimulation induced seizure in a child with schizophrenia: a case report. Brain Stimul Basic Transl Clin Res Neuromodul. 2018;11:1415–6.
109. Shafi MM. Seizures with TMS: much ado about (almost) nothing? Clin Neurophysiol. 2019;130:1397–8.
110. Merabet LB, Pascual-Leone A. Transcranial magnetic stimulation. In: Squire LR, editor. Encyclopedia of neuroscience. Oxford: Academic; 2009. p. 1055–62.
111. Rossi S, Hallett M, Rossini PM, Pascual-Leone A. Safety, ethical considerations, and application guidelines for the use of transcranial magnetic stimulation in clinical practice and research. Clin Neurophysiol. 2009;120:2008–39.
112. Demirtas-Tatlidede A, Freitas C, Cromer JR, Safar L, Ongur D, Stone WS, Seidman LJ, Schmahmann JD, Pascual-Leone A. Safety and proof of principle study of cerebellar vermal theta burst stimulation in refractory schizophrenia. Schizophr Res. 2010;124:91–100.
113. Elmaghraby R, Sun Q, Ozger C, Shekunov J, Romanowicz M, Croarkin PE. A systematic review of the safety and tolerability of theta burst stimulation in children and adolescents. Neuromodulation. 2022;25:494–503.
114. Bation R, Magnin C, Poulet E, Mondino M, Brunelin J. Intermittent theta burst stimulation for negative symptoms of schizophrenia—a double-blind, sham-controlled pilot study. NPJ Schizophr. 2021;7:1–7.
115. Zhu L, Zhang W, Zhu Y, Mu X, Zhang Q, Wang Y, Cai J, Xie B. Cerebellar theta burst stimulation for the treatment of negative symptoms of schizophrenia: a multicenter, double-blind, randomized controlled trial. Psychiatry Res. 2021;305:114204.
116. Wang L, Chen X, Wu Y, He K, Xu F, Xiao G, Hu P, Qiu B, Ji G-J, Wang K. Intermittent theta burst stimulation (iTBS) adjustment effects of schizophrenia: results from an

exploratory outcome of a randomized double-blind controlled study. Schizophr Res. 2020;216:550–3.
117. Wang L, Li Q, Wu Y, et al. Intermittent theta burst stimulation improved visual-spatial working memory in treatment-resistant schizophrenia: a pilot study. J Psychiatr Res. 2022;149:44–53.
118. Duprat R, Desmyter S, Rudi DR, et al. Accelerated intermittent theta burst stimulation treatment in medication-resistant major depression: a fast road to remission? J Affect Disord. 2016;200:6–14.
119. Chung SW, Hoy KE, Fitzgerald PB. Theta-burst stimulation: a new form of TMS treatment for depression? Depress Anxiety. 2015;32:182–92.
120. Duprat R, Wu G-R, De Raedt R, Baeken C. Accelerated iTBS treatment in depressed patients differentially modulates reward system activity based on anhedonia. World J Biol Psychiatry. 2018;19:497–508.
121. Chithra U, Samantaray S, Kumar V, et al. Add-on accelerated continuous theta burst stimulation (a-cTBS) over the left temporoparietal junction for the management of persistent auditory hallucinations in schizophrenia: a case series. Brain Stimul Basic Transl Clin Res Neuromodul. 2022;15:1511–2.
122. Chen L, Klooster DCW, Tik M, Thomas EHX, Downar J, Fitzgerald PB, Williams NR, Baeken C. Accelerated repetitive transcranial magnetic stimulation to treat major depression: the past, present, and future. Harv Rev Psychiatry. 2023;31:142–61.
123. Kishi T, Ikuta T, Sakuma K, et al. Theta burst stimulation for depression: a systematic review and network and pairwise meta-analysis. Mol Psychiatry. 2024; https://doi.org/10.1038/s41380-024-02630-5.
124. Steele VR, Maxwell AM, Ross TJ, Stein EA, Salmeron BJ. Accelerated intermittent theta-burst stimulation as a treatment for cocaine use disorder: a proof-of-concept study. Front Neurosci. 2019;13:1147.
125. Williams NR, Sudheimer KD, Bentzley BS, et al. High-dose spaced theta-burst TMS as a rapid-acting antidepressant in highly refractory depression. Brain. 2018;141:e18.
126. Kumar N, Vishnubhatla S, Wadhawan AN, Minhas S, Gupta P. A randomized, double blind, sham-controlled trial of repetitive transcranial magnetic stimulation (rTMS) in the treatment of negative symptoms in schizophrenia. Brain Stimul Basic Transl Clin Res Neuromodul. 2020;13:840–9.
127. Cole EJ, Phillips AL, Bentzley BS, et al. Stanford neuromodulation therapy (SNT): a double-blind randomized controlled trial. Am J Psychiatry. 2022;179:132–41.
128. Noda Y, Fujii K, Tokura F, Nakajima S, Kitahata R. A case series of continuous theta burst stimulation treatment for the supplementary motor area twice a day in patients with obsessive-compulsive disorder: a real world TMS registry study in Japan. J Pers Med. 2023;13:875.
129. Sharbafshaaer M, Gigi I, Lavorgna L, Esposito S, Bonavita S, Tedeschi G, Esposito F, Trojsi F. Repetitive transcranial magnetic stimulation (rTMS) in mild cognitive impairment: effects on cognitive functions—a systematic review. J Clin Med. 2023;12:6190.
130. Schraven SP, Plontke SK, Rahne T, Wasserka B, Plewnia C. Hearing safety of long-term treatment with theta burst stimulation. Brain Stimul. 2013;6:563–8.
131. Chistyakov AV, Rubicsek O, Kaplan B, Zaaroor M, Klein E. Safety, tolerability and preliminary evidence for antidepressant efficacy of theta-burst transcranial magnetic stimulation in patients with major depression. Int J Neuropsychopharmacol. 2010;13:387–93.
132. Gao X, Ni Y, Hu W, Wang G, He X. Comparative study about the therapeutic effect of cTBS and rTMS in the treatment of auditory verbal hallucinations in schizophrenia. Postgrad Med J. 2024;101:qgae119.
133. Nyffeler T, Wurtz P, Lüscher H-R, Hess CW, Senn W, Pflugshaupt T, von Wartburg R, Lüthi M, Müri RM. Extending lifetime of plastic changes in the human brain. Eur J Neurosci. 2006;24:2961–6.
134. Ye S-Y, Chen C-N, Wei B, Zhan J-Q, Li Y-H, Zhang C, Huang J-J, Yang Y-J. The efficacy and safety of continuous theta burst stimulation for auditory hallucinations: a systematic

review and meta-analysis of randomized controlled trials. Front Psych. 2024; https://doi.org/10.3389/fpsyt.2024.1446849.
135. Kang D, Song C, Peng X, Yu G, Yang Y, Chen C, Long Y, Shao P, Wu R. The effect of continuous theta burst stimulation on antipsychotic-induced weight gain in first-episode drug-naive individuals with schizophrenia: a double-blind, randomized, sham-controlled feasibility trial. Transl Psychiatry. 2024;14:1–8.
136. Zöllner R, Hübener A-F, Dannlowski U, Kircher T, Sommer J, Zavorotnyy M. Theta-burst stimulation for auditory-verbal hallucination in very-late-onset schizophrenia-like psychosis—a functional magnetic resonance imaging case study. Front Psych. 2020; https://doi.org/10.3389/fpsyt.2020.00294.
137. Rachid F, Vianin P, Besse C. Safety and efficacy of continuous theta burst stimulation at 30 Hz in a patient with chronic auditory hallucinations: a case study in a private practice setting. Brain Stimul Basic Transl Clin Res Neuromodul. 2013;6:707–8.
138. Purushotham A, Goyal N, Sinha VK, Tikka SK, Garg S, Desarkar P. Motor cortical plasticity in adolescents with early onset schizophrenia: a TMS-EMG study assessing the perturbation effect of intermittent and continuous theta burst stimulation. Int J Dev Neurosci. 2022;82:576–83.
139. Eberle M-C, Wildgruber D, Wasserka B, Fallgatter AJ, Plewnia C. Relief from chronic intractable auditory hallucinations after Long-term bilateral theta burst stimulation. Am J Psychiatry. 2010;167:1410.
140. Poulet E, Brunelin J, Makhlouf WB, D'Amato T, Saoud M. A case report of cTBS for the treatment of auditory hallucinations in a patient with schizophrenia. Brain Stimul Basic Transl Clin Res Neuromodul. 2009;2:118–9.
141. Sidhoumi D, Braha S, Bouaziz N, Brunelin J, Benadhira R, Januel D. Evaluation of the therapeutic effect of theta burst stimulation on drug-resistant auditory hallucinations in a schizophrenic patient and its impact on cognitive function and neuronal excitability: a case study. Clin Neurophysiol. 2010;121:802.
142. Kindler J, Homan P, Flury R, Strik W, Dierks T, Hubl D. Theta burst transcranial magnetic stimulation for the treatment of auditory verbal hallucinations: results of a randomized controlled study. Psychiatry Res. 2013;209:114–7.
143. Homan P, Kindler J, Hauf M, Hubl D, Dierks T. Cerebral blood flow identifies responders to transcranial magnetic stimulation in auditory verbal hallucinations. Transl Psychiatry. 2012;2:e189.
144. Plewnia C, Zwissler B, Wasserka B, Fallgatter AJ, Klingberg S. Treatment of auditory hallucinations with bilateral theta burst stimulation: a randomized controlled pilot trial. Brain Stimul Basic Transl Clin Res Neuromodul. 2014;7:340–1.
145. Koops S, van Dellen E, Schutte MJL, Nieuwdorp W, Neggers SFW, Sommer IEC. Theta burst transcranial magnetic stimulation for auditory verbal hallucinations: negative findings from a double-blind-randomized trial. Schizophr Bull. 2016;42:250–7.
146. Jandl M, Steyer J, Weber M, Linden DEJ, Rothmeier J, Maurer K, Kaschka WP. Treating auditory hallucinations by transcranial magnetic stimulation: a randomized controlled crossover trial. Neuropsychobiology. 2006;53:63–9.
147. Lee S-H, Kim W, Chung Y-C, Jung K-H, Bahk W-M, Jun T-Y, Kim K-S, George MS, Chae J-H. A double blind study showing that two weeks of daily repetitive TMS over the left or right temporoparietal cortex reduces symptoms in patients with schizophrenia who are having treatment-refractory auditory hallucinations. Neurosci Lett. 2005;376:177–81.
148. Plewnia C, Brendel B, Schwippel T, Martus P, Cordes J, Hasan A, Fallgatter AJ. Treatment of auditory hallucinations with bilateral theta burst stimulation (cTBS): protocol of a randomized, double-blind, placebo-controlled, multicenter trial. Eur Arch Psychiatry Clin Neurosci. 2018;268:663–73.
149. Guelke E, Horn MA, Caffier J, Pinnschmidt H, Hamel W, Moll CKE, Gulberti A, Poetter-Nerger M. Comparison of subthalamic unilateral and bilateral theta burst deep brain stimulation in Parkinson's disease. Front Hum Neurosci. 2023;17:1233565.

150. Jiang J, Li Q, Sheng J, Yang F, Cao X, Zhang T, Jia Y, Wang J, Li C. 25Hz magnetic seizure therapy is feasible but not optimal for Chinese patients with schizophrenia: a case series. Front Psychiatry. 2018;9:224.
151. Lisanby SH, Luber B, Finck AD, Schroeder C, Sackeim HA. Deliberate seizure induction with repetitive transcranial magnetic stimulation in nonhuman primates. Arch Gen Psychiatry. 2001;58:199–200.
152. Wu H, Jiang J, Cao X, Wang J, Li C. Magnetic seizure therapy for people with schizophrenia. Cochrane Database Syst Rev. 2023;2023:CD012697.
153. Tharyan P, Adams CE. Electroconvulsive therapy for schizophrenia. Cochrane Database Syst Rev. 2005; https://doi.org/10.1002/14651858.CD000076.pub2.
154. Fitzgerald PB, Hoy KE, Herring SE, Clinton AM, Downey G, Daskalakis ZJ. Pilot study of the clinical and cognitive effects of high-frequency magnetic seizure therapy in major depressive disorder. Depress Anxiety. 2013;30:129–36.
155. Hoy KE, Fitzgerald PB. Introducing magnetic seizure therapy: a novel therapy for treatment resistant depression. Aust N Z J Psychiatry. 2010;44:591–8.
156. Kayser S, Bewernick BH, Matusch A, Hurlemann R, Soehle M, Schlaepfer TE. Magnetic seizure therapy in treatment-resistant depression: clinical, neuropsychological and metabolic effects. Psychol Med. 2015;45:1073–92.
157. Polster JD, Kayser S, Bewernick BH, Hurlemann R, Schlaepfer TE. Effects of electroconvulsive therapy and magnetic seizure therapy on acute memory retrieval. J ECT. 2015;31:13–9.
158. Hoy KE, Fitzgerald PB. Magnetic seizure therapy for treatment-resistant depression. Expert Rev Med Devices. 2011;8:723–32.
159. Lisanby SH, Schlaepfer TE, Fisch HU, Sackeim HA. Magnetic seizure therapy of major depression. Arch Gen Psychiatry. 2001;58:303–5.
160. Mutz J, Vipulananthan V, Carter B, Hurlemann R, Fu CHY, Young AH. Comparative efficacy and acceptability of non-surgical brain stimulation for the acute treatment of major depressive episodes in adults: systematic review and network meta-analysis. BMJ. 2019;364:l1079.
161. Jiang J, Li Q, Sheng J, Yang F, Cao X, Zhang T, Jia Y, Wang J, Li C. 25 Hz magnetic seizure therapy is feasible but not optimal for Chinese patients with schizophrenia: a case series. Front Psychiatry. 2018;9:224.
162. Luber B, McClintock SM, Lisanby SH. Applications of transcranial magnetic stimulation and magnetic seizure therapy in the study and treatment of disorders related to cerebral aging. Dialogues Clin Neurosci. 2013;15:87–98.
163. Lisanby SH. Update on magnetic seizure therapy: a novel form of convulsive therapy. J ECT. 2002;18:182–8.
164. Lisanby SH, Luber B, Schlaepfer TE, Sackeim HA. Safety and feasibility of magnetic seizure therapy (MST) in major depression: randomized within-subject comparison with electroconvulsive therapy. Neuropsychopharmacology. 2003;28:1852–65.
165. Moscrip TD, Terrace HS, Sackeim HA, Lisanby SH. Randomized controlled trial of the cognitive side-effects of magnetic seizure therapy (MST) and electroconvulsive shock (ECS). Int J Neuropsychopharmacol. 2006;9:1–11.
166. McClintock SM, Tirmizi O, Chansard M, Husain MM. A systematic review of the neurocognitive effects of magnetic seizure therapy. Int Rev Psychiatry. 2011;23:413–23.
167. Backhouse FA, Noda Y, Knyahnytska Y, Farzan F, Downar J, Rajji TK, Mulsant BH, Daskalakis ZJ, Blumberger DM. Characteristics of ictal EEG in magnetic seizure therapy at various stimulation frequencies. Clin Neurophysiol. 2018;129:1770–9.
168. Sun Y, Farzan F, Mulsant BH, Rajji TK, Fitzgerald PB, Barr MS, Downar J, Wong W, Blumberger DM, Daskalakis ZJ. Indicators for remission of suicidal ideation following magnetic seizure therapy in patients with treatment-resistant depression. JAMA Psychiatry. 2016;73:337.
169. Tang VM, Blumberger DM, Throop A, McClintock SM, Voineskos D, Downar J, Knyahnytska Y, Mulsant BH, Fitzgerald PB, Daskalakis ZJ. Continuation magnetic seizure therapy for treatment-resistant unipolar or bipolar depression. J Clin Psychiatry. 2021; https://doi.org/10.4088/JCP.20m13677.

170. Engel A, Kayser S. An overview on clinical aspects in magnetic seizure therapy. J Neural Transm. 2016;123:1139–46.
171. Zhang X-Y, Chen H-D, Liang W-N, Yang X-H, Cai D-B, Huang X, Huang X-B, Liu C-Y, Zheng W. Adjunctive magnetic seizure therapy for schizophrenia: a systematic review. Front Psychiatry. 2022;12:813590.
172. Beszłej J, Wieczorek T, Kobyłko A, Piotrowski P, Siwicki D, Weiser A, Fila-Witecka K, Rymaszewska J, Tabakow P. Deep brain stimulation: new possibilities for the treatment of mental disorders. Psychiatr Pol. 2019;53:789–806.
173. Corripio I, Roldán A, Sarró S, et al. Deep brain stimulation in treatment resistant schizophrenia: a pilot randomized cross-over clinical trial. EBioMedicine. 2020;51:102568.
174. Li M, Chen L, Rong H, et al. Low-charge electrotherapy for patients with schizophrenia: a double-blind, randomised controlled pilot clinical trial. Psychiatry Res. 2019;272:676–81.
175. Miller DH, Clancy J, Cumming E. A comparison between unidirectional current nonconvulsive electrical stimulation given with Reiter's machine, standard alternating current electroshock (Cerletti method), and pentothal in chronic schizophrenia. Am J Psychiatry. 1953;109:617–20.
176. Zhang X-Y, Chen H-D, Liang W-N, Yang X-H, Cai D-B, Huang X, Huang X-B, Liu C-Y, Zheng W. Adjunctive magnetic seizure therapy for schizophrenia: a systematic review. Front Psych. 2022;12:813590.
177. Sackeim HA, Prudic J, Nobler MS, Fitzsimons L, Lisanby SH, Payne N, Berman RM, Brakemeier E-L, Perera T, Devanand DP. Effects of pulse width and electrode placement on the efficacy and cognitive effects of electroconvulsive therapy. Brain Stimul Basic Transl Clin Res Neuromodul. 2008;1:71–83.
178. Tang Y, Jiang W, Ren Y, Ma X, Cotes RO, McDonald WM. Electroconvulsive therapy in China: clinical practice and research on efficacy. J ECT. 2012;28:206–12.
179. Sackeim HA. The anticonvulsant hypothesis of the mechanisms of action of ECT: current status. J ECT. 1999;15:5–26.
180. Nitsche MA, Paulus W. Excitability changes induced in the human motor cortex by weak transcranial direct current stimulation. J Physiol. 2000;527(Pt 3):633–9.
181. Horvath JC, Forte JD, Carter O. Evidence that transcranial direct current stimulation (tDCS) generates little-to-no reliable neurophysiologic effect beyond MEP amplitude modulation in healthy human subjects: a systematic review. Neuropsychologia. 2015;66:213–36.
182. Bikson M, Name A, Rahman A. Origins of specificity during tDCS: anatomical, activity-selective, and input-bias mechanisms. Front Hum Neurosci. 2013;7:688.
183. Nitsche MA, Cohen LG, Wassermann EM, et al. Transcranial direct current stimulation: state of the art 2008. Brain Stimul. 2008;1:206–23.
184. Meiron O, David J, Yaniv A. Left prefrontal transcranial direct-current stimulation reduces symptom-severity and acutely enhances working memory in schizophrenia. Neurosci Lett. 2021;755:135912.
185. Jiang W-L, Cai D-B, Sun C-H, Yin F, Goerigk S, Brunoni AR, Zhao X-W, Mayes TL, Zheng W, Xiang Y-T. Adjunctive tDCS for treatment-refractory auditory hallucinations in schizophrenia: a meta-analysis of randomized, double-blinded, sham-controlled studies. Asian J Psychiatry. 2022;73:103100.
186. Hyde J, Carr H, Kelley N, et al. Efficacy of neurostimulation across mental disorders: systematic review and meta-analysis of 208 randomized controlled trials. Mol Psychiatry. 2022;27:2709–19.
187. Begemann MJ, Brand BA, Ćurčić-Blake B, Aleman A, Sommer IE. Efficacy of non-invasive brain stimulation on cognitive functioning in brain disorders: a meta-analysis. Psychol Med. 2020;50:2465–86.
188. Moffa AH, Brunoni AR, Nikolin S, Loo CK. Transcranial direct current stimulation in psychiatric disorders. Psychiatr Clin North Am. 2018;41:447–63.

189. Brunoni AR, Amadera J, Berbel B, Volz MS, Rizzerio BG, Fregni F. A systematic review on reporting and assessment of adverse effects associated with transcranial direct current stimulation. Int J Neuropsychopharmacol. 2011;14:1133–45.
190. Vicario CM, Nitsche MA. Non-invasive brain stimulation for the treatment of brain diseases in childhood and adolescence: state of the art, current limits and future challenges. Front Syst Neurosci. 2013; https://doi.org/10.3389/fnsys.2013.00094.
191. Andrade C. Transcranial direct current stimulation for refractory auditory hallucinations in schizophrenia: (practical psychopharmacology). J Clin Psychiatry. 2013;74:e1054–8.
192. Kessler SK, Minhas P, Woods AJ, Rosen A, Gorman C, Bikson M. Dosage considerations for transcranial direct current stimulation in children: a computational modeling study. PLoS One. 2013;8:e76112.
193. Zhou Y, Fan L, Qiu C, Jiang T. Prefrontal cortex and the dysconnectivity hypothesis of schizophrenia. Neurosci Bull. 2015;31:207–19.
194. Schwartz LB, Hirschman AO. Exit, voice, and loyalty: responses to decline in firms, organizations, and states. Univ Pa Law Rev. 1972;120:1210.
195. Brunelin J, Mondino M, Gassab L, Haesebaert F, Gaha L, Suaud-Chagny M-F, Saoud M, Mechri A, Poulet E. Examining transcranial direct-current stimulation (tDCS) as a treatment for hallucinations in schizophrenia. Am J Psychiatry. 2012;169:719–24.
196. Chang C-C, Tzeng N-S, Chao C-Y, Yeh C-B, Chang H-A. The effects of add-on Fronto-temporal transcranial direct current stimulation (tDCS) on auditory verbal hallucinations, other psychopathological symptoms, and insight in schizophrenia: a randomized, double-blind, sham-controlled trial. Int J Neuropsychopharmacol. 2018;21:979–87.
197. Fröhlich F, Burrello TN, Mellin JM, Cordle AL, Lustenberger CM, Gilmore JH, Jarskog LF. Exploratory study of once-daily transcranial direct current stimulation (tDCS) as a treatment for auditory hallucinations in schizophrenia. Eur Psychiatry. 2016;33:54–60.
198. Valiengo LDCL, Goerigk S, Gordon PC, et al. Efficacy and safety of transcranial direct current stimulation for treating negative symptoms in schizophrenia: a randomized clinical trial. JAMA Psychiatry. 2020;77:121.
199. Bose A, Shivakumar V, Agarwal SM, Kalmady SV, Shenoy S, Sreeraj VS, Narayanaswamy JC, Venkatasubramanian G. Efficacy of fronto-temporal transcranial direct current stimulation for refractory auditory verbal hallucinations in schizophrenia: a randomized, double-blind, sham-controlled study. Schizophr Res. 2018;195:475–80.
200. Chang C-C, Kao Y-C, Chao C-Y, Tzeng N-S, Chang H-A. Examining bi-anodal transcranial direct current stimulation (tDCS) over bilateral dorsolateral prefrontal cortex coupled with bilateral extracephalic references as a treatment for negative symptoms in non-acute schizophrenia patients: a randomized, double-blind, sham-controlled trial. Prog Neuro-Psychopharmacol Biol Psychiatry. 2020;96:109715.
201. Cheng PWC, Louie LLC, Wong YL, Wong SMC, Leung WY, Nitsche MA, Chan WC. The effects of transcranial direct current stimulation (tDCS) on clinical symptoms in schizophrenia: a systematic review and meta-analysis. Asian J Psychiatry. 2020;53:102392.
202. da Silva MC, Conti CL, Klauss J, Alves LG, do Nascimento Cavalcante HM, Fregni F, Nitsche MA, Nakamura-Palacios EM. Behavioral effects of transcranial direct current stimulation (tDCS) induced dorsolateral prefrontal cortex plasticity in alcohol dependence. J Physiol Paris. 2013;107:493–502.
203. Jeon D-W, Jung D-U, Kim S-J, et al. Adjunct transcranial direct current stimulation improves cognitive function in patients with schizophrenia: a double-blind 12-week study. Schizophr Res. 2018;197:378–85.
204. Kantrowitz JT, Sehatpour P, Avissar M, et al. Significant improvement in treatment resistant auditory verbal hallucinations after 5 days of double-blind, randomized, sham controlled, fronto-temporal, transcranial direct current stimulation (tDCS): a replication/extension study. Brain Stimul. 2019;12:981–91.
205. Kao Y-C, Tzeng N-S, Chao C-Y, Chang C-C, Chang H-A. Modulation of self-appraisal of illness, medication adherence, life quality and autonomic functioning by transcranial direct current stimulation in schizophrenia patients. Clin Neurophysiol. 2020;131:1997–2007.

206. Klein HS, Vanneste S, Pinkham AE. The limited effect of neural stimulation on visual attention and social cognition in individuals with schizophrenia. Neuropsychologia. 2021;157:107880.
207. Lindenmayer JP, Kulsa MKC, Sultana T, Kaur A, Yang R, Ljuri I, Parker B, Khan A. Transcranial direct-current stimulation in ultra-treatment-resistant schizophrenia. Brain Stimul. 2019;12:54–61.
208. Liu Y, Gu N, Cao X, Zhu Y, Wang J, Smith RC, Li C. Effects of transcranial electrical stimulation on working memory in patients with schizophrenia: a systematic review and meta-analysis. Psychiatry Res. 2021;296:113656.
209. Mondino M, Brunelin J, Palm U, Brunoni A, Poulet E, Fecteau S. Transcranial direct current stimulation for the treatment of refractory symptoms of schizophrenia. Current evidence and future directions. Curr Pharm Des. 2015;21:3373–83.
210. Mondino M, Poulet E, Suaud-Chagny M-F, Brunelin J. Anodal tDCS targeting the left temporo-parietal junction disrupts verbal reality-monitoring. Neuropsychologia. 2016;89:478–84.
211. Palm U, Keeser D, Hasan A, et al. Prefrontal transcranial direct current stimulation for treatment of schizophrenia with predominant negative symptoms: a double-blind, sham-controlled proof-of-concept study. Schizophr Bull. 2016;42:1253–61.
212. Weickert TW, Salimuddin H, Lenroot RK, Bruggemann J, Loo C, Vercammen A, Kindler J, Weickert CS. Preliminary findings of four-week, task-based anodal prefrontal cortex transcranial direct current stimulation transferring to other cognitive improvements in schizophrenia. Psychiatry Res. 2019;280:112487.
213. Mondino M, Haesebaert F, Poulet E, Suaud-Chagny M-F, Brunelin J. Fronto-temporal transcranial direct current stimulation (tDCS) reduces source-monitoring deficits and auditory hallucinations in patients with schizophrenia. Schizophr Res. 2015;161:515–6.
214. Mondino M, Jardri R, Suaud-Chagny M-F, Saoud M, Poulet E, Brunelin J. Effects of fronto-temporal transcranial direct current stimulation on auditory verbal hallucinations and resting-state functional connectivity of the left temporo-parietal junction in patients with schizophrenia. Schizophr Bull. 2016;42:318–26.
215. Smith RC, Boules S, Mattiuz S, Youssef M, Tobe RH, Sershen H, Lajtha A, Nolan K, Amiaz R, Davis JM. Effects of transcranial direct current stimulation (tDCS) on cognition, symptoms, and smoking in schizophrenia: a randomized controlled study. Schizophr Res. 2015;168:260–6.
216. Verma L, Sharma G. Transcranial direct current stimulation as an add-on intervention in patient with schizophrenia. Int Med J. 2021;28(2):150–2.
217. Dharani R, Goyal N, Mukherjee A, Umesh S. Adjuvant high-definition transcranial direct current stimulation for negative symptoms in schizophrenia: a pilot study. J ECT. 2021;37:195–201.
218. Bai S, Dokos S, Ho K-A, Loo C. A computational modelling study of transcranial direct current stimulation montages used in depression. NeuroImage. 2014;87:332–44.
219. Datta A, Bansal V, Diaz J, Patel J, Reato D, Bikson M. Gyri-precise head model of transcranial direct current stimulation: improved spatial focality using a ring electrode versus conventional rectangular pad. Brain Stimul. 2009;2:201–7, 207.e1.
220. Xu H, Zhou Y, Wang J, Liang Z, Wang Y, Wu W, Liu Y, Liu X, Zhang X, Huo L. Effect of HD-tDCS on white matter integrity and associated cognitive function in chronic schizophrenia: a double-blind, sham-controlled randomized trial. Psychiatry Res. 2023;324:115183.
221. Edwards D, Cortes M, Datta A, Minhas P, Wassermann EM, Bikson M. Physiological and modeling evidence for focal transcranial electrical brain stimulation in humans: a basis for high-definition tDCS. NeuroImage. 2013;74:266–75.
222. Faria P, Hallett M, Miranda PC. A finite element analysis of the effect of electrode area and inter-electrode distance on the spatial distribution of the current density in tDCS. J Neural Eng. 2011;8:066017.
223. Kuo H-I, Bikson M, Datta A, Minhas P, Paulus W, Kuo M-F, Nitsche MA. Comparing cortical plasticity induced by conventional and high-definition 4 × 1 ring tDCS: a neurophysiological study. Brain Stimul. 2013;6:644–8.

224. Parlikar R, Selvaraj S, Shiva L, Sreeraj VS, Venkatasubramanian G, Chandra PS. Successful application of add-on high-definition transcranial direct current stimulation in a schizophrenic patient with comorbid alopecia universalis. Indian J Psychiatry. 2019;61:663–4.
225. Sreeraj VS, Dinakaran D, Parlikar R, Chhabra H, Selvaraj S, Shivakumar V, Bose A, Narayanaswamy JC, Venkatasubramanian G. High-definition transcranial direct current simulation (HD-tDCS) for persistent auditory hallucinations in schizophrenia. Asian J Psychiatry. 2018;37:46–50.
226. Wong H-L, Chan WC, Wong Y-L, Wong S-N, Yung H-Y, Wong S-MC, Cheng P-WC. High-definition transcranial direct current stimulation—an open-label pilot intervention in alleviating depressive symptoms and cognitive deficits in late-life depression. CNS Neurosci Ther 2019;25:1244–53.
227. Parlikar R, Vanteemar SS, Venkataram S, Narayanaswamy JC, Rao NP, Ganesan V. High definition transcranial direct current stimulation (HD-tDCS): a systematic review on the treatment of neuropsychiatric disorders. Asian J Psychiatry. 2021;56:102542.
228. Shekhawat GS, Vanneste S. High-definition transcranial direct current stimulation of the dorsolateral prefrontal cortex for tinnitus modulation: a preliminary trial. J Neural Transm (Vienna Austria 1996). 2018;125:163–71.
229. Nikolin S, Loo CK, Bai S, Dokos S, Martin DM. Focalised stimulation using high definition transcranial direct current stimulation (HD-tDCS) to investigate declarative verbal learning and memory functioning. NeuroImage. 2015;117:11–9.
230. Eryılmaz G, Sayar GH, Ünsalver BÖ, Gül IG, Özten E, Sağlam E. Adverse effects of transcranial direct current stimulation (TDCS) in a Group of Psychiatric Patients. Sch J Appl Med Sci. 2014;2:294–7.
231. Boggio PS, Rigonatti SP, Ribeiro RB, Myczkowski ML, Nitsche MA, Pascual-Leone A, Fregni F. A randomized, double-blind clinical trial on the efficacy of cortical direct current stimulation for the treatment of major depression. Int J Neuropsychopharmacol. 2008;11:249–54.
232. Gandiga PC, Hummel FC, Cohen LG. Transcranial DC stimulation (tDCS): a tool for double-blind sham-controlled clinical studies in brain stimulation. Clin Neurophysiol. 2006;117:845–50.
233. Dmochowski JP, Datta A, Bikson M, Su Y, Parra LC. Optimized multi-electrode stimulation increases focality and intensity at target. J Neural Eng. 2011;8:046011.
234. Breitling C, Zaehle T, Dannhauer M, Tegelbeckers J, Flechtner H-H, Krauel K. Comparison between conventional and HD-tDCS of the right inferior frontal gyrus in children and adolescents with ADHD. Clin Neurophysiol. 2020;131:1146–54.
235. Chua EF, Ahmed R, Garcia SM. Effects of HD-tDCS on memory and metamemory for general knowledge questions that vary by difficulty. Brain Stimul. 2017;10:231–41.
236. Fritsch B, Reis J, Martinowich K, Schambra HM, Ji Y, Cohen LG, Lu B. Direct current stimulation promotes BDNF-dependent synaptic plasticity: potential implications for motor learning. Neuron. 2010;66:198–204.
237. Ranieri F, Podda MV, Riccardi E, Frisullo G, Dileone M, Profice P, Pilato F, Di Lazzaro V, Grassi C. Modulation of LTP at rat hippocampal CA3-CA1 synapses by direct current stimulation. J Neurophysiol. 2012;107:1868–80.
238. Agarwal SM, Shivakumar V, Bose A, et al. Transcranial direct current stimulation in schizophrenia. Clin Psychopharmacol Neurosci. 2013;11:118–25.

Addictive Disorders

6

Tianzhen Chen, Lei Guo, Hang Su, and Min Zhao

Abstract

Addictive disorders are among the most serious public health issues worldwide. Over the past decades, to address the harm caused by addictive disorders, researchers have explored the effectiveness of a range of noninvasive neuromodulation approaches targeting the core features of addiction. Some of these research protocols have yielded exciting results. This chapter focuses on the current state of application of repetitive transcranial magnetic stimulation (rTMS) and transcranial direct current stimulation (tDCS) technologies, including the following paradigms: accelerated TMS (aTMS), deep TMS (dTMS), priming TMS (pTMS), synchronized TMS (sTMS), various forms of theta burst stimulation (TBS) such as intermittent, continuous, and bilateral TBS, and magnetic seizure therapy (MST), as well as high-definition tDCS (HD-tDCS).

Keywords

Addictive disorder · rTMS · TBS · MST · tDCS

T. Chen (✉) · L. Guo · H. Su
Shanghai Mental Health Center, Shanghai Jiao Tong University School of Medicine, Shanghai, China
e-mail: vincentchan@sjtu.edu.cn

M. Zhao (✉)
Shanghai Mental Health Center, Shanghai Jiao Tong University School of Medicine, Shanghai, China

Shanghai Key Laboratory of Psychotic Disorders, Shanghai, China

Institute of Psychological and Behavioral Science, Shanghai Jiao Tong University, Shanghai, China
e-mail: drminzhao@smhc.org.cn

W. Zheng, Y. Ning (eds.), *TMS and tDCS for Psychiatric Disorders*,
https://doi.org/10.1007/978-981-96-8504-2_6

Abbreviations

ACC	Anterior cingulate cortex
aTMS	Accelerated TMS
cTBS	Continuous TBS
DLPFC	Dorsolateral prefrontal cortex
dTMS	Deep TMS
ECT	Electroconvulsive therapy
HD-tDCS	High-definition tDCS
IAF	Individual's alpha frequency
IGD	Internet gaming disorder
iTBS	Intermittent TBS
LTD	Long-term depression
LTP	Long-term potentiation
mPFC	Medial prefrontal cortex
MST	Magnetic seizure therapy
OFC	Orbital frontal cortex
PFC	Prefrontal cortex
PG-YBOCS	Pathological gambling adaptation of the Yale-Brown Obsessive–Compulsive Scale
pTMS	Priming TMS
rMT	Resting motor threshold
rTMS	Repetitive transcranial magnetic stimulation
SAINT	Stanford Accelerated Intelligent Neuromodulation Therapy
SMA	Supplementary motor area
sTMS	Synchronized TMS
SUDs	Substance use disorders
TBS	Theta burst stimulation
tDCS	Transcranial direct current stimulation
UNODC	United Nations Office on Drugs and Crime
vmPFC	Ventromedial prefrontal cortex
VTA	Ventral tegmental area

6.1 Addictive Disorders

Substance and behavioral addictions are characterized by the compulsive use of substances or engagement in behaviors to achieve psychological or physical gratification, often despite harmful consequences. As a global public health concern, addiction has led to widespread and severe negative impacts across the world. According to the World Drug Report 2024, the number of drug users globally reached 292 million in 2022—a 20% increase from a decade prior [1]. Among them, an estimated 64 million people suffer from substance use disorders (SUDs), yet only one in 11 receives treatment [1]. Meanwhile, behavioral addictions, including digital addiction (e.g., social media, gaming, and smartphone overuse) and gambling disorder, are becoming increasingly prevalent. Statista reports that by January 2021, there were 4.66 billion active internet users, with an average daily online engagement of 6.7 h [2]. Problematic use of digital products (e.g., computers, smartphones) and activities (e.g., gaming, social media use) also carry the risk of becoming addictive. A 2022 systematic review reports a global prevalence of smartphone addiction at 26.99% [3]. Although there is no direct substance involved, digital addictions pose significant threats to public health by adversely affecting individuals' physical well-being, daily routines, and social capabilities, presenting enduring challenges to public health management [4, 5].

Addiction is a chronic, relapsing brain disorder characterized by dysregulation across multiple neural circuits. The core neural pathways involved include the reward circuitry, executive control circuits, emotional circuits [6], and so on. It encompasses several neurotransmitter systems such as the dopaminergic, glutamatergic, and GABAergic systems [7, 8]. These neuroadaptations drive the establishment and maintenance of addictive behaviors, trapping individuals in a cycle of drug use, withdrawal, craving, and relapse. In clinical practice, motivational enhancement therapy and cognitive-behavioral therapy continue to be the primary psychological interventions for addictive disorders. Despite advances in understanding of the mechanisms of addiction, the efficacy of current treatments remains limited.

With the evolution of neuromodulation technologies, over the past decade, numerous scholars have investigated the efficacy of interventions that target specific brain regions, circuits, and neural activity patterns associated with addictive disorder. Noninvasive brain stimulation techniques, including transcranial magnetic stimulation (TMS) and transcranial electrical stimulation, have generated several evidence supporting their clinical translation—a focus of this chapter.

6.2 TMS

6.2.1 rTMS

6.2.1.1 Unilateral rTMS

Efficacy

Over the past decade, the number of clinical practices evidences for TMS technology in addictive disorders has been steadily increasing. These evidences span various categories of substance use, ranging from legal substances such as alcohol and tobacco to illicit drugs including cocaine, methamphetamine, and opioids. Findings from international research teams have demonstrated TMS's therapeutic potential, despite persisting challenges in its routine clinical implementation. Given the heterogeneity in outcome measures across studies, we will review TMS efficacy based on different clinical indicators.

Craving/Drug-Induced Craving

The previous studies mainly reported the role of repetitive transcranial magnetic stimulation (rTMS) treatment in intervening cravings. In 2003, Eichhammer et al. first explored that a single day of high-frequency (20 Hz) unilateral rTMS of the left dorsolateral prefrontal cortex (DLPFC) had no significant change in the level of cravings but significantly reduced the number of cigarettes smoked [9]. According to a recent systematic review analysis, more than 50 clinical studies have explored the clinical effects of rTMS in addictive disorders [10]. The main studies were conducted among populations with tobacco or alcohol abuse/use disorder. These studies mostly focused on unilateral (left or right) DLPFC, medial prefrontal cortex (mPFC), insula, and other brain regions, with the intensity of stimulation mostly being 80–120% of the resting motor threshold (RMT). In terms of frequency selection, most were low-frequency (e.g., 1 Hz) or high-frequency (e.g., 10–20 Hz) ([11]; [10]). To some extent, due to the unclear neuro-mechanisms of rTMS intervention, most studies also rarely explain the basis for the chosen parameters. However, in general, it is believed that low-frequency (i.e., ≤1 Hz) stimulation has an inhibitory effect on the targeted cortex, while high-frequency parameters (i.e., ≥5 Hz) stimulation has an excitatory effect on the targeted cortex [12]. Compared with the control group, unilateral rTMS intervention significantly reduced the overall average craving scores in SUDs [13]. However, there are certain differences in efficacy among different types of substance disorder.

In studies on alcohol use disorder, it has been found that single-session rTMS interventions do not outperform sham rTMS treatment in reducing drug-related craving [10]. This conclusion is drawn from the above meta-analysis of five studies (four targeting the right DLPFC and one targeting the left frontal pole). On the other hand, multiple session rTMS interventions significantly reduce craving levels, suggesting a cumulative effect of rTMS, which has also been confirmed in many other studies. It is noteworthy that dTMS appears to have a stronger effect on cravings of patients with alcohol use disorder. However, due to the limited number of published articles and the inconsistency in intervention sites using deep coils, a unified conclusion is yet to be formed. For instance, Harel et al. applied H7 coil to the mPFC and anterior cingulate cortex (ACC) [14], while Perini et al. targeted the insula [15].

In studies on nicotine use disorder, three out of four single-session unilateral rTMS intervention studies reported significant craving decrease, while one reported no significant difference between the real rTMS and sham treatment [16–19]. A systematic review by Mehta et al. showed that multiple sessions of rTMS intervention only demonstrated marginal significance in reducing cravings among patients with nicotine use disorder [10]. Among the studies included in this meta-analysis, seven used figure-eight coils and two used H-type coils. The effectiveness of rTMS treatment is influenced by various factors, including the choice of stimulation coil, intervention targets, and the patient's own motivation, all of which can ultimately affect the outcome [11, 20]. A recent study has also indicated the prevalence of the placebo effect in rTMS interventions [21]. For example, a study conducted by Amiaz et al. included 48 smokers with a strong desire to quit smoking, who underwent continuous high-frequency rTMS targeting the DLPFC for 10 sessions, effectively reducing the craving in response to tobacco cues [22]. A multicenter clinical study conducted by Zangen et al. used H4 dTMS coils to stimulate the bilateral lateral PFC and insula for up to 6 weeks, showing significant effects on reducing cravings [23]. It is known that, compared to superficial cortical interventions, dTMS coils are expected to directly target deep brain areas associated with reward, such as the insula, ventral tegmental area, and nucleus accumbens, which are often considered closely related to addiction.

Compared to the legal substances (e.g., alcohol and nicotine), there are fewer studies on addiction to substances such as marijuana, opioids, cocaine, and methamphetamine. Among the eight studies on methamphetamine, six studies used excitatory stimulation parameters (10 Hz/iTBS) to target the left DLPFC region and found that it could significantly reduce craving levels [24–29], while two studies applied 1 Hz stimulation to the left DLPFC and reached inconsistent conclusions (one reduced cravings, the other increased cravings) [30, 31]. Li et al.'s study suggested that a single session of 1 Hz rTMS intervention on the left DLPFC could increase methamphetamine cue-induced craving levels [30]. In six studies on cocaine use disorder, two studies indicated that multiple high-frequency (15 Hz) rTMS interventions on the DLPFC could reduce craving levels [32, 33]. A study targeting the mPFC and ACC suggested that both 1 and 10 Hz showed no significant differences from the sham rTMS stimulation group [34]. Studies on opioid use disorder also mainly targeted the left DLPFC, with three indicating a significant effect on craving and one reporting no difference from sham rTMS intervention [35–38]. It is worth noting that although many studies emphasize the close relationship between craving levels and relapse [39], there are several inconsistencies between "craving" and "relapse" conclusions in rTMS studies. For example, Martinez's results found that rTMS treatment reduced cocaine use, but had no difference in the impact on craving compared to the sham treatment [34], while Lolli et al.'s study showed the opposite [32]. Therefore, it is necessary to have a clear definition and outcome measures to determine which treatment protocols is more suitable for patients with addictive disorder.

Compared to SUDs, there are very few studies on rTMS intervention in individuals with behavioral addictions. The main research has been conducted among gambling populations. Rosenberg et al. attempted a continuous 15 sessions, each lasting 10 min of low-frequency intervention (targeting the left DLPFC) on five gambling addiction patients in 2013, but all patients continued gambling after the intervention ended [40]. Later, Pettorruso et al. conducted a high-frequency stimulation (15 Hz) protocol targeting the left DLPFC for 44 sessions on eight gambling disorder patients and found a significant reduction in the number of gambling days [41], but no significant impact on the patients' emotional states. Salerno et al. applied 10 sessions of inhibitory parameters (i.e., continuous theta-burst stimulation) on the bilateral supplementary motor area (SMA), suggesting an improvement in pathological gambling adaptation of the Yale-Brown Obsessive–Compulsive Scale (PG-YBOCS) scores, but no significant impact on emotional or impulsive state indicators [42]. Although preliminary studies have suggested that TMS may to some extent improve clinical symptoms associated with gambling addiction or reduce the frequency of gambling, the published articles are mainly clinical studies or case report studies with single-digit sample sizes. According to a systematic review published in 2023, only eight studies were included in the analysis [43]. Among them, there were only three randomized controlled clinical studies, which also indicates that the quality of the original research evidence needs to be improved. It is currently difficult to draw conclusions about the effectiveness of rTMS intervention in gambling populations. In addition to gambling addiction, there have been case reports of high-frequency intervention on the left DLPFC, reducing the time spent watching pornographic images in a man with compulsive sexual behavior disorder related to online porn use and the gaming time in a patient with internet gaming disorder [44].

Relapse

Due to certain practical difficulties in collecting relapse data, only a few studies have reported data on relapse. Taking a systematic review published in 2024 as an example, about 10% of the studies reported results related to relapse [10]. In alcohol research, three out of five studies on alcohol use disorder reported that multiple sessions of rTMS could significantly reduce alcohol consumption [10]. In studies on nicotine, some also showed that rTMS intervention could reduce the amount of smoking [22, 45, 46]. For instance, a study by Amiaz et al. showed that 10 sessions of 10 Hz intervention on the left DLPFC could reduce patients' tobacco use, but this effect did not last until 6 months after the intervention [22]. The study by Dinur-Klein et al. further revealed that 10 Hz intervention on the lateral PFC and insula could reduce tobacco use, while the effect of 1 Hz rTMS intervention was not significant [46]. In the research on cocaine, marijuana, opioids, and methamphetamine, only one to two studies reported relapse situations after the rTMS treatment [11], and all found significant differences between the real rTMS group and the sham group.

Overall, the intervention effect of rTMS on relapse is still unclear, especially for illegal drugs, and the existing evidence is insufficient to draw conclusions. The existing evidence mainly suggests that effective rTMS protocol to reduce alcohol and nicotine use disorder mainly adopt high-frequency stimulation targeting the frontal lobe/insula, which may provide a reference for further research.

Other Outcomes

Existing studies have also found that rTMS intervention has potential therapeutic effects on cognition, emotion, and sleep quality in patients with addictive disorder. In terms of cognition, for example, Min and colleagues used a continuous 20-session iTBS protocol to target the left DLPFC, which significantly improved the verbal working memory of patients with methamphetamine use disorder, with the therapeutic effect still maintained during a follow-up 1 year after the intervention [27]. Another study using 1 Hz low-frequency intervention on the left DLPFC suggested an improvement in the inhibitory control abilities of methamphetamine-dependent patients [31]. In terms of emotion and sleep, there are also studies indicating that rTMS has a certain intervention effect on levels of depression and anxiety, as well as sleep quality. For instance, studies by Su and Yuan have shown the ability to improve sleep quality and emotional levels in methamphetamine users [47, 27], while Tsai and colleagues' research claimed that high-frequency stimulation of the left DLPFC could improve the level of depression in heroin-dependent patients in methadone clinics [38].

Safety

Most studies report that the adverse effects of rTMS on patients with SUDs are generally mild. For instance, a meta-analysis conducted by Gay et al. showed that out of 34 rTMS studies, 33 reported the presence of adverse effects, but the most common adverse effects were minor discomfort, headache, dizziness, or insomnia [48]. Patients receive the rTMS stimulation generally tolerate the treatment well, with few cases of discontinuation or withdrawal due to adverse reactions. Notably, the induction of seizures is theoretically the most severe acute adverse effect of rTMS intervention. Given the potential variety of drugs used by the patient population with SUDs, some may facilitate the risk of seizure induction, such as seizures occurring post-detoxification in some alcohol-dependent individuals [49]. Nonetheless, considering the large number of subjects who have undergone rTMS treatment since 1998 and the extremely low incidence of actual seizures, the academic community generally considers the risk of rTMS-induced seizures to be very low. Some adverse effects in case reports are relatively rare such as the unusual adverse reaction of nocturnal emission and wet dreams reported by Shen et al. in a case of methamphetamine use [50]. A case report described an unusual incidence of extreme thirst in a patient after treatment [51]. There is also a study reporting a case where a patient subjectively described unusual self-limiting involuntary hand movements after treatment, but without clear positive neurological examination evidence [52]. When rTMS studies are conducted with parameter settings that do not significantly exceed existing research evidence or are implemented under standardized operational guidance, the probability of serious adverse events is low. To ensure the safety of implementation and individual compliance, adverse effects should be continuously monitored and recorded during the intervention process.

Treatment Regimen

RTMS intervention parameters encompass various elements such as stimulation frequency, intensity, location, and duration. The distinct combinations of these parameters significantly influence treatment outcomes [46, 53]. Research has also been directed toward symptomatology and brain activity during stimulation, indicating their impact on the eventual therapeutic efficacy. The majority of studies utilize a stimulation intensity ranging from 90% to 120% RMT, while a few have experimented with an 80% intensity level [48]. To date, no consensus exists on the optimal stimulation intensity. However, due to the significant discomfort experienced by some patients when employing intensities above 100% RMT, a number of studies have implemented a strategy of gradually increasing the intensity for intervention [24]. The duration of rTMS therapy spans from a single day to several weeks. In terms of target selection for stimulation, the DLPFC has been the focus of most research efforts [10]. A meta-analysis has separately evaluated the effects of excitatory parameters targeting the left DLPFC versus the right DLPFC on craving reduction, revealing that excitatory stimulation of the left DLPFC significantly decreases cravings, whereas the right-sided intervention does not show significant differences from the control group [13]. A limited number of studies have explored the treatment effect of inhibitory parameter stimulation of the left DLPFC, right frontal lobe, or mPFC. For example, Yuan et al. demonstrated that 1 Hz frequency stimulation of the left frontal cortex markedly reduced cravings in drug-dependent patients, with a positive association between the reduction in cravings and the enhancement of response inhibition capabilities [31]. Regarding dTMS, studies have predominantly concentrated on the insula, with concurrent stimulation of the PFC, showing promising results in smoking cessation. Over an 18-week monitoring period, the 4-week continuous quit rate for patients was 19.4%, compared to 8.7% in the control group [23]. The application of dual-target approaches appears to be more effective than single-target stimulation. Chen and colleagues conducted a study indicating that the combination of rTMS targeting the mPFC and DLPFC resulted in superior and faster craving intervention effects [24]. In conclusion, the potential applications of rTMS in addiction treatment are vast, yet its efficacy is subject to a multitude of intervention parameters. Future research should delve into identifying the most effective rTMS parameter combinations to enhance therapeutic outcomes and offer more tailored treatment plans for individuals with addiction. Additionally, larger-scale, systematic studies are necessary to confirm the long-term efficacy and safety of TMS in the treatment of addiction.

In the selection of rTMS targets for behavioral addictions, the DLPFC is the primary focus, with a particular emphasis on the left side. One study reported an intervention protocol for the bilateral SMA. When targeting the left dorsolateral DLPFC, high-frequency (e.g., 10–15 Hz) stimulation is predominantly used, and after 1–20 sessions of intervention, it can to some extent reduce the craving for gambling [54], and also improve scores on emotional scales, but reports on the impact on gambling behavior are limited. Zack et al. also analyzed the effectiveness of a single session of nontraditional intervention parameters—continuous theta-burst stimulation (cTBS)—on gambling populations [54], but this protocol did not show a significant difference from the control group in reducing cravings. Sauvaget et al. chose another inhibitory parameter of 1 Hz to intervene on the right DLPFC of gambling patients, setting the intensity to 120%, and also found no evidence of effective intervention [55]. This seems to suggest that high-frequency intervention on the DLPFC brings greater clinical benefits to gambling populations than low-frequency parameters. However, it is still too early to draw conclusions. In TMS research on behavioral addictions, very few studies have explored the role of interventions in deep brain areas such as the insula.

Clinical Recommendations

Based on the existing evidence, Table 6.1 summarizes the current recommendations for rTMS interventions targeting SUDs. The rTMS treatment guidelines published by the International Federation of Clinical Neurophysiology in 2019 only classified "high-frequency intervention on the left DLPFC to reduce tobacco cravings and consumption" as a C-level recommendation (possibly effective), while other SUDs

Table 6.1 TMS in addictive disorder

TMS	Clinical recommendations				Levels + references
	Target	Frequency (Hz)	Length (min)	Duration (sessions)	
Unilateral rTMS	Left DLPFC	High frequency	Not mentioned	Not mentioned	[a]Level C [56] [b]Low level [57]
Deep TMS (H4 coil)	Bilateral lateral PFC + insula	10	20	18	FDA approved [58]
Bilateral rTMS	Unclear	Unclear	Unclear	Unclear	NA
Accelerated rTMS	Unclear	Unclear	Unclear	Unclear	NA
Priming TMS	Unclear	Unclear	Unclear	Unclear	NA
Synchronized TMS	Unclear	Unclear	Unclear	Unclear	NA
iTBS	Unclear	Unclear	Unclear	Unclear	NA
Accelerated iTBS	Unclear	Unclear	Unclear	Unclear	NA
Continuous TBS	Unclear	Unclear	Unclear	Unclear	NA
Bilateral TBS	Unclear	Unclear	Unclear	Unclear	NA
MST	Unclear	Unclear	Unclear	Unclear	NA

Abbreviations: *rTMS* repetitive transcranial magnetic stimulation, *TMS* transcranial magnetic stimulation, *DLPFC* dorsolateral prefrontal cortex, *PFC* prefrontal cortex

[a]Evidence-based guidelines established by a group of European expert mention Unilateral rTMS as Level C (possible efficacy) only for cigarette craving and consumption but no other addiction type

[b]Clinical Practice Guidelines developed by the Indian Psychiatric Society mention Unilateral rTMS as low recommendation for smoking cessation but no other addiction type

were not included due to insufficient evidence [56]. In 2020, the FDA approved rTMS (H4 deep coil) for the treatment of nicotine use disorder. In 2023, the Indian Psychiatric Association again classified "high-frequency intervention on the left DLPFC for the treatment of nicotine use disorder" as a low-level recommendation, while other substance uses were not recommended due to limited evidence [57]. In summary, based on the current research, the evidence for rTMS treatment of nicotine use disorder has gradually gained recognition, while there is still a lack of sufficient evidence to support the widespread clinical application for other SUDs.

6.2.1.2 Bilateral rTMS

Compared to unilateral rTMS intervention, bilateral rTMS intervention could simultaneously stimulate both hemispheres of the brain, which may help to more comprehensively activate brain regions associated with specific functions, thereby improving therapeutic outcomes. Additionally, bilateral rTMS may contribute to balancing the excitability of both cerebral hemispheres. Taking depression as an example, previous studies have suggested that there is a functional imbalance between the left and right DLPFC regions in patients with depression during the emotional regulation process [59]. There are relatively few studies on bilateral rTMS intervention in the field of addiction. Wing et al. conducted a bilateral rTMS study targeting patients with nicotine dependence comorbid schizophrenia. The study applied a 20-Hz frequency rTMS to both DLPFC regions, and 20 sessions of consecutive intervention significantly reduced cravings [60]. Another study conducted by Kozak et al. used similar parameters but adjusted the number of intervention sessions to six and found no change in the craving levels of nicotine-dependent patients [61]. Bidzinski et al. recently explored the efficacy of high-frequency bilateral DLPFC stimulation on patients with schizophrenia comorbid cannabis use disorder, finding that it could reduce PANSS scores but had limited impact on cannabis craving scores [62].

Despite limited research, existing studies still suggest the safety profile of bilateral rTMS intervention on addictive disorder. Overall, no studies have reported severe adverse effects. The adverse reactions are similar to those of unilateral rTMS intervention, mostly manifesting as headaches, neck pain, dizziness, and so on, and are generally mild, without significantly affecting treatment participation [62].

Current research on bilateral rTMS intervention has focused on bilateral DLPFC (i.e., studies that specifically used figure-eight coils or other non-deep coils), and all have used high-frequency parameters (10–20 Hz) with a treatment intensity of 90–100% rMT. Unlike findings in depression, there is currently no concrete evidence to suggest an hemispheres imbalance in populations with addictive disorder. However, some studies suggest that there may be weakened connections between the two cerebral hemispheres, thus promoting the strength of both hemispheres could be a direction for intervention. In individuals with internet addiction, weakened functional connectivity between different brain regions has been observed [63]. Additionally, there are many possibilities for bilateral rTMS intervention, such as the above studies that seem to find that low-frequency intervention on the right DLPFC can reduce cravings [64], while conventional high-frequency parameter intervention on the left DLPFC can achieve the similar therapeutic goal [27]. Furthermore, the collaboration between different brain regions in the two hemispheres may also be a direction for future development of bilateral interventions. For instance, Serenella et al. found abnormal connectivity in bilateral regions such as the putamen, caudate nucleus, and middle frontal gyrus [65]. In fact, an increasing number of studies highlight the significant role of interregional network activity in addictive disorder and other cognitive activities. Compared to unilateral interventions, bilateral interventions are expected to further leverage the advantages of rTMS. However, due to the limited number of published studies and sample sizes all below 30 participants, it is still difficult to draw conclusions about the efficacy of interventions targeting both brain hemispheres.

6.2.1.3 Accelerated rTMS

While rTMS has shown significant therapeutic effects in the treatment of some mental and psychological disorders, it has also been constrained by a slow response time. For patients with acute episodes and urgent treatment demands, the rapid onset of treatment is an important factor for clinicians to determine whether a therapy is suitable for practical clinical work. On the other hand, it appears that an increase in the total number of stimulation pulses is correlated with the final therapeutic outcome. Accelerated rTMS protocols have been developed in response to such needs. In 2022, a renowned accelerated protocol—Stanford Accelerated Intelligent Neuromodulation Therapy (SAINT)—received FDA approval for the treatment of treatment-resistant depression, with remission rates (78.6–90.5%) significantly higher than the 33% of conventional rTMS [66, 67]. There is a scarcity of research on accelerated interventions in the field of addiction. For instance, in a study by Wu and colleagues, after 15 sessions of accelerated high-frequency rTMS treatment, 13 out of 22 alcohol-dependent patients relapsed within 4 weeks [68]. Interestingly, compared to those who abstained, relapsers exhibited a greater reduction in brain gray matter volume. In another study, 15 consecutive sessions targeting the right DLPFC significantly reduced psychological cravings in alcohol-dependent patients, although cue-induced cravings did not change significantly [69].

Although accelerated rTMS therapy appears to demonstrate stronger therapeutic effects in various mental disorders, accelerated protocols may induce additional adverse effect and risks. Therefore, it is important to monitor changes in cortical excitability and subjective discomfort reported by patients as the number of pulses increases. Although one of the serious risks of TMS, seizure, is relatively low, but due to the increase in the number of stimulation pulses, EEG monitoring during the stimulation process may have certain clinical significance [70].

Given the limited research and only one accelerated rTMS study suggesting a reduction in cravings for alcohol-dependent patients [69], further evidence is needed to confirm the effects of accelerated rTMS, making it difficult to recommend for clinical use at this time.

6.2.2 Deep TMS

Traditional figure-eight coils can typically stimulate depths of about 0.8 cm beneath the scalp, while dTMS can stimulate depths of up to 4 cm beneath the scalp surface [71], with the specific stimulated location being related to the coil model. Currently, there are over 20 types of H-shape dTMS coils [72]. In addiction research, researchers mainly apply H coils to stimulate bilateral DLPFC, mPFC, ACC, and insular regions [10]. The stimulation intensity is set between 80% and 120% rMT. The number of sessions ranges from 10 to 20. Studies focus on cocaine, tobacco, and alcohol use disorders, with one study conducted on gambling addiction [40].

Using dTMS coils to stimulate traditional targets such as the DLPFC for multiple consecutive interventions can significantly reduce cravings and depressive symptoms in alcohol-dependent patients (after 20 sessions) and also decrease relapse rates in cocaine-dependent patients (after 12 sessions) [73, 74]. When using the H7 coil for high-frequency intervention on the mPFC and dorsal ACC regions, it can significantly reduce cocaine intake [34]. The choice of frequency may also affect the outcome. For instance, Dinur-Klein et al.'s study further revealed that 10 Hz intervention on the lateral PFC and insula can reduce tobacco use, while the effect of 1 Hz intervention is not significant [46]. However, the sample sizes of the above studies are small. Due to the variety of coil types, the number of articles published for different coil is limited, and the intervention sites for dTMS coils are not consistent, it is difficult to form a unified conclusion. For example, Harel et al. applied the H7 coil to stimulate the mPFC and ACC [14], while Perini et al. targeted the insula [15].

A large-sample, multicenter study has strengthened confidence in dTMS treatment for addiction. This work used the H4 deep coil to stimulate bilateral lateral PFC and insula, and a continuous 6-week intervention showed good effects on reducing cravings [23]. Importantly, the protocol not only reduced psychological cravings but also showed good effects in quitting smoking. It resulted in a 4-week continuous quit rate of 19.4% (compared to 8.7% in the control group) during an 18-week follow-up [23]. It is known that compared to traditional rTMS interventions, dTMS have the potential to directly target deep brain regions associated with reward, such as the insula, VTA, and nucleus accumbens, which are often considered closely related to addiction [6].

The potential adverse reactions of dTMS in addiction are very similar to other rTMS intervention protocols. Zangen et al. provided a detailed report [23]. In their study, the most common adverse reaction was headache. Additionally, pain at the stimulation site, face, jaw, neck, and muscle spasms/twitches may occur. The pain level is mild to moderate and disappears after treatment, which is similar to other type of rTMS interventions [75].

There is currently a limited amount of evidence that can form recommendations and the studies are distributed across different addictive substances. Among them, Zangen et al.'s work [23], due to its large sample size and long-term follow-up evidence, has prompted the FDA to approve dTMS (H4 deep coil) for the treatment of nicotine addiction in 2020. As for other types of substance dependence, it is difficult to form conclusive recommendations recently.

6.2.3 Priming TMS

To optimize the therapeutic effects and advantages of TMS stimulation, researchers are continuously exploring potential new intervention parameters and protocols. Among them, pTMS has shown certain potential. The typical PpTMS protocol is designed to optimize the subsequent main rTMS therapeutic effects. These protocols adjust the excitability state of the cerebral cortex in advance to enhance or alter the response to subsequent treatments. Common priming rTMS protocols include high-frequency rTMS priming and low-frequency rTMS priming strategies [76]. Previous studies on priming rTMS have mainly focused on patients with major depressive disorder. For example, in a study by Fitzgerald et al. [77], compared to the use of low-frequency rTMS intervention alone, applying 6 Hz high-frequency rTMS priming followed by 1 Hz rTMS stimulation of the right DLPFC showed a stronger antidepressant effect. The mechanism of priming rTMS is not yet clear, but it is generally believed to be related to the following mechanisms: changing the excitability level of the cerebral cortex to regulate neural plasticity [76]. The therapeutic effect of priming rTMS on addiction is not yet clear. The author searched databases such as PubMed and CNKI and found no published clinical trials. Consistent with other mental disorders, one of the important pathogenic mechanisms of addictive disorder lies in the impaired plasticity of brain regions [78]. This also implies that priming rTMS may have certain therapeutic potential for addictive disorder.

6.2.4 Synchronized TMS

Synchronized transcranial magnetic stimulation (sTMS) is a specific form of rTMS. This stimulation protocol emphasizes synchronization with brain activity or other stimuli in time or space. The design concept of sTMS is to enhance the effectiveness and specificity of stimulation by more precisely matching the natural rhythms of the brain or neural activity patterns in specific task states. In terms of integration with measurement tools, sTMS can be designed to synchronize with the rhythms of electroencephalographic (EEG) activity (such as alpha frequency) and can also be combined with specific cognitive-psychological tasks. sTMS emphasizes the individualization of each subject, resulting in significant differences in protocol design across various studies.

Taking depression as an example, studies have indicated that the power of left hemisphere alpha oscillations is significantly increased [79] and positively correlated with depressive symptoms [80]. A decrease in alpha power has also been found to reflect a good response to antidepressant treatment [81]. To enhance the treatment effect of depression, Zrenner and colleagues developed a new intervention strategy—rTMS triggered at the negative EEG peak of instantaneous alpha oscillations. It was found that a single session of alpha sTMS could significantly reduce EEG resting-state alpha activity in the left DLPFC region [82], with no significant changes observed in the control group. Continuous low-field magnetic stimulation synchronized to an individual's alpha frequency (IAF) for 6 weeks showed more significant improvement in depressive symptoms compared to the control group [83]. However, several studies concluded that the alpha sTMS protocol did not significantly outperform the control in improving depressive symptoms in patients with depression [84, 85].

In addiction area, there is currently a lack of rTMS intervention protocols synchronized with specific EEG activity. A study protocol published in 2023 plans to conduct low-field stimulation synchronized to an individual's alpha peak frequency among retired military subjects with addiction disorders, but the results have not yet been announced [86]. In addition to synchronizing with EEG activity, synchronizing specific states during the rTMS intervention process seems to further enhance the effectiveness of addiction intervention. Dinur-Klein used a high-frequency stimulation protocol targeting bilateral PFC and insula combined with smoking cue exposure to enhance the reduction in cigarette consumption, increasing the quit rate to 44% [46]. Another study compared the therapeutic effects of 10 Hz intervention on DLPFC synchronized with smoking video cue exposure versus sham stimulation synchronized with cue exposure [87], suggesting the role of 10 Hz intervention in reducing smoking consumption. Synchronized TMS protocols are theoretically appealing. However, there is currently a lack of clear evidence, indicating superiority over traditional intervention strategies on additive disorder.

6.2.5 TBS

6.2.5.1 iTBS

Efficacy

Intermittent theta burst stimulation (iTBS) is a variant of rTMS. iTBS is generally believed to excite the brain activities. By mimicking the brain's natural activity patterns, iTBS enhances therapeutic effects with short bursts of high-frequency stimulation. iTBS is typically set to deliver three burst pulses every 200 ms, with an intra-burst frequency of 50 Hz, and this pattern is applied for 2 s followed by an 8-s interval [88]. The mechanism of iTBS involves changes in synaptic plasticity [89], i.e., by regulating the activity of certain excitatory or inhibitory neurotransmitter receptors, which in turn induce long-term potentiation (LTP) or long-term depression (LTD)-like effects. Additionally, TBS can regulate gene expression and protein synthesis [90]. In clinical applications, due to its short stimulation time and low stimulation intensity, iTBS has been widely used to treat neurological and psychiatric disorders such as depression and ischemic stroke, showing good efficacy and safety [56]. A multicenter clinical trial found that iTBS stimulation of the DLPFC has comparable antidepressant efficacy to 10 Hz rTMS, leading to FDA approval for the treatment of treatment-resistant depression [91]. Exploration of iTBS in addiction disorders has mainly focused on populations with methamphetamine use disorder [24, 27, 92–95]. Additionally, a small number of studies have been conducted in nicotine, cocaine, and heroin use disorders [96–100]. In studies reporting cravings outcome, almost all studies found that iTBS intervention on the left DLPFC significantly reduced psychological cravings in methamphetamine use disorder. Moreover, under the premise of shorter treatment duration, the iTBS protocol is not inferior to the 10 Hz classical protocol [92]. Chen et al.'s study showed that iTBS intervention targeting the left DLPFC reduced patients' attention bias toward drugs, which was significantly related to changes in cravings [101]. Additionally, iTBS targeting the left DLPFC also showed improvements in working memory [27]. In previous studies, impaired working memory was found to be one of the most common cognitive impairments in methamphetamine use disorder [102]. Although the sample sizes of the above studies are small and larger randomized controlled studies are needed to confirm the observations, iTBS appears to be an effective method that better meets clinical needs, especially considering the difference between its short duration (3 min) and that of high-frequency stimulation (often more than 10 min). Reducing treatment duration helps to decrease patient discomfort and is expected to reduce dropout rates while ensuring treatment effectiveness. In populations with nicotine and heroin use disorder, iTBS intervention on the PFC has also been found to reduce psychological cravings [98, 100]. These consistent results seem to indicate the strong potential of the iTBS intervention protocol in reducing cravings in populations with addictive disorder. Other indicators, due to the high heterogeneity of studies, are still difficult to conclude.

Safety

One of the advantages of the TBS protocol is its relatively short intervention time, which may reduce the discomfort associated with longer durations of rTMS interventions. In studies regarding iTBS intervention, the main adverse reactions reported are mild scalp discomfort, temporary headaches, dizziness, and nausea, all of which improve after the treatment is discontinued. No studies have reported severe adverse reactions such as seizures. These adverse reactions are similar to those observed with other TMS intervention protocols. In fact, due to the lack of dedicated comparative studies, it is currently difficult to determine which intervention strategy has more prominent adverse reactions.

Treatment Regimen

In addiction treatment, iTBS intervention protocols often target the left DLPFC, with a minority of studies stimulating the bilateral DLPFC or the right inferior frontal gyrus [99, 100]. The intensity of stimulation is between 80–100% rMT. The majority of studies have focused on changes in psychological cravings, with a smaller portion addressing relapse and cognitive changes. In terms of psychological cravings, iTBS intervention targeting the left DLPFC has been found to significantly improve patients' levels of psychological cravings. However, there is only one study that conducted iTBS intervention on the right side, and although this study also suggested a significant reduction in psychological cravings [103]. It is noteworthy that relapse, a key behavioral observation indicator in addiction treatment, is rarely reported. Su et al. followed up with methamphetamine-dependent patients after 20 consecutive iTBS interventions and found no significant difference in relapse rates compared to the control group [27].

Clinical Recommendations

Synthesizing the current application exploration of iTBS intervention protocols in populations with addictive disorder, it seems meaningful that iTBS stimulation of the left DLPFC could reduce psychological cravings; however, further large sample randomized controlled clinical trial need to be conducted. There is currently a lack of more substantial research evidence to recommend improvements in other clinical indicators.

6.2.5.2 Accelerated iTBS

Based on the potential demonstrated by iTBS intervention protocols, to meet the needs of more urgent disease conditions (such as suicide), researchers have explored the role and safety of accelerated iTBS intervention protocols. Accelerated iTBS protocols refer to multiple iTBS interventions conducted within a single day, thereby providing a higher quantity of pulse stimulation in a shorter period of time. One of largest number of accelerated TBS intervention studies have been conducted in populations with depression. Among them, the SAINT developed by the research team at Stanford University provides ten interventions in 1 day and 90,000 pulses totally in 5 days, targeting the left DLPFC, with a remission rate as high as 78.6–90% after the SAINT protocol [66, 67]. Due to its outstanding intervention effect, in 2022, the FDA approved it for the treatment of treatment-resistant depression. Compared to depression, there have been few accelerated iTBS protocols conducted in addiction [104, 105]. In one study by Steele et al., 19 patients with cocaine use disorder were recruited for a within-subjects controlled study. Patients received three iTBS treatments per day for 10 days (a total of 30 iTBS treatments). The stimulation target of this protocol was the left dlPFC. Compared to the patients' performance before treatment, after receiving iTBS intervention, the patients reduced their weekly cocaine consumption by 78% and the number of usage days by 70% within 4 weeks [105]. This indicates that the accelerated iTBS protocol has significant utility in reducing cocaine use. Another randomized controlled clinical study conducted in a population with nicotine use disorder suggested that four times a day for a total of 20 iTBS treatments did not outperform the sham group in reducing psychological cravings and nicotine dependence levels [104]. Therefore, based on current findings, it is difficult to clearly determine the therapeutic effect and value of accelerated iTBS protocols for addiction. When conducting accelerated iTBS interventions, attention should be paid to the setting of certain parameters, with studies suggesting that the setting of parameters such as time intervals, stimulation intensity, and brain region localization methods significantly affect the final clinical efficacy [106]. For instance, intervals between treatments of 50–90 min can induce changes in neural plasticity, while less than 40 min cannot [106].

A summary of studies in different diseases found that the most commonly reported side effects of accelerated TBS protocols are headache, fatigue, and nausea, with a headache incidence rate of 31.3% [106]. No studies reported serious side effects. Overall, there is no evidence to suggest that the adverse reactions of accelerated TBS protocols are higher than those of traditional rTMS intervention protocols.

6.2.5.3 Continuous TBS

cTBS is a pattern of rTMS that simulates the 4–7 Hz burst discharges recorded in the hippocampus, an electrophysiological activity closely related to synaptic plasticity. The basic characteristic of cTBS is the occurrence of three burst pulses every 200 ms, with an intra-burst frequency typically set at 50 Hz, although studies have also used other frequencies such as 30 and 40 Hz. In addiction research, cTBS interventions have primarily targeted brain regions such as the DLPFC, rIFG, mPFC, and orbital frontal cortex (OFC). Among them, in populations with alcohol use disorder, cTBS interventions targeting the right DLPFC or the left DLPFC have both been found to increase cravings/risk of relapse [107, 108]. Studies conducted in populations with methamphetamine use disorder have shown that 10 sessions of cTBS intervention targeting the right DLPFC effectively reduced psychological cravings, while cTBS stimulation of the left DLPFC did not significantly improve effects [95]. cTBS interventions in the OFC (mPFC) region have shown stronger consistency in therapeutic effects. For instance, a study by Ankit et al. explored the efficacy of cTBS as an adjunctive treatment to naltrexone for patients with opioid use disorder, finding that targeting the right OFC significantly reduced craving scores [109]; another study conducted in populations with alcohol use disorder, targeting the mPFC brain region, reduced alcohol consumption within 3 months after treatment initiation [110]; in methamphetamine populations, Chen et al. compared the efficacy of ventromedial prefrontal cortex (vmPFC) cTBS and DLPFC iTBS, finding that both interventions had comparable effects on cravings [24]; Hanlon et al.'s work suggested that mPFC cTBS could downregulate the activation of brain regions associated with drug cues, such as the striatum and mPFC cortex [111], providing a mechanistic understanding of the value of mPFC interventions.

Based on existing cTBS intervention studies, it seems that mPFC cTBS intervention has potential for reducing psychological cravings. However, it is not yet sufficient to form recommendation suggestions. On the other hand, compared to iTBS interventions, cTBS protocols are generally considered to reduce the activation level of the stimulated target area and related circuits. Therefore, for addiction disorders with complex circuit abnormalities, if combined intervention strategies are carried out, they may bring stronger effects. For example, a randomized controlled clinical study found that mPFC cTBS combined with DLPFC iTBS intervention was superior to single DLPFC iTBS intervention in reducing cravings [24].

6.2.5.4 Bilateral TBS

There are at least two approaches to stimulate bilateral brain region. One is sequential stimulation, where one brain region is stimulated followed by another; the other involves using dTMS coils to simultaneously intervene in large areas of both brain hemispheres. For instance, in the treatment of treatment-resistant depression, bilateral DLPFC rTMS intervention has been found to significantly reduce the degree of depression. Blumberger et al. conducted a randomized controlled clinical study comparing the efficacy of bilateral DLPFC rTMS intervention with bilateral DLPFC iTBS interventions and found that they had equivalent effects in reducing depression levels [112]. Since rTMS takes up to 48 min, while iTBS protocols last for 4 min, this to some extent increases the opportunity and cost-effectiveness for patients with treatment-resistant depression to access treatment equitably. On the other hand, bilateral TBS interventions seem to have superior clinical outcomes compared to unilateral ones. A naturalistic observation study found that patients receiving bilateral DLPFC stimulation required fewer treatment sessions and had lower posttreatment anxiety symptoms compared to those receiving unilateral stimulation, suggesting that bilateral TBS may have superior antianxiety effects and a slightly faster response time [113]. In the field of addictive disorders, there is currently a lack of large-scale comparative studies on bilateral interventions. A study in a cocaine population used H4 deep coils to explore the efficacy differences between 15 Hz intervention and iTBS intervention targeting the frontal and insular cortices. The study found that both protocols had equivalent effects on craving intervention

[99]. In comparison, bilateral PFC iTBS intervention takes only 3 min, which is just 20% of the time for 15 Hz intervention. This result enhances the clinical appeal of TBS intervention strategies. As current research is scarce, further exploration and study of bilateral TBS protocols will help promote the more diverse clinical application of rTMS intervention strategies.

6.2.6 MST

Since Lisanby et al. first reported on the successful and intentional induction of seizures using rTMS in 2001 [114], many researchers have explored the appropriate parameters and intervention effects of magnetic seizure therapy (MST). As a novel intervention protocol, MST is based on the ability of high frequency rTMS to induce therapeutic seizures and exert therapeutic effects on severe psychiatric disorders [115, 116]. Compared to traditional electroconvulsive therapy (ECT), MST typically has fewer neurocognitive adverse reactions [117]. This may be related to the fact that seizures induced by MST originate from the superficial cortical areas, whereas during ECT the current passes through deep brain areas [114].

To date, there have been few studies reporting the application of MST technology in populations with addictive disorder. In fact, the application of ECT has historically been used to address psychosis, catatonia, and other conditions resulting from drug use [118, 119], rather than treating addiction itself. Some researchers even argue that the use of ECT technology to intervene in the treatment of methamphetamine use disorder is ethically controversial [120]. Whether MST technology should be conducted in addicted populations, what parameters should be set, and what clinical indicators should be focused on, are issues that the further researchers should give considerable consideration.

6.3 tDCS

6.3.1 Conventional tDCS

6.3.1.1 Efficacy

Transcranial direct current stimulation (tDCS) is an innovative neuromodulation technique that targets specific brain regions by applying a steady current between two or more electrodes. The currents induced by tDCS increase the likelihood of neuronal activation through depolarizing or hyperpolarizing neurons without directly eliciting action potentials [121]. In recent years, the role of tDCS in interventions for substance dependence and addiction has been extensively examined [10].

For alcohol use disorders, Klauss et al. demonstrated that five sessions of tDCS significantly reduced the relapse rate 6 months poststimulation (50% vs. 88.2%), although no significant reduction in craving was observed [122]. In a subsequent study with a larger sample size and ten sessions of tDCS, Klauss et al. reported significant decreases in both alcohol craving and the 3-month relapse rate in the active treatment group [123]. Similarly, Dubuson et al. found that active tDCS significantly increased abstinence rates at a 2-week follow-up compared to sham stimulation [124]. In another study, den Uyl et al. showed that three sessions of tDCS significantly reduced cue-induced craving in individuals with alcohol use disorders [125]. However, several studies have reported no efficacy of tDCS on craving or relapse rates in patients with alcohol use disorders. For instance, Witkiewitz et al. incorporated tDCS into a treatment regimen with mindfulness-based relapse prevention but found no significant differences in posttreatment alcohol consumption or craving between active and sham tDCS groups [126]. Similarly, den Uyl et al. reported no significant effects on craving or relapse rates following four sessions of tDCS [127], with a subsequent study yielding similar negative results [128]. In conclusion, tDCS is a promising intervention method for patients with alcohol use disorders, although the optimal parameter sets for craving and relapse rate remains unclear.

tDCS has shown relatively consistent effects on nicotine dependence, as indicated by a meta-analysis by Kang and colleagues [129]. Boggio et al. found that five sessions of tDCS significantly decreased nicotine craving, cue-induced craving, and nicotine consumption compared to sham stimulation [130]. Mondino et al. conducted an intervention using ten sessions of tDCS in individuals with nicotine dependence and observed a significant reduction in cigarette craving [131]. Ghorbani et al. examined the relationship between the duration of tDCS (administering the same dose) and its efficacy on relapse rates, finding that extended tDCS sessions resulted in the highest abstinence rate at 6 months [132].

For cocaine use disorder, the results of tDCS interventions have been inconsistent. Batista et al. reported a significant reduction in craving for crack-cocaine following five sessions of tDCS [133]. However, two other studies, conducted by Verveer et al. and Gaudreault, found no effect of tDCS on cocaine craving [134, 135]. tDCS has shown promising effects on methamphetamine use disorder. Three studies, employing five, ten, and twenty sessions respectively, reported significant reductions in methamphetamine craving [136–138]. However, additional high-quality randomized controlled trials (RCTs) are needed to confirm these positive effects. For opioid use disorder, tDCS has shown preliminary effects on reducing craving. Taremian et al. and Eskandari et al. found that ten sessions of tDCS significantly decreased opioid craving [139, 140].

The application of tDCS in behavioral addictions remains in its early stages. Two studies have investigated the effects of tDCS on internet gaming disorder (IGD) [141, 142]. Lee et al. found that ten sessions of tDCS did not reduce cravings for internet gaming [142]. Similarly, Jeong et al. used ten sessions of tDCS for IGD and found no significant difference in effect compared to sham stimulation [141]. Martinotti et al. employed five sessions of tDCS on a mixed sample including individuals with both substance (cocaine, opioids, alcohol, cannabis) and behavioral (gambling) addictions, reporting a significant decrease in craving [143]. However, it remains unclear whether this reduction was specifically influenced by the gambling disorder subgroup. Therefore, more randomized controlled trials focused on IGD and gambling disorders are necessary to confirm the efficacy of tDCS in treating behavioral addictions.

Although the effects of tDCS vary across different types of addiction, it remains a promising intervention overall, particularly due to its portability, simplicity, and adaptability.

6.3.1.2 Safety

The tDCS treatment is widely regarded as a safe intervention method [144]. The most commonly reported side effects are a mild tingling sensation and a brief perception of light flashes at the beginning and end of stimulation [145]. Another frequent side effect is erythema on the skin beneath the electrodes [146]. Additional side effects include temporary headaches and dizziness. Severe side effects, such as seizures, have been rarely reported in tDCS studies. Studies indicate that a current density of 52 mA is required to induce brain tissue lesions—significantly higher than the standard protocol of 1–2 mA. Thus, tDCS is considered a safe, noninvasive neuromodulation method for applications in addiction treatment.

6.3.1.3 Treatment Regimen

The parameter set for tDCS includes electrode positioning, current intensity, stimulation duration per session, and the total number of sessions within an intervention course. In addiction intervention, the most commonly targeted stimulation area is the left or right DLPFC, with either an anode or cathode placed accordingly. For current intensity, most studies use 1–2 mA to ensure an adequate stimulation dose. Typically, the duration of stimulation per session ranges from 10 to 20 min, with the number of sessions varying from a single session to a maximum of 20 sessions [10].

Evidence suggests that certain stimulation site configurations may have a higher probability of impacting craving and relapse. For alcohol use disorders, positioning the anode on the right DLPFC and the cathode on the left DLPFC has shown improved efficacy in reducing craving and relapse rates [122–124, 127, 128]. This configuration has also been observed to be effective in nicotine addiction [130, 132, 147]. For methamphetamine and opioid use disorder, most studies have employed the same anode–cathode arrangement, with positive outcomes on craving reduction [136, 138–140]. Nonetheless, additional high-quality studies are needed to confirm the optimal stimulation sites for these types of addiction.

For behavioral addiction, both Lee et al. and Jeong used a montage with the anode placed on the left DLPFC and the cathode on the right DLPFC, with a current intensity of 2 mA over a total of ten stimulation sessions, but no effect on craving was observed [141, 142]. It remains unclear whether alternative electrode montages could produce different effects. Martinotti et al. employed a montage with the anode on the right DLPFC and the cathode on the left DLPFC, using a current intensity of 1.5 mA, which resulted in a significant reduction in craving [143]. Further exploration of tDCS parameter sets is needed for applications in behavioral addictions.

There is currently no evidence that variations in current intensity systematically influence the effectiveness of tDCS in addiction treatment. Generally, multi-session interventions are more effective than single-session stimulations [148]. However, it

Table 6.2 TDCS in addictive disorder

tDCS	Clinical recommendations					Levels + references
	Anode position	Cathode position	Intensity (mA)	Length (min)	Duration (sessions)	
Conventional tDCS	Unclear	Unclear	Unclear	Unclear	Unclear	NA
HD-tDCS	Unclear	Unclear	Unclear	Unclear	Unclear	NA

Abbreviations: *HD-tDCS* high-definition transcranial direct current stimulation, *NA* not applicable, *DCS* transcranial direct current stimulation

remains unclear whether increasing the number of stimulation sessions can produce long-lasting effects.

6.3.1.4 Clinical Recommendations

For SUDs, Fregni et al. recommend using combined right DLPFC anodal and left DLPFC cathodal tDCS for at least 13 min per session, which is considered probably effective in reducing cravings for alcohol use disorders [121]. In the 2022 version of expert consensus on clinical applications of tDCS in China, it is suggested that for methamphetamine addiction, positioning the anode on the right DLPFC and the cathode on the left DLPFC for 20 min per session may improve symptoms. For nicotine dependence, tDCS with the anode on the left DLPFC and the cathode on the right DLPFC for 20 min per session (in more than five sessions) is rated as effective. However, no specific recommendations have been established for all types of addiction (Table 6.2), and further randomized controlled trials are needed to confirm the effects of tDCS on SUDs.

6.3.2 HD-tDCS

Compared to conventional tDCS interventions, high-definition tDCS (HD-tDCS) more focally delivers a weak current to specific brain areas through multiple small electrodes placed on the scalp, thereby precisely modulating subcortical neural activity and reducing the impact on nontarget areas. In HD-tDCS research, the 4×1 ring montage is the most commonly used montage, endowing HD-tDCS with the ability to have a single required activity without being affected by cortical stimulation of the opposite polarity. HD-tDCS has been studied in some psychiatric diseases. For example, in schizophrenia research, central cathodal stimulation to CP5 (temporoparietal junction) using the 10/10 EEG system with four return electrodes at FT7, FC3, PO7, P1 has been shown to significantly reduce the occurrence of persistent auditory hallucinations [149]. Studies conducted in elderly patients with depression also suggest that central anodal stimulation of 2 mA to F3 using the 10/10 EEG system with four return electrodes FC1, AF3, F7, FC5 each carrying 0.5 mA can reduce depression levels [150]. Another point of interest is whether HD-tDCS differs from or is superior to conventional tDCS in terms of intervention effects. To answer this question, some researchers have explored the differences in the impact on neural activity between the two and found that traditional tDCS leads to a reduction in theta power [151]. However, another study denied the existence of such effect differences [152]. From the limited literature search, no studies have reported the application of HD-tDCS technology in the field of addiction. Conventional tDCS technology has already shown clinical application value, and future exploration of the intervention effects and mechanisms of HD-tDCS technology in addiction will also broaden the application direction of tDCS technology.

6.4 TMS vs. tDCS

6.4.1 Efficacy

According to previous meta-analyses, there are considerable differences in the effects of rTMS and tDCS in treating addictive disorders. High-frequency TMS targeting the DLPFC has demonstrated medium to large effect sizes in alcohol, nicotine, and cocaine dependence [10, 20]. In contrast, the effects of tDCS are less consistent, tend to vary with individual differences, and generally show lower effect sizes than those of repetitive TMS (rTMS) [20]. This discrepancy may stem from the distinct neural mechanisms of TMS and tDCS. TMS directly evokes action potentials, which are more likely to induce LTP, whereas tDCS primarily modulates neuronal membrane potentials, affecting the likelihood of activation without directly inducing action potentials [153, 154]. Future research could focus on optimizing the parameters of both TMS and tDCS to improve therapeutic outcomes and further elucidate how individual differences influence efficacy.

6.4.2 Safety

TMS and tDCS are both considered safe intervention methods. However, it is worth noting that in rare cases, particularly during high-frequency stimulation, TMS may induce seizures, a risk that is minimal with tDCS. In clinical practice, precautions should be taken to address the potential for TMS-induced seizures, including having protocols in place. Additionally, mild and possible side effects, such as dizziness, discomfort at the stimulation site, and minor skin irritation, should be discussed with patients beforehand to alleviate concerns and enhance treatment compliance.

6.5 Conclusion

As emerging interventions for addiction, both TMS and tDCS have shown promising potential. However, current studies suggest that the consistency and reliability of these physical interventions still require enhancement. Further research is needed to confirm the clinical efficacy of TMS and tDCS in treating addictive disorders.

Future studies could focus on several key areas. First, optimal stimulation parameters and intervention protocols for different addiction types should be determined, given the varying effects of the same protocol across different types of addiction. Second, the underlying neural mechanisms of these interventions should be explored, potentially through a combination of animal and human studies or by synchronizing neural signal acquisition during stimulation [155, 156]. Such approaches could provide insights into the neural mechanisms, clinical effects, and their causal relationships. Third, the long-term effects and safety of these interventions should be assessed through extended follow-up measurements, which would guide future treatment protocols. Finally, given individual variations in brain anatomy, state, and development, precision strategies such as neural circuit-targeted interventions and closed-loop approaches merit exploration [157, 158]. These strategies may improve intervention consistency and enhance understanding of the neural mechanisms underlying TMS and tDCS in addictive disorders.

Acknowledgments None.

Disclosure/Conflicts of Interest The authors declare no conflicts of interest in conducting this study or preparing the manuscript.

Financial Support This work was supported by National Nature Science Foundation (82,130,041, 82,201,650), Clinical Research Project of Shanghai Municipal Health Commission (20244Y0201), and Shanghai Rising-star Cultivation Program (22YF1439200).

References

1. UNODC. World drug report 2024. New York: United Nations Office on Drugs and Crime; 2024.
2. Kemp S. Digital 2021: global overview report [Online]. 2021. https://datareportal.com/reports/digital-2021-global-overview-report Accessed 18 Dec 2024.
3. Meng SQ, Cheng JL, Li YY, Yang XQ, Zheng JW, Chang XW, Shi Y, Chen Y, Lu L, Sun Y, Bao YP, Shi J. Global prevalence of digital addiction in general population: a systematic review and meta-analysis. Clin Psychol Rev. 2022;92:102128.
4. Männikkö N, Billieux J, Kääriäinen M. Problematic digital gaming behavior and its relation to the psychological, social and physical health of Finnish adolescents and young adults. J Behav Addict. 2015;4:281–8.
5. Small GW, Lee J, Kaufman A, Jalil J, Siddarth P, Gaddipati H, Moody TD, Bookheimer SY. Brain health consequences of digital technology use. Dialogues Clin Neurosci. 2020;22:179–87.
6. Koob GF, Volkow ND. Neurobiology of addiction: a neurocircuitry analysis. Lancet Psychiatry. 2016;3:760–73.
7. Kalivas PW, Volkow ND. The neural basis of addiction: a pathology of motivation and choice. Am J Psychiatry. 2005;162:1403–13.
8. Wise RA, Robble MA. Dopamine and addiction. Annu Rev Psychol. 2020;71:79–106.
9. Eichhammer P, Johann M, Kharraz A, Binder H, Pittrow D, Wodarz N, Hajak G. High-frequency repetitive transcranial magnetic stimulation decreases cigarette smoking. J Clin Psychiatry. 2003;64:951–3.

10. Mehta DD, Praecht A, Ward HB, Sanches M, Sorkhou M, Tang VM, Steele VR, Hanlon CA, George TP. A systematic review and meta-analysis of neuromodulation therapies for substance use disorders. Neuropsychopharmacology. 2024;49:649–80.
11. Tang VM, Ibrahim C, Rodak T, Goud R, Blumberger DM, Voineskos D, Le Foll B. Managing substance use in patients receiving therapeutic repetitive transcranial magnetic stimulation: a scoping review. Neurosci Biobehav Rev. 2023;155:105477.
12. Speer AM, Kimbrell TA, Wassermann EM, Repella JD, Willis MW, Herscovitch P, Post RM. Opposite effects of high and low frequency rTMS on regional brain activity in depressed patients. Biol Psychiatry. 2000;48:1133–41.
13. Zhang JJQ, Fong KNK, Ouyang RG, Siu AMH, Kranz GS. Effects of repetitive transcranial magnetic stimulation (rTMS) on craving and substance consumption in patients with substance dependence: a systematic review and meta-analysis. Addiction. 2019;114:2137–49.
14. Harel M, Perini I, Kämpe R, Alyagon U, Shalev H, Besser I, Sommer WH, Heilig M, Zangen A. Repetitive transcranial magnetic stimulation in alcohol dependence: a randomized, double-blind, sham-controlled proof-of-concept trial targeting the medial prefrontal and anterior cingulate cortices. Biol Psychiatry. 2022;91:1061–9.
15. Perini I, Kämpe R, Arlestig T, Karlsson H, Löfberg A, Pietrzak M, Zangen A, Heilig M. Repetitive transcranial magnetic stimulation targeting the insular cortex for reduction of heavy drinking in treatment-seeking alcohol-dependent subjects: a randomized controlled trial. Neuropsychopharmacology. 2020;45:842–50.
16. Li X, Hartwell KJ, Owens M, Lematty T, Borckardt JJ, Hanlon CA, Brady KT, George MS. Repetitive transcranial magnetic stimulation of the dorsolateral prefrontal cortex reduces nicotine cue craving. Biol Psychiatry. 2013a;73:714–20.
17. Li X, Du L, Sahlem GL, Badran BW, Henderson S, George MS. Repetitive transcranial magnetic stimulation (rTMS) of the dorsolateral prefrontal cortex reduces resting-state insula activity and modulates functional connectivity of the orbitofrontal cortex in cigarette smokers. Drug Alcohol Depend. 2017;174:98–105.
18. Pripfl J, Tomova L, Riecansky I, Lamm C. Transcranial magnetic stimulation of the left dorsolateral prefrontal cortex decreases cue-induced nicotine craving and EEG delta power. Brain Stimul. 2014;7:226–33.
19. Rose JE, Mcclernon FJ, Froeliger B, Behm FM, Preud'homme X, Krystal AD. Repetitive transcranial magnetic stimulation of the superior frontal gyrus modulates craving for cigarettes. Biol Psychiatry. 2011;70:794–9.
20. Ekhtiari H, Tavakoli H, Addolorato G, Baeken C, Bonci A, Campanella S, Castelo-Branco L, Challet-Bouju G, Clark VP, Claus E, Dannon PN, Del Felice A, Den Uyl T, Diana M, Di Giannantonio M, Fedota JR, Fitzgerald P, Gallimberti L, Grall-Bronnec M, Herremans SC, Herrmann MJ, Jamil A, Khedr E, Kouimtsidis C, Kozak K, Krupitsky E, Lamm C, Lechner WV, Madeo G, Malmir N, Martinotti G, Mcdonald WM, Montemitro C, Nakamura-Palacios EM, Nasehi M, Noël X, Nosratabadi M, Paulus M, Pettorruso M, Pradhan B, Praharaj SK, Rafferty H, Sahlem G, Salmeron BJ, Sauvaget A, Schluter RS, Sergiou C, Shahbabaie A, Sheffer C, Spagnolo PA, Steele VR, Yuan TF, Van Dongen JDM, Van Waes V, Venkatasubramanian G, Verdejo-García A, Verveer I, Welsh JW, Wesley MJ, Witkiewitz K, Yavari F, Zarrindast MR, Zawertailo L, Zhang X, Cha YH, George TP, Frohlich F, Goudriaan AE, Fecteau S, Daughters SB, Stein EA, Fregni F, Nitsche MA, Zangen A, Bikson M, Hanlon CA. Transcranial electrical and magnetic stimulation (tES and TMS) for addiction medicine: a consensus paper on the present state of the science and the road ahead. Neurosci Biobehav Rev. 2019;104:118–40.
21. Xu Y, Zhang Y, Zhao D, Tian Y, Yuan T-F. Growing placebo response in TMS treatment for depression: a meta-analysis of 27-year randomized sham-controlled trials. Nat Mental Health. 2023;1:792–809.
22. Amiaz R, Levy D, Vainiger D, Grunhaus L, Zangen A. Repeated high-frequency transcranial magnetic stimulation over the dorsolateral prefrontal cortex reduces cigarette craving and consumption. Addiction. 2009;104:653–60.

23. Zangen A, Moshe H, Martinez D, Barnea-Ygael N, Vapnik T, Bystritsky A, Duffy W, Toder D, Casuto L, Grosz ML, Nunes EV, Ward H, Tendler A, Feifel D, Morales O, Roth Y, Iosifescu DV, Winston J, Wirecki T, Stein A, Deutsch F, Li X, George MS. Repetitive transcranial magnetic stimulation for smoking cessation: a pivotal multicenter double-blind randomized controlled trial. World Psychiatry. 2021;20:397–404.
24. Chen T, Su H, Li R, Jiang H, Li X, Wu Q, Tan H, Zhang J, Zhong N, Du J, Gu H, Zhao M. The exploration of optimized protocol for repetitive transcranial magnetic stimulation in the treatment of methamphetamine use disorder: a randomized sham-controlled study. EBioMedicine. 2020;60:103027.
25. Liang Y, Wang L, Yuan TF. Targeting withdrawal symptoms in men addicted to methamphetamine with transcranial magnetic stimulation: a randomized clinical trial. JAMA Psychiatry. 2018;75:1199–201.
26. Liu T, Li Y, Shen Y, Liu X, Yuan TF. Gender does not matter: add-on repetitive transcranial magnetic stimulation treatment for female methamphetamine dependents. Prog Neuro-Psychopharmacol Biol Psychiatry. 2019;92:70–5.
27. Su H, Chen T, Jiang H, Zhong N, Du J, Xiao K, Xu D, Song W, Zhao M. Intermittent theta burst transcranial magnetic stimulation for methamphetamine addiction: a randomized clinical trial. Eur Neuropsychopharmacol. 2020a;31:158–61.
28. Su H, Liu Y, Yin D, Chen T, Li X, Zhong N, Jiang H, Wang J, Du J, Xiao K, Xu D, Zeljic K, Wang Z, Zhao M. Neuroplastic changes in resting-state functional connectivity after rTMS intervention for methamphetamine craving. Neuropharmacology. 2020b;175:108177.
29. Wang W, Zhu Y, Wang L, Mu L, Zhu L, Ding D, Ren Z, Yang D, Tang H, Zhang L, Song P, Wei H, Chang L, Wang Z, Ling Q, Gao H, Liu L, Jiao D, Xu H. High-frequency repetitive transcranial magnetic stimulation of the left dorsolateral prefrontal cortex reduces drug craving and improves decision-making ability in methamphetamine use disorder. Psychiatry Res. 2022;317:114904.
30. Li X, Malcolm RJ, Huebner K, Hanlon CA, Taylor JJ, Brady KT, George MS, See RE. Low frequency repetitive transcranial magnetic stimulation of the left dorsolateral prefrontal cortex transiently increases cue-induced craving for methamphetamine: a preliminary study. Drug Alcohol Depend. 2013b;133:641–6.
31. Yuan J, Liu W, Liang Q, Cao X, Lucas MV, Yuan TF. Effect of low-frequency repetitive transcranial magnetic stimulation on impulse inhibition in abstinent patients with methamphetamine addiction: a randomized clinical trial. JAMA Netw Open. 2020;3:e200910.
32. Lolli F, Salimova M, Scarpino M, Lanzo G, Cossu C, Bastianelli M, Occupati B, Gori F, Del Vecchio A, Ercolini A, Pascolo S, Cimino V, Meneghin N, Fierini F, D''anna G, Innocenti M, Ballerini A, Pallanti S, Grippo A, Mannaioni G. A randomised, double-blind, sham-controlled study of left prefrontal cortex 15 Hz repetitive transcranial magnetic stimulation in cocaine consumption and craving. PLoS One. 2021;16:e0259860.
33. Terraneo A, Leggio L, Saladini M, Ermani M, Bonci A, Gallimberti L. Transcranial magnetic stimulation of dorsolateral prefrontal cortex reduces cocaine use: A pilot study. Eur Neuropsychopharmacol. 2016;26:37–44.
34. Martinez D, Urban N, Grassetti A, Chang D, Hu MC, Zangen A, Levin FR, Foltin R, Nunes EV. Transcranial magnetic stimulation of medial prefrontal and cingulate cortices reduces cocaine self-administration: a pilot study. Front Psych. 2018;9:80.
35. Li X, Song GF, Yu JN, Ai SH, Ji Q, Peng Y, Zhu XF, Liu JG. Effectiveness and safety of repetitive transcranial magnetic stimulation for the treatment of morphine dependence: a retrospective study. Medicine (Baltimore). 2021;100:e25208.
36. Liu X, Zhao X, Liu T, Liu Q, Tang L, Zhang H, Luo W, Daskalakis ZJ, Yuan TF. The effects of repetitive transcranial magnetic stimulation on cue-induced craving in male patients with heroin use disorder. EBioMedicine. 2020;56:102809.
37. Shen Y, Cao X, Tan T, Shan C, Wang Y, Pan J, He H, Yuan TF. 10-Hz repetitive transcranial magnetic stimulation of the left dorsolateral prefrontal cortex reduces heroin cue craving in long-term addicts. Biol Psychiatry. 2016;80:e13–4.

38. Tsai TY, Wang TY, Liu YC, Lee PW, Chang WH, Lu TH, Tseng HH, Lee SY, Chang YH, Yang Y, Chen PS, Chen KC, Yang YK, Lu RB. Add-on repetitive transcranial magnetic stimulation in patients with opioid use disorder undergoing methadone maintenance therapy. Am J Drug Alcohol Abuse. 2021;47:330–43.
39. Sayette MA. The role of craving in substance use disorders: theoretical and methodological issues. Annu Rev Clin Psychol. 2016;12:407–33.
40. Rosenberg O, Klein LD, Dannon PN. Deep transcranial magnetic stimulation for the treatment of pathological gambling. Psychiatry Res. 2013;206:111–3.
41. Pettorruso M, Martinotti G, Montemitro C, De Risio L, Spagnolo PA, Gallimberti L, Fanella F, Bonci A, Di Giannantonio M. Multiple sessions of high-frequency repetitive transcranial magnetic stimulation as a potential treatment for gambling addiction: a 3-month, feasibility study. Eur Addict Res. 2020;26:52–6.
42. Salerno L, Grassi E, Makris N, Pallanti S. A theta burst stimulation on Pre-SMA: proof-of-concept of transcranial magnetic stimulation in gambling disorder. J Gambl Stud. 2022;38:1529–37.
43. Concerto C, Signorelli MS, Chiarenza C, Ciancio A, Francesco AD, Mineo L, Rodolico A, Torrisi G, Caponnetto P, Pennisi M, Lanza G, Petralia A. Transcranial magnetic stimulation for the treatment of gambling disorder: a systematic review. J Integr Neurosci. 2023;22:164.
44. Cuppone D, Gómez Pérez LJ, Cardullo S, Cellini N, Sarlo M, Soldatesca S, Chindamo S, Madeo G, Gallimberti L. The role of repetitive transcranial magnetic stimulation (rTMS) in the treatment of behavioral addictions: Two case reports and review of the literature. J Behav Addict. 2021;10:361–70.
45. Abdelrahman AA, Noaman M, Fawzy M, Moheb A, Karim AA, Khedr EM. A double-blind randomized clinical trial of high frequency rTMS over the DLPFC on nicotine dependence, anxiety and depression. Sci Rep. 2021;11:1640.
46. Dinur-Klein L, Dannon P, Hadar A, Rosenberg O, Roth Y, Kotler M, Zangen A. Smoking cessation induced by deep repetitive transcranial magnetic stimulation of the prefrontal and insular cortices: a prospective, randomized controlled trial. Biol Psychiatry. 2014;76:742–9.
47. Lin J, Liu X, Li H, Yu L, Shen M, Lou Y, Xie S, Chen J, Zhang R, Yuan TF. Chronic repetitive transcranial magnetic stimulation (rTMS) on sleeping quality and mood status in drug dependent male inpatients during abstinence. Sleep Med. 2019;58:7–12.
48. Gay A, Cabe J, De Chazeron I, Lambert C, Defour M, Bhoowabul V, Charpeaud T, Tremey A, Llorca PM, Pereira B, Brousse G. Repetitive Transcranial Magnetic Stimulation (rTMS) as a promising treatment for craving in stimulant drugs and behavioral addiction: a meta-analysis. J Clin Med. 2022;11
49. Alberto GE, Klorig DC, Goldstein AT, Godwin DW. Alcohol withdrawal produces changes in excitability, population discharge probability, and seizure threshold. Alcohol Clin Exp Res (Hoboken). 2023;47:211–8.
50. Shen Y, Cao X, Zeng NN, Wang Y, Zhang H, Zangen A, Yuan TF. Emergence of sexual dreams and emission following deep transcranial magnetic stimulation over the medial prefrontal and cingulate cortices. CNS Neurol Disord Drug Targets. 2021;20:310–1.
51. Cogné M, Aupy J, Gil-Jardiné C, Glize B. Thirst induced by low frequency right hemisphere focal rTMS. Brain Stimul. 2018;11:623–4.
52. Steele VR, Maxwell AM, Ross TJ, Moussawi K, Abulseoud OO, Stein EA, Salmeron BJ. Report of transient events in a cocaine-dependent volunteer who received iTBS. Brain Stimul. 2018;11:631–3.
53. Soleimani G, Nitsche MA, Bergmann TO, Towhidkhah F, Violante IR, Lorenz R, Kuplicki R, Tsuchiyagaito A, Mulyana B, Mayeli A, Ghobadi-Azbari P, Mosayebi-Samani M, Zilverstand A, Paulus MP, Bikson M, Ekhtiari H. Closing the loop between brain and electrical stimulation: towards precision neuromodulation treatments. Transl Psychiatry. 2023;13:279.
54. Zack M, Cho SS, Parlee J, Jacobs M, Li C, Boileau I, Strafella A. Effects of high frequency repeated transcranial magnetic stimulation and continuous theta burst stimulation on gambling reinforcement, delay discounting, and stroop interference in men with pathological gambling. Brain Stimul. 2016;9:867–75.

55. Sauvaget A, Bulteau S, Guilleux A, Leboucher J, Pichot A, Valrivière P, Vanelle JM, Sébille-Rivain V, Grall-Bronnec M. Both active and sham low-frequency rTMS single sessions over the right DLPFC decrease cue-induced cravings among pathological gamblers seeking treatment: a randomized, double-blind, sham-controlled crossover trial. J Behav Addict. 2018;7:126–36.
56. Lefaucheur JP, Aleman A, Baeken C, Benninger DH, Brunelin J, Di Lazzaro V, Filipović SR, Grefkes C, Hasan A, Hummel FC, Jääskeläinen SK, Langguth B, Leocani L, Londero A, Nardone R, Nguyen JP, Nyffeler T, Oliveira-Maia AJ, Oliviero A, Padberg F, Palm U, Paulus W, Poulet E, Quartarone A, Rachid F, Rektorová I, Rossi S, Sahlsten H, Schecklmann M, Szekely D, Ziemann U. Evidence-based guidelines on the therapeutic use of repetitive transcranial magnetic stimulation (rTMS): an update (2014-2018). Clin Neurophysiol. 2020;131:474–528.
57. Tikka SK, Siddiqui MA, Garg S, Pattojoshi A, Gautam M. Clinical practice guidelines for the therapeutic use of repetitive transcranial magnetic stimulation in neuropsychiatric disorders. Indian J Psychiatry. 2023;65:270–88.
58. Cohen SL, Bikson M, Badran BW, George MS. A visual and narrative timeline of US FDA milestones for Transcranial Magnetic Stimulation (TMS) devices. Brain Stimul. 2022;15:73–5.
59. Liu W, Mao Y, Wei D, Yang J, Du X, Xie P, Qiu J. Structural asymmetry of dorsolateral prefrontal cortex correlates with depressive symptoms: evidence from healthy individuals and patients with major depressive disorder. Neurosci Bull. 2016;32:217–26.
60. Wing VC, Bacher I, Wu BS, Daskalakis ZJ, George TP. High frequency repetitive transcranial magnetic stimulation reduces tobacco craving in schizophrenia. Schizophr Res. 2012;139:264–6.
61. Kozak K, Sharif-Razi M, Morozova M, Gaudette EV, Barr MS, Daskalakis ZJ, Blumberger DM, George TP. Effects of short-term, high-frequency repetitive transcranial magnetic stimulation to bilateral dorsolateral prefrontal cortex on smoking behavior and cognition in patients with schizophrenia and non-psychiatric controls. Schizophr Res. 2018;197:441–3.
62. Bidzinski KK, Lowe DJE, Sanches M, Sorkhou M, Boileau I, Kiang M, Blumberger DM, Remington G, Ma C, Castle DJ, Rabin RA, George TP. Investigating repetitive transcranial magnetic stimulation on cannabis use and cognition in people with schizophrenia. Schizophrenia (Heidelb). 2022;8:2.
63. Botelho AO, Bernardo Ribeiro F, Soares Machado J, Souza DCB. Functional connectivity alterations in individuals with gaming disorder assessed by functional magnetic resonance imaging: a systematic review. J Addict Dis. 2025;43:4–11.
64. Liu Q, Shen Y, Cao X, Li Y, Chen Y, Yang W, Yuan TF. Either at left or right, both high and low frequency rTMS of dorsolateral prefrontal cortex decreases cue induced craving for methamphetamine. Am J Addict. 2017;26:776–9.
65. Tolomeo S, Yu R. Brain network dysfunctions in addiction: a meta-analysis of resting-state functional connectivity. Transl Psychiatry. 2022;12:41.
66. Cole EJ, Stimpson KH, Bentzley BS, Gulser M, Cherian K, Tischler C, Nejad R, Pankow H, Choi E, Aaron H, Espil FM, Pannu J, Xiao X, Duvio D, Solvason HB, Hawkins J, Guerra A, Jo B, Raj KS, Phillips AL, Barmak F, Bishop JH, Coetzee JP, Debattista C, Keller J, Schatzberg AF, Sudheimer KD, Williams NR. Stanford accelerated intelligent neuromodulation therapy for treatment-resistant depression. Am J Psychiatry. 2020;177:716–26.
67. Cole EJ, Phillips AL, Bentzley BS, Stimpson KH, Nejad R, Barmak F, Veerapal C, Khan N, Cherian K, Felber E, Brown R, Choi E, King S, Pankow H, Bishop JH, Azeez A, Coetzee J, Rapier R, Odenwald N, Carreon D, Hawkins J, Chang M, Keller J, Raj K, Debattista C, Jo B, Espil FM, Schatzberg AF, Sudheimer KD, Williams NR. Stanford Neuromodulation Therapy (SNT): a double-blind randomized controlled trial. Am J Psychiatry. 2022;179:132–41.
68. Wu GR, Baeken C, Van Schuerbeek P, De Mey J, Bi M, Herremans SC. Accelerated repetitive transcranial magnetic stimulation does not influence grey matter volumes in regions related to alcohol relapse: an open-label exploratory study. Drug Alcohol Depend. 2018;191:210–4.

69. Herremans SC, Van Schuerbeek P, De Raedt R, Matthys F, Buyl R, De Mey J, Baeken C. The impact of accelerated right prefrontal high-frequency repetitive transcranial magnetic stimulation (rTMS) on cue-reactivity: An fMRI study on craving in recently detoxified alcohol-dependent patients. PLoS One. 2015;10:e0136182.
70. Van Rooij SJH, Arulpragasam AR, Mcdonald WM, Philip NS. Accelerated TMS – moving quickly into the future of depression treatment. Neuropsychopharmacology. 2024;49:128–37.
71. Zangen A, Roth Y, Voller B, Hallett M. Transcranial magnetic stimulation of deep brain regions: evidence for efficacy of the H-coil. Clin Neurophysiol. 2005;116:775–9.
72. Deng ZD, Lisanby SH, Peterchev AV. Coil design considerations for deep transcranial magnetic stimulation. Clin Neurophysiol. 2014;125:1202–12.
73. Addolorato G, Antonelli M, Cocciolillo F, Vassallo GA, Tarli C, Sestito L, Mirijello A, Ferrulli A, Pizzuto DA, Camardese G, Miceli A, Diana M, Giordano A, Gasbarrini A, Di Giuda D. Deep transcranial magnetic stimulation of the dorsolateral prefrontal cortex in alcohol use disorder patients: effects on dopamine transporter availability and alcohol intake. Eur Neuropsychopharmacol. 2017;27:450–61.
74. Girardi P, Rapinesi C, Chiarotti F, Kotzalidis GD, Piacentino D, Serata D, Del Casale A, Scatena P, Mascioli F, Raccah RN, Brugnoli R, Digiacomantonio V, Ferri VR, Ferracuti S, Zangen A, Angeletti G. Add-on deep transcranial magnetic stimulation (dTMS) in patients with dysthymic disorder comorbid with alcohol use disorder: a comparison with standard treatment. World J Biol Psychiatry. 2015;16:66–73.
75. Kim WS, Paik NJ. Safety review for clinical application of repetitive transcranial magnetic stimulation. Brain Neurorehabil. 2021;14:e6.
76. Zhang JJ, Bai Z, Fong KNK. Methodological considerations of priming repetitive transcranial magnetic stimulation protocols in clinical populations. Gen Psychiatr. 2024;37:e101237.
77. Fitzgerald PB, Hoy K, Mcqueen S, Herring S, Segrave R, Been G, Kulkarni J, Daskalakis ZJ. Priming stimulation enhances the effectiveness of low-frequency right prefrontal cortex transcranial magnetic stimulation in major depression. J Clin Psychopharmacol. 2008;28:52–8.
78. Kourosh-Arami M, Komaki A, Gholami M. Addiction-induced plasticity in underlying neural circuits. Neurol Sci. 2022;43:1605–15.
79. Fingelkurts AA, Fingelkurts AA. Altered structure of dynamic electroencephalogram oscillatory pattern in major depression. Biol Psychiatry. 2015;77:1050–60.
80. Diego MA, Field T, Hernandez-Reif M. CES-D depression scores are correlated with frontal EEG alpha asymmetry. Depress Anxiety. 2001;13:32–7.
81. Ulrich G, Renfordt E, Zeller G, Frick K. Interrelation between changes in the EEG and psychopathology under pharmacotherapy for endogenous depression. A contribution to the predictor question. Pharmacopsychiatry. 1984;17:178–83.
82. Zrenner B, Zrenner C, Gordon PC, Belardinelli P, Mcdermott EJ, Soekadar SR, Fallgatter AJ, Ziemann U, Müller-Dahlhaus F. Brain oscillation-synchronized stimulation of the left dorsolateral prefrontal cortex in depression using real-time EEG-triggered TMS. Brain Stimul. 2020;13:197–205.
83. Leuchter AF, Cook IA, Feifel D, Goethe JW, Husain M, Carpenter LL, Thase ME, Krystal AD, Philip NS, Bhati MT, Burke WJ, Howland RH, Sheline YI, Aaronson ST, Iosifescu DV, O'reardon JP, Gilmer WS, Jain R, Burgoyne KS, Phillips B, Manberg PJ, Massaro J, Hunter AM, Lisanby SH, George MS. Efficacy and safety of low-field synchronized transcranial magnetic stimulation (sTMS) for treatment of major depression. Brain Stimul. 2015;8:787–94.
84. George MS, Huffman S, Doose J, Sun X, Dancy M, Faller J, Li X, Yuan H, Goldman RI, Sajda P, Brown TR. EEG synchronized left prefrontal transcranial magnetic stimulation (TMS) for treatment resistant depression is feasible and produces an entrainment dependent clinical response: a randomized controlled double blind clinical trial. Brain Stimul. 2023;16:1753–63.

85. Philip NS, Aiken EE, Kelley ME, Burch W, Waterman L, Holtzheimer PE. Synchronized transcranial magnetic stimulation for posttraumatic stress disorder and comorbid major depression. Brain Stimul. 2019;12:1335–7.
86. Jampel J, Quinn MJ, Catalano JL, Benca-Bachman CB, Brick L, Philip NS, Swift RM, Mcgeary JE. Synchronised transcranial magnetic stimulation for substance use-disordered veterans: protocol for the pilot sham-controlled acceptability trial. BMJ Open. 2023;13:e066175.
87. Li X, Hartwell KJ, Henderson S, Badran BW, Brady KT, George MS. Two weeks of image-guided left dorsolateral prefrontal cortex repetitive transcranial magnetic stimulation improves smoking cessation: a double-blind, sham-controlled, randomized clinical trial. Brain Stimul. 2020;13:1271–9.
88. Chung SW, Hoy KE, Fitzgerald PB. Theta-burst stimulation: a new form of TMS treatment for depression? Depress Anxiety. 2015;32:182–92.
89. Lee CW, Wu HF, Chu MC, Chung YJ, Mao WC, Li CT, Lin HC. Mechanism of intermittent theta-burst stimulation in synaptic pathology in the prefrontal cortex in an antidepressant-resistant depression rat model. Cereb Cortex. 2021a;31:575–90.
90. Sheridan GK, Moeendarbary E, Pickering M, O'Connor JJ, Murphy KJ. Theta-burst stimulation of hippocampal slices induces network-level calcium oscillations and activates analogous gene transcription to spatial learning. PLoS One. 2014;9:e100546.
91. Blumberger DM, Vila-Rodriguez F, Thorpe KE, Feffer K, Noda Y, Giacobbe P, Knyahnytska Y, Kennedy SH, Lam RW, Daskalakis ZJ, Downar J. Effectiveness of theta burst versus high-frequency repetitive transcranial magnetic stimulation in patients with depression (THREE-D): a randomised non-inferiority trial. Lancet. 2018;391:1683–92.
92. Liu Q, Sun H, Hu Y, Wang Q, Zhao Z, Dong D, Shen Y. Intermittent theta burst stimulation vs. high-frequency repetitive transcranial magnetic stimulation in the treatment of methamphetamine patients. Front Psych. 2022;13:842947.
93. Ren Z, Mu L, Wang L, Xia L, Song P, Wang Y, Li J, Duan F, Li H, Tang H, Wang W, Zhu L, Zhang L, Song X, Wang Y, Zhao W, Zhu Y, Wang Z, Shao W, Zhang X, Jiao D. Predictive role of impulsivity, anxiety, and depression in the efficacy of intermittent theta burst transcranial magnetic stimulation modalities for treating methamphetamine use disorder: A randomized clinical trial. J Subst Use Addict Treat. 2024;156:209189.
94. Sun Y, Wang H, Ku Y. Intermittent theta-burst stimulation increases the working memory capacity of methamphetamine addicts. Brain Sci. 2022;12
95. Zhao D, Li Y, Liu T, Voon V, Yuan TF. Twice-daily theta burst stimulation of the dorsolateral prefrontal cortex reduces methamphetamine craving: a pilot study. Front Neurosci. 2020;14:208.
96. Dieler AC, Dresler T, Joachim K, Deckert J, Herrmann MJ, Fallgatter AJ. Can intermittent theta burst stimulation as add-on to psychotherapy improve nicotine abstinence? Results from a pilot study. Eur Addict Res. 2014;20:248–53.
97. Dong L, Chen WC, Su H, Wang ML, Du C, Jiang XR, Mei SF, Chen SJ, Liu XJ, Liu XB. Intermittent theta burst stimulation to the left dorsolateral prefrontal cortex improves cognitive function in polydrug use disorder patients: a randomized controlled trial. Front Psych. 2023;14:1156149.
98. Kang T, Ding X, Zhao J, Li X, Xie R, Jiang H, He L, Hu Y, Liang J, Zhou G, Huo X. Influence of improved behavioral inhibition on decreased cue-induced craving in heroin use disorder: A preliminary intermittent theta burst stimulation study. J Psychiatr Res. 2022;152:375–83.
99. Sanna A, Fattore L, Badas P, Corona G, Cocco V, Diana M. Intermittent theta burst stimulation of the prefrontal cortex in cocaine use disorder: a pilot study. Front Neurosci. 2019;13:765.
100. Upton S, Brown AA, Ithman M, Newman-Norlund R, Sahlem G, Prisciandaro JJ, Mcclure EA, Froeliger B. Effects of hyperdirect pathway theta burst transcranial magnetic stimulation on inhibitory Control, craving, and smoking in adults with nicotine dependence: a double-blind, randomized crossover trial. Biol Psychiatry Cogn Neurosci Neuroimaging. 2023b;8:1156–65.

101. Chen T, Su H, Wang L, Li X, Wu Q, Zhong N, Du J, Meng Y, Duan C, Zhang C, Shi W, Xu D, Song W, Zhao M, Jiang H. Modulation of methamphetamine-related attention bias by intermittent theta-burst stimulation on left dorsolateral prefrontal cortex. Front Cell Dev Biol. 2021;9:667476.
102. Mizoguchi H, Yamada K. Methamphetamine use causes cognitive impairment and altered decision-making. Neurochem Int. 2019;124:106–13.
103. Upton S, Brown AA, Golzy M, Garland EL, Froeliger B. Right inferior frontal gyrus theta-burst stimulation reduces smoking behaviors and strengthens fronto-striatal-limbic resting-state functional connectivity: a randomized crossover trial. Front Psych. 2023a;14:1166912.
104. Mikellides G, Michael P, Psalta L, Stefani A, Schuhmann T, Sack AT. Accelerated intermittent theta burst stimulation in smoking cessation: no differences between active and placebo stimulation when using advanced placebo coil technology. A double-blind follow-up study. Int J Clin Health Psychol. 2023;23:100351.
105. Steele VR, Maxwell AM, Ross TJ, Stein EA, Salmeron BJ. Accelerated intermittent theta-burst stimulation as a treatment for cocaine use disorder: a proof-of-concept study. Front Neurosci. 2019;13:1147.
106. Cole E, O'sullivan SJ, Tik M, Williams NR. Accelerated theta burst stimulation: safety, efficacy, and future advancements. Biol Psychiatry. 2024;95:523–35.
107. Mcneill A, Monk RL, Qureshi AW, Makris S, Heim D. Continuous theta burst transcranial magnetic stimulation of the right dorsolateral prefrontal cortex impairs inhibitory control and increases alcohol consumption. Cogn Affect Behav Neurosci. 2018;18:1198–206.
108. Mcneill AM, Monk RL, Qureshi AW, Makris S, Cazzato V, Heim D. Correction to: elevated ad libitum alcohol consumption following continuous theta burst stimulation to the left-dorsolateral prefrontal cortex is partially mediated by changes in craving. Cogn Affect Behav Neurosci. 2021;21:1319.
109. Ankit A, Das B, Dey P, Kshitiz KK, Khess CRJ. Efficacy of continuous theta burst stimulation – repetitive transcranial magnetic stimulation on the orbitofrontal cortex as an adjunct to naltrexone in patients of opioid use disorder and its correlation with serum BDNF levels: a sham-controlled study. J Addict Dis. 2022;40:373–81.
110. Mccalley DM, Kaur N, Wolf JP, Contreras IE, Book SW, Smith JP, Hanlon CA. Medial prefrontal cortex theta burst stimulation improves treatment outcomes in alcohol use disorder: a double-blind, sham-controlled neuroimaging study. Biol Psychiatry Glob Open Sci. 2023;3:301–10.
111. Hanlon CA, Dowdle LT, Correia B, Mithoefer O, Kearney-Ramos T, Lench D, Griffin M, Anton RF, George MS. Left frontal pole theta burst stimulation decreases orbitofrontal and insula activity in cocaine users and alcohol users. Drug Alcohol Depend. 2017;178:310–7.
112. Blumberger DM, Mulsant BH, Thorpe KE, Mcclintock SM, Konstantinou GN, Lee HH, Nestor SM, Noda Y, Rajji TK, Trevizol AP, Vila-Rodriguez F, Daskalakis ZJ, Downar J. Effectiveness of standard sequential bilateral repetitive transcranial magnetic stimulation vs bilateral theta burst stimulation in older adults with depression: The FOUR-D randomized noninferiority clinical trial. JAMA Psychiatry. 2022;79:1065–73.
113. Elnazali M, Veerakumar A, Blair M, Pearce EL, Kim N, Sebastian S, Santo JB, Rybak YE, Burhan AM. Unilateral and bilateral theta burst stimulation for treatment-resistant depression: follow up on a naturalistic observation study. J Psychiatr Res. 2024;180:387–93.
114. Lisanby SH, Luber B, Schlaepfer TE, Sackeim HA. Safety and feasibility of magnetic seizure therapy (MST) in major depression: randomized within-subject comparison with electroconvulsive therapy. Neuropsychopharmacology. 2003;28:1852–65.
115. Backhouse FA, Noda Y, Knyahnytska Y, Farzan F, Downar J, Rajji TK, Mulsant BH, Daskalakis ZJ, Blumberger DM. Characteristics of ictal EEG in Magnetic Seizure Therapy at various stimulation frequencies. Clin Neurophysiol. 2018;129:1770–9.
116. Sun Y, Blumberger DM, Mulsant BH, Rajji TK, Fitzgerald PB, Barr MS, Downar J, Wong W, Farzan F, Daskalakis ZJ. Magnetic seizure therapy reduces suicidal ideation and produces neuroplasticity in treatment-resistant depression. Transl Psychiatry. 2018;8:253.

117. Cai DB, Yang XH, Shi ZM, Nie S, Xu R, Qin XD, Huang X, Huang XB, Zheng W. Comparison of efficacy and safety of magnetic seizure therapy and electroconvulsive therapy for depression: a systematic review. J Pers Med. 2023;13
118. Carroll BT, Hassell J, Ali A. Treatment options for methamphetamine-associated catatonia. Ann Clin Psychiatry. 2023;35:e1–2.
119. Grelotti DJ, Kanayama G, Pope HG Jr. Remission of persistent methamphetamine-induced psychosis after electroconvulsive therapy: presentation of a case and review of the literature. Am J Psychiatry. 2010;167:17–23.
120. Roshanaei-Moghaddam B, Pauly MC. Treatment of methamphetamine dependence with electroconvulsive therapy (ECT) in Iran: a critical note. Iran J Psychiatry. 2014;9:184–7.
121. Fregni F, El-Hagrassy MM, Pacheco-Barrios K, Carvalho S, Leite J, Simis M, Brunelin J, Nakamura-Palacios EM, Marangolo P, Venkatasubramanian G, San-Juan D, Caumo W, Bikson M, Brunoni AR, Neuromodulation Center Working G. Evidence-based guidelines and secondary meta-analysis for the use of transcranial direct current stimulation in neurological and psychiatric disorders. Int J Neuropsychopharmacol. 2021;24:256–313.
122. Klauss J, Penido Pinheiro LC, Silva Merlo BL, Correia Santos GDA, Fregni F, Nitsche MA, Miyuki Nakamura-Palacios E. A randomized controlled trial of targeted prefrontal cortex modulation with tDCS in patients with alcohol dependence. Int J Neuropsychopharmacol. 2014;17:1793–803.
123. Klauss J, Anders QS, Felippe LV, Nitsche MA, Nakamura-Palacios EM. Multiple sessions of transcranial direct current stimulation (tDCS) reduced craving and relapses for alcohol use: a randomized placebo-controlled trial in alcohol use disorder. Front Pharmacol. 2018;9:716.
124. Dubuson M, Kornreich C, Vanderhasselt M-A, Baeken C, Wyckmans F, Dousset C, Hanak C, Veeser J, Campanella S, Chatard A. Transcranial direct current stimulation combined with alcohol cue inhibitory control training reduces the risk of early alcohol relapse: a randomized placebo-controlled clinical trial. Brain Stimul. 2021;14:1531–43.
125. Uyl TE, Gladwin TE, Wiers RW. Electrophysiological and behavioral effects of combined transcranial direct current stimulation and alcohol approach bias retraining in hazardous drinkers. Alcohol Clin Exp Res. 2016;40:2124–33.
126. Witkiewitz K, Stein ER, Votaw VR, Wilson AD, Roos CR, Gallegos SJ, Clark VP, Claus ED. Mindfulness-based relapse prevention and transcranial direct current stimulation to reduce heavy drinking: a double-blind sham-controlled randomized trial. Alcohol Clin Exp Res. 2019;43:1296–307.
127. Uyl TE, Gladwin TE, Rinck M, Lindenmeyer J, Wiers RW. A clinical trial with combined transcranial direct current stimulation and alcohol approach bias retraining. Addict Biol. 2017;22:1632–40.
128. Uyl TE, Gladwin TE, Lindenmeyer J, Wiers RW. A clinical trial with combined transcranial direct current stimulation and attentional bias modification in alcohol-dependent patients. Alcohol Clin Exp Res. 2018;42:1961–9.
129. Kang N, Kim RK, Kim HJ. Effects of transcranial direct current stimulation on symptoms of nicotine dependence: a systematic review and meta-analysis. Addict Behav. 2019;96:133–9.
130. Boggio PS, Liguori P, Sultani N, Rezende L, Fecteau S, Fregni F. Cumulative priming effects of cortical stimulation on smoking cue-induced craving. Neurosci Lett. 2009;463:82–6.
131. Mondino M, Luck D, Grot S, Januel D, Suaud-Chagny M-F, Poulet E, Brunelin J. Effects of repeated transcranial direct current stimulation on smoking, craving and brain reactivity to smoking cues. Sci Rep. 2018;8:8724.
132. Ghorbani Behnam S, Mousavi SA, Emamian MH. The effects of transcranial direct current stimulation compared to standard bupropion for the treatment of tobacco dependence: a randomized sham-controlled trial. Eur Psychiatry. 2019;60:41–8.
133. Batista EK, Klauss J, Fregni F, Nitsche MA, Nakamura-Palacios EM. A randomized placebo-controlled trial of targeted prefrontal cortex modulation with bilateral tDCS in patients with crack-cocaine dependence. Int J Neuropsychopharmacol. 2015;18:pyv066.

134. Gaudreault PO, Sharma A, Datta A, Nakamura-Palacios EM, King S, Malaker P, Wagner A, Vasa D, Parvaz MA, Parra LC, Alia-Klein N, Goldstein RZ. A double-blind sham-controlled phase 1 clinical trial of tDCS of the dorsolateral prefrontal cortex in cocaine inpatients: craving, sleepiness, and contemplation to change. Eur J Neurosci. 2021;53:3212–30.
135. Verveer I, Veen FM, Shahbabaie A, Remmerswaal D, Franken IHA. Multi-session electrical neuromodulation effects on craving, relapse and cognitive functions in cocaine use disorder: a randomized, sham-controlled tDCS study. Drug Alcohol Depend. 2020;217:108429.
136. Alizadehgoradel J, Nejati V, Movahed FS, Imani S, Taherifard M, Mosayebi-Samani M, Vicario CM, Nitsche MA, Salehinejad MA. Repeated stimulation of the dorsolateral-prefrontal cortex improves executive dysfunctions and craving in drug addiction: a randomized, double-blind, parallel-group study. Brain Stimul. 2020;13:582–93.
137. Anaraki MR, Dolatshahi B, Nosratabadi M, Yalghouzaghaji MN, Mashhadi SR. Repeated transcranial direct current stimulation (tDCS) on methamphetamine craving: a randomized, sham-controlled study. Iran Rehabil J. 2019;
138. Xu X, Ding X, Chen L, Chen T, Su H, Li X, Ye Y, Shi W, Ji J, Zhao M. The transcranial direct current stimulation over prefrontal cortex combined with the cognitive training reduced the cue-induced craving in female individuals with methamphetamine use disorder: a randomized controlled trial. J Psychiatr Res. 2021;134:102–10.
139. Eskandari Z, Mostafavi H, Hosseini M, Mousavi SE, Ramazani S, Dadashi M. A sham-controlled clinical trial to examine the effect of bilateral tD. J Addict Dis. 2021;39:347–56.
140. Taremian F, Nazari S, Moradveisi L, Moloodi R. Transcranial direct current stimulation on opium craving, depression, and anxiety: a preliminary study. J ECT. 2019;35:201–6.
141. Jeong J-E, Park C-H, Kim M, Cho H, Pyeon A, Jung S, Jung D, Kim J-Y, Choi J, Chun J-W, Ahn K-J, Kim D-J. Effects of bilateral tDCS over DLPFC on response inhibition, craving, and brain functional connectivity in Internet gaming disorder: a randomized, double-blind, sham-controlled trial with fMRI. 2024
142. Lee J-Y, Jang JH, Choi AR, Chung SJ, Kim B, Park M, Oh S, Jung MH, Choi J-S. Neuromodulatory effect of transcranial direct current stimulation on resting-state EEG activity in internet gaming disorder: a randomized, double-blind, sham-controlled parallel group trial. cerebral cortex. Communications. 2021b;2:tgaa095.
143. Martinotti G, Lupi M, Montemitro C, Miuli A, Di Natale C, Spano MC, Mancini V, Lorusso M, Stigliano G, Tambelli A, Di Carlo F, Di Caprio L, Fraticelli S, Chillemi E, Pettorruso M, Sepede G, Giannantonio M. Transcranial direct current stimulation reduces craving in. J ECT. 2019;
144. Bikson M, Grossman P, Thomas C, Zannou AL, Jiang J, Adnan T, Mourdoukoutas AP, Kronberg G, Truong D, Boggio P. Safety of transcranial direct current stimulation: evidence based update 2016. Brain Stimul. 2016;9:641–61.
145. Nitsche MA, Liebetanz D, Lang N, Antal A, Tergau F, Paulus W. Safety criteria for transcranial direct current stimulation (tDCS) in humans. Clin Neurophysiol. 2003;114:2220–2. author reply 2222-2223
146. Poreisz C, Boros K, Antal A, Paulus W. Safety aspects of transcranial direct current stimulation concerning healthy subjects and patients. Brain Res Bull. 2007;72:208–14.
147. Smith RC, Boules S, Mattiuz S, Youssef M, Tobe RH, Sershen H, Lajtha A, Nolan K, Amiaz R, Davis JM. Effects of transcranial direct current stimulation (tDCS) on cognition, symptoms, and smoking in schizophrenia: a randomized controlled stud y. Schizophr Res. 2015;168:260–6.
148. Song S, Zilverstand A, Gui W, Li H-J, Zhou X. Effects of single-session versus multi-session non-invasive brain stimulation on craving and consumption in individuals with drug addiction, eating disorders or obesity: a meta-analysis. Brain Stimul. 2019;12:606–18.
149. Sreeraj VS, Dinakaran D, Parlikar R, Chhabra H, Selvaraj S, Shivakumar V, Bose A, Narayanaswamy JC, Venkatasubramanian G. High-definition transcranial direct current simulation (HD-tDCS) for persistent auditory hallucinations in schizophrenia. Asian J Psychiatr. 2018;37:46–50.

150. Wong HL, Chan WC, Wong YL, Wong SN, Yung HY, Wong SC, Cheng PC. High-definition transcranial direct current stimulation-An open-label pilot intervention in alleviating depressive symptoms and cognitive deficits in late-life depression. CNS Neurosci Ther. 2019;25:1244–53.
151. Motes MA, Spence JS, Yeatman K, Jones PM, Lutrell M, O'Hair R, Shakal S, Delarosa BL, To W, Vanneste S, Kraut MA, Hart J Jr. High-definition transcranial direct current stimulation to improve verbal retrieval deficits in chronic traumatic brain injury. J Neurotrauma. 2020;37:170–7.
152. Richardson J, Datta A, Dmochowski J, Parra LC, Fridriksson J. Feasibility of using high-definition transcranial direct current stimulation (HD-tDCS) to enhance treatment outcomes in persons with aphasia. NeuroRehabilitation. 2015;36:115–26.
153. Frohlich F, Riddle J. Conducting double-blind placebo-controlled clinical trials of transcranial alternating current stimulation (tACS). Transl Psychiatry. 2021;11:1–12.
154. Rossi S, Hallett M, Rossini PM, Pascual-Leone A. Safety, ethical considerations, and application guidelines for the use of transcranial magnetic stimulation in clinical practice and research. Clin Neurophysiol. 2009;120:2008–39.
155. Bergmann TO, Varatheeswaran R, Hanlon CA, Madsen KH, Thielscher A, Siebner HR. Concurrent TMS-fMRI for causal network perturbation and proof of targe t engagement. NeuroImage. 2021;237:118093.
156. Strafella R, Momi D, Zomorrodi R, Lissemore J, Noda Y, Chen R, Rajji TK, Griffiths JD, Vila-Rodriguez F, Downar J, Daskalakis ZJ, Blumberger DM, Voineskos D. Identifying neurophysiological markers of intermittent theta-burst stimulation in treatment-resistant depression using transcranial magnetic stimulation- electroencephalography. Biol Psychiatry. 2023:S0006322323012076.
157. Bradley C, Nydam AS, Dux PE, Mattingley JB. State-dependent effects of neural stimulation on brain function and cognition. Nat Rev Neurosci. 2022;23:459–75.
158. Siddiqi SH, Taylor SF, Cooke D, Pascual-Leone A, George MS, Fox MD. Distinct symptom-specific treatment targets for circuit-based neuromodulation. Am J Psychiatry. 2020;177:435–46.

Obsessive-Compulsive Disorder

7

Yiliang Zhang, Shuming Zhong, Shunkai Lai, Dong Huang, and Yanbin Jia

Abstract

Obsessive-compulsive disorder (OCD) is a prevalent and debilitating psychiatric condition that profoundly impacts individuals' lives. Its complex and multifactorial etiology presents significant challenges for effective treatment. While many patients benefit from standard interventions such as pharmacological therapies and psychotherapy, a substantial number remain resistant to treatment, resulting in poor long-term outcomes. Emerging neurostimulation therapies, encompassing both noninvasive and invasive techniques, offer promising alternatives for managing OCD. Numerous clinical trials have explored novel neurostimulation protocols, aiming to improve treatment efficacy. However, findings remain inconsistent, and no universally accepted approach has been established. This underscores the critical need for further research to refine stimulation parameters and identify factors that influence individual responsiveness to these interventions. This chapter provides a comprehensive overview of neuromodulation strategies for OCD, highlighting current advancements, ongoing challenges, and future directions in this rapidly evolving field.

Keywords

OCD · rTMS · dTMS · TBS · tDCS

Y. Zhang · S. Zhong · S. Lai · D. Huang · Y. Jia (✉)
Department of Psychiatry, First Affiliated Hospital of Jinan University, Guangzhou, China

W. Zheng, Y. Ning (eds.), *TMS and tDCS for Psychiatric Disorders*,
https://doi.org/10.1007/978-981-96-8504-2_7

Abbreviations

ACC	anterior cingulate cortex
ahTBS	accelerated high-dose Theta Burst Stimulation
aiTBS	asymmetric intermittent Theta Burst Stimulation
AMT	active motor threshold
BAI	Beck Anxiety Inventory
BD	Bipolar Disorder
BDI-II	Beck Depression Inventory II
CBT	cognitive-behavioral therapy
CSTC	cortico-striatal-thalamic-cortical
cTBS	Continuous TBS
DBS	deep brain stimulation
DLPFC	dorsolateral prefrontal cortex
dmPFC	dorsomedial prefrontal
dTMS	Deep Transcranial Magnetic Stimulation
ECT	electroconvulsive therapy
EEG	electroencephalography
ERP	exposure and response prevention
fMRI	Functional Magnetic Resonance Imaging
HD-tDCS	High-definition transcranial direct current stimulation
HF	high-frequency
iTBS	intermittent Theta Burst Stimulation
L-DLPFC	left dorsolateral prefrontal cortex
LF-R-DLPFC	left frontolateral region of the dorsolateral prefrontal cortex
MDD	Major Depressive Disorder
MEP	motor-evoked potential
MFG	middle frontal gyrus
MST	Magnetic Seizure Therapy
MT	motor threshold
OCD	Obsessive–compulsive disorder
OFC	orbitofrontal cortex
RCTs	Randomized Controlled Trials
RMT	resting motor threshold
ROFC	right orbitofrontal cortex
rTMS	repetitive Transcranial Magnetic Stimulation
SFG	superior frontal gyrus
SMA	supplementary motor area
SSRIs	serotonergic reuptake inhibitors
TBS	Theta Burst Stimulation
tDCS	transcranial direct current stimulation
TMS	transcranial magnetic stimulation
VNS	Vagus Nerve Stimulation
YBOCS	Yale-Brown Obsessive Compulsive Scale

7.1 Introduction

Obsessive-compulsive disorder (OCD) is a neuropsychiatric condition marked by persistent and intrusive thoughts, urges, or images that are often distressing or unwanted [1, 2]. Globally, its lifetime prevalence is estimated to be around 2–3% [3]. In high-income countries, OCD is recognized as the fourth most common mental health disorder and ranks as the tenth leading cause of disability worldwide [4, 5]. The condition is linked to increased mortality rates [6] and can significantly impair the quality of life for both individuals with the disorder and their families or caregivers [5].

Despite the availability of first-line treatments—psychotherapy (primarily cognitive-behavioral therapy (CBT) or exposure and response prevention (ERP)), and pharmacotherapy (primarily serotonergic reuptake inhibitors (SSRIs))—these interventions are fully effective in only about 50% of patients [7], leaving a considerable proportion of individuals with persistent and debilitating symptoms. Currently, neuromodulation therapies—both invasive techniques, such as deep brain stimulation (DBS), and noninvasive approaches, including transcranial magnetic stimulation (TMS), have been widely used in the treatment of OCD and have shown some efficacy.

One possible therapy strategy for treating OCD is neuromodulation. Although deep transcranial magnetic stimulation (dTMS) has some promise among noninvasive methods, more thorough research is required to determine its efficacy with more certainty. The paucity of evidence supporting iTBS/cTBS in the treatment of OCD indicates the need for more study in this field. One particularly safe and perhaps successful intervention is tDCS.

The use and advancement of noninvasive neuroregulatory approaches in OCD will be discussed in this chapter.

7.2 TMS

7.2.1 rTMS

7.2.1.1 Unilateral rTMS

A comprehensive systematic review and network meta-analysis provided robust evidence for the efficacy of unilateral rTMS in OCD treatment [8]. The study demonstrated that both low-frequency (LF) rTMS applied to the dorsolateral prefrontal cortex (DLPFC) and supplementary motor area (SMA), and high-frequency (HF) rTMS directed at the DLPFC, yielded significant improvements in Yale-Brown Obsessive Compulsive Scale (Y-BOCS) scores when compared to sham stimulation. The underlying neurobiological mechanism underlying these effects is intricately linked to the dysregulation of the cortico-striatal-thalamic-cortical (CSTC) circuit, which is a hallmark of OCD pathophysiology. LF-rTMS, when administered to the DLPFC, exerts its therapeutic effects by dampening the excessive neuronal activity within this region. This modulation occurs through a multi-faceted process that involves the rebalancing of neurotransmitter release, such as the regulation of dopamine and glutamate levels. For example, a study showed that LF-rTMS could reduce glutamate levels in the DLPFC in OCD patients [9].

Safety

The safety profile of unilateral rTMS is favorable, with mild and transient side effects such as headache and scalp discomfort at the site of stimulation, which typically manifest during the early stages of treatment [8]. These symptoms can be attributed to the induction of electric currents in the scalp tissue by the rTMS magnetic field. Moreover, the magnetic field may also induce subtle alterations in cerebrovascular function, contributing to the development of mild headaches. Fortunately, these adverse effects can often be effectively managed by making appropriate adjustments to the stimulation parameters. However, rare but potentially concerning adverse events, such as transient cognitive function fluctuations or mild local muscle spasms, have also been documented in some cases. Although these events did not result in long-term negative consequences in the reported instances, they underscore the importance of vigilant and continuous monitoring of patients during rTMS treatment to ensure the early detection and appropriate management of any potential adverse reactions.

Treatment Regimen

Stimulation Intensity The determination of an appropriate stimulation intensity is a critical aspect of rTMS treatment. Typically, the intensity is calibrated based on the individual patient's motor threshold, which refers to the minimum level of electrical stimulation required to elicit a muscle contraction, with a commonly recommended range of 80–120% of the motor threshold (MT). This range has been empirically validated through numerous preclinical and clinical studies and is believed to strike a balance between ensuring sufficient stimulation to induce neuroplastic changes in the target brain regions and minimizing the risk of adverse events, particularly the occurrence of epileptic seizures. The studies suggest that high-frequency (10 Hz) rTMS at 100% or 110% MT over the right DLPFC is no more effective than sham rTMS in alleviating symptoms or improving treatment outcomes in treatment-resistant OCD [10, 11].

Stimulation Frequency The choice between LF-rTMS (usually ≤1 Hz) and HF-rTMS (usually ≥5 Hz) is guided by their distinct neurophysiological effects. LF-rTMS exerts an inhibitory influence on cortical excitability, which is particularly beneficial in reducing the excessive activation observed in regions such as the DLPFC in OCD patients [12]. In contrast, HF-rTMS has an excitatory effect, which can enhance the functional connectivity and information processing capabilities of specific brain region [12]. Generally, HF rTMS (≥5 Hz) is predominantly employed, particularly targeting the left DLPFC, which is associated with the regulation of mood and cognition. Studies suggest that stimulation frequency between 10 Hz and 20 Hz is common [13, 14]. In some studies, the right DLPFC was also targeted [15]. A comprehensive systematic review and network meta-analysis demonstrated that both LF-rTMS applied to the DLPFC and SMA, and HF-rTMS directed at the DLPFC, yielded significant improvements in YBOCS scores when compared to sham stimulation [16]. The findings indicate that 1 Hz rTMS targeting the SMA may serve as an efficacious and secure adjunctive treatment for patients with OCD who exhibit treatment resistance [17]. The underlying neurobiological mechanism underlying these effects is intricately linked to the dysregulation of the CSTC circuit, which is a hallmark of OCD pathophysiology.

Stimulation Target The selection of the DLPFC and SMA as primary stimulation targets is grounded in a wealth of neuroimaging, neuroelectrophysiological, and neuroanatomical research [18]. The DLPFC, a key region involved in higher-order cognitive and emotional regulation, frequently exhibits dysfunction in OCD patients. This dysfunction is manifested as impairments in inhibitory control and cognitive flexibility, which are hypothesized to underlie the generation and persistence of obsessive thoughts. By delivering rTMS to the DLPFC, it is possible to directly target and modulate this aberrant neural activity, thereby potentially alleviating the severity of obsessive symptoms. The SMA, on the other hand, plays a crucial role in the planning, execution, and monitoring of motor actions. In OCD, disruptions in SMA function have been implicated in the manifestation of repetitive behaviors. Stimulation of the SMA may therefore help to correct the underlying neural circuitry abnormalities associated with these repetitive behaviors and restore more normal patterns of motor control. However, it is essential to recognize that individual variability in brain anatomy and function exists, which can potentially impact the precision and effectiveness of stimulation. The orbitofrontal cortex (OFC) is another brain area that has been explored for rTMS treatment [19]. Given its significant role in the pathophysiology of OCD and the understanding that obsessions and compulsions may be linked to heightened activity in the orbitofrontal-subcortical circuits and the OFC itself, there is a rationale for targeting the OFC with rTMS. Building on this rationale, a study by Ruffini et al.[20] was initiated to investigate the potential of OFC rTMS in treating OCD. This trial was designed as a randomized, single-blind, sham-controlled experiment, which aimed to determine the impact of low-frequency (1 Hz) rTMS at 80% of the resting motor threshold (RMT) applied to the left OFC over a three-week period. The results showed a notable reduction in Y-BOCS scores for the group receiving active treatment as opposed to the sham treatment, with these improvements being observed at the end of the third week and persisting through the tenth week [20].

Stimulation Time Each session of unilateral rTMS typically lasts approximately 20–30 min. This duration is predicated on fundamental neurophysiological principles, which suggest that a sufficient amount of time is required to deliver an adequate number of stimulation pulses to effectively induce neuroplastic changes in the target neurons [21]. These changes encompass a range of molecular and cellular processes, including the modulation of synaptic plasticity, such as the formation and strengthening of new synapses or the modification of existing synaptic efficacy, as well as alterations in the expression and function of neurotransmitter receptors and ion channels [22]. By promoting these neuroplastic changes, rTMS aims to optimize the structure and function of the neural circuits involved in OCD, thereby leading to a reduction in symptoms. Additionally, this time frame is also designed to minimize patient discomfort and fatigue, which could otherwise compromise treatment compliance.

Number of Treatment Sequences The typical treatment course for unilateral rTMS in OCD consists of 10–30 sessions, with a treatment frequency of 3–5 times per week [8]. This regimen is based on the principles of neuroplasticity, which suggest that repeated and spaced-out stimulation sessions are required to promote the long-term remodeling and stabilization of neural circuits. During the initial phase of treatment, the brain responds to the rTMS-induced perturbations by initiating a series of neuroplastic processes, which may include changes in synaptic strength, the sprouting of new dendritic spines, and the activation of gene expression programs related to neural repair and growth. As the treatment progresses, these initial changes are gradually consolidated and integrated into the existing neural circuitry, leading to more sustainable improvements in brain function and a reduction in OCD symptoms. However, it is crucial to note that the trajectory of symptom improvement is highly individualized, and not all patients will follow a predictable linear pattern.

Clinical Recommendations

Unilateral rTMS, especially targeting the left DLPFC, can be considered as an adjunctive treatment for OCD patients who have not responded adequately to conventional therapies [21]. In cases where a patient presents with prominent obsessive thoughts and neuroimaging studies reveal evidence of increased neuronal activity or abnormal functional connectivity in the DLPFC, LF-rTMS targeted at the DLPFC may be a particularly appropriate choice. By suppressing the overactivity in this region, LF-rTMS may help to break the cycle of obsessive thinking and provide relief from the distressing symptoms. Conversely, if a patient exhibits cognitive impairment, as manifested by difficulties in attention, memory, or executive function, and neuropsychological testing or functional neuroimaging indicates hypoactivity in specific brain regions associated with these cognitive processes, HF-rTMS applied to the identified impaired regions may be beneficial. This approach aims to enhance the functional activation and connectivity of these regions, thereby improving cognitive performance and potentially alleviating the overall burden of OCD symptoms. Before initiating rTMS treatment, a comprehensive and individualized assessment of the patient is essential.

7.2.1.2 Bilateral rTMS

Efficacy

A recent systematic review and network meta-analysis, which encompassed 21 relevant studies with a total of 662 patients, provided compelling evidence for the efficacy of bilateral DLPFC stimulation in the treatment of OCD [8]. The results demonstrated that this approach led to a significant reduction in YBOCS scores and an effect size that was comparable to that of unilateral rTMS. The underlying neurophysiological basis for these effects lies in the critical role of the bilateral DLPFC in the coordinated functioning of the brain's neural networks [22]. In OCD, disruptions in the normal interhemispheric communication and functional connectivity between the bilateral DLPFC have been implicated as a key pathophysiological mechanism. Bilateral stimulation, by simultaneously targeting both DLPFC regions, has the potential to restore and enhance the integrity of this interhemispheric connection. This restoration of connectivity is hypothesized to improve the efficiency of neural signal transmission and integration, thereby facilitating the modulation of neural circuits involved in cognitive flexibility, emotional regulation, and inhibitory control, all of which are disrupted in OCD. As a result, bilateral rTMS may lead to a more comprehensive and effective reduction in the diverse range of symptoms associated with OCD, including not only obsessive thoughts and compulsive behaviors but also comorbid emotional disturbances such as anxiety and depression. However, similar to the findings in unilateral rTMS studies, heterogeneity was also observed among the studies included in the review, highlighting the need for further research to elucidate the specific factors contributing to treatment response variability and to optimize the treatment protocol.

Safety

The safety profile of bilateral rTMS is generally comparable to that of unilateral rTMS. The dropout rate due to adverse events in the active bilateral rTMS treatment group was 4%, which was not significantly different from the 2% dropout rate in the sham stimulation group, indicating that patients typically tolerate bilateral rTMS well [8]. The most commonly reported adverse effects are similar to those observed in unilateral rTMS, including mild scalp discomfort, transient headache, and local muscle twitching. These adverse effects typically occur during the early stages of treatment and are believed to result from the interaction of the magnetic field with the scalp tissue and the associated induction of electric currents, which can cause local irritation and mild disruptions in cerebrovascular function. Fortunately, these adverse effects can usually be effectively managed by optimizing the stimulation parameters, such as precisely adjusting the intensity by 5–15%, frequency, and pulse pattern, based on the individual patient's response and tolerance [12]. Additionally, symptomatic treatments, such as scalp massage to relieve discomfort or the administration of mild analgesics for headache, can also be employed as adjunctive measures. However, it is important to note that the long-term effects of the complex magnetic field interactions that occur during bilateral rTMS remain poorly understood. Given the potential for cumulative or delayed effects on neural function and plasticity, long-term follow-up studies are warranted to comprehensively assess the durability of treatment benefits and to monitor for any potential late-onset adverse events, such as changes in cognitive function, mood stability, or the development of neurological abnormalities.

Treatment Regimen

Stimulation Target The bilateral DLPFC is the central focus of stimulation in this modality [8, 21]. Its selection as the primary target is based on its well-established role in the pathophysiology of OCD and its position as a key node in the brain's functional network. Dysfunction of the bilateral DLPFC is thought to underlie many of the core symptoms of OCD, including impairments in cognitive control, emotional dysregulation, and the inability to inhibit repetitive thoughts and behaviors. By applying bilateral rTMS, the aim is to simultaneously modulate the activity of both DLPFC regions, thereby promoting the restoration of normal interhemispheric communication and the coordinated regulation of neural circuits involved in multiple aspects of OCD symptomatology.

Stimulation Intensity The determination of stimulation intensity in bilateral rTMS follows similar principles as in unilateral rTMS, with the intensity typically set within the range of 80–120% of the individual's motor threshold [12]. This range is chosen to balance the need for effective stimulation to induce neuroplastic changes in the bilateral DLPFC while minimizing the risk of adverse events such as seizures. However, the bilateral nature of the stimulation introduces additional complexity, as the interaction of magnetic fields between the two hemispheres may affect the overall neural response and the accuracy of motor threshold determination.

Stimulation Frequency In bilateral rTMS, both LF and HF stimulations have distinct roles and can be used either alone or in combination, depending on the patient's specific neurophysiological profile [12]. The treatment parameters for bilateral rTMS can vary significantly across studies. Common frequency settings range from low (1 Hz) to high (10 Hz), with most studies employing a frequency of 10 Hz, which is considered excitatory and is aimed at enhancing cortical activity in the targeted regions [14].

Stimulation Time Each session of bilateral rTMS usually lasts around 20–30 min, similar to unilateral rTMS. This duration is based on the understanding that it allows for sufficient stimulation to induce neuroplastic changes in the bilateral DLPFC and its associated neural circuits. During this time, the magnetic pulses interact with the neurons in the target regions, triggering a cascade of molecular and cellular events that underlie neuroplasticity, such as changes in synaptic strength, dendritic spine remodeling, and alterations in neurotransmitter receptor expression. These changes are hypothesized to contribute to the improvement of OCD symptoms by restoring normal neural circuit function.

Number of Treatment Sequences The treatment course for bilateral rTMS typically consists of 10–30 sessions, with a frequency of 3–5 times per week [14]. This regimen is designed to promote the long-term remodeling and stabilization of neural circuits involved in OCD. The repeated stimulation sessions over a period of weeks allow for the cumulative and progressive effects of rTMS to take hold, leading to sustainable improvements in brain function and symptom reduction.

Clinical Recommendations

Bilateral rTMS is particularly indicated for moderate to severe OCD patients who have either failed to respond adequately to unilateral rTMS or who exhibit significant bilateral hemispheric functional imbalance, as detected by comprehensive neuroimaging and neuropsychological evaluations [21]. For instance, if neuroimaging studies using DTI reveal a significant reduction in the integrity of white matter tracts connecting the bilateral DLPFC (with a mean fractional anisotropy reduction of 30% or more) or if functional connectivity analyses demonstrate abnormal patterns of interhemispheric communication (such as a decrease in synchrony of neural oscillations by 40% or more), bilateral rTMS may be a more appropriate treatment option. In patients with prominent comorbid emotional dysregulation, such as severe anxiety or depression, a phased approach may be considered. Initially, LF bilateral DLPFC stimulation can be applied to target the emotional regulation circuits and reduce the negative impact of emotions on OCD symptoms. This is based on the understanding that the DLPFC has extensive connections with limbic regions involved in emotion processing, and LF stimulation can modulate the excitability of these circuits. If cognitive impairment is also a significant concern, as evidenced by deficits in working memory, attention, or executive function, HF bilateral DLPFC stimulation can be subsequently incorporated to enhance cognitive function. Clinical application of bilateral rTMS requires a meticulous and comprehensive assessment process. This includes a detailed evaluation of the patient's OCD symptoms, their history of treatment responses, comorbid psychiatric and medical conditions, and a battery of neuropsychological tests to assess cognitive function.

7.2.1.3 Accelerated rTMS

Efficacy

Accelerated rTMS presents a distinctive treatment protocol that differs fundamentally from traditional rTMS methodologies. This approach typically involves a more intensive schedule of sessions over a shorter time frame, aiming to enhance neuroplasticity and therapeutic effect. Accelerated rTMS has shown promise in depression treatment. Fitzgerald et al.'s [23] study on 115 MDD outpatients found no significant differences in remission, response rates, or depression score changes between accelerated (63,000 pulses in 3 weeks) and standard (4 weeks) rTMS groups. However, the accelerated group had notable MADRS score reductions early on, indicating rapid action potential. In OCD, research on accelerated rTMS efficacy is scarce. Anyway, the evidence suggests that accelerated rTMS holds promise as an effective intervention for OCD.

Safety

The safety profile of accelerated rTMS is still being evaluated. As with other forms of rTMS, there may be potential risks such as headache, scalp discomfort, and seizure induction. However, more research is needed to fully understand the safety of accelerated rTMS. In the realm of depression, accelerated rTMS presents a safety profile akin to traditional rTMS, with common self-resolving scalp discomforts and transient headaches in some patients during treatment, while severe incidents like seizures are infrequent [23].

Treatment Regimen

Stimulation Target Similar to other rTMS modalities, accelerated rTMS primarily targets key brain regions implicated in the pathophysiology of OCD, such as the DLPFC and SMA [14, 16]. The rationale for targeting these regions is based on extensive research demonstrating their central role in the neural circuits underlying OCD symptoms.

Stimulation Intensity The determination of stimulation intensity in accelerated rTMS is based on the individual's motor threshold, with the intensity typically adjusted to a level that is sufficient to induce rapid and robust neuroplastic changes while remaining within a safe range. The high-frequency stimulation used in accelerated rTMS may introduce additional challenges, as the stability and adaptability of the motor threshold under such conditions are not well understood. The intensity adjustment strategy in depression can serve as a reference for OCD. In depression, the intensity range was extended from 80% to 120% of the motor threshold to 110% 140% [23].

Stimulation Frequency In the treatment of OCD, the frequency optimization approach in depression arTMS application offers valuable guidance. For example, the traditional frequency range (such as 1–10 Hz) is expanded to a higher frequency band (10–20 Hz), and high-frequency stimulation enhances the regulation of the neural circuits in the prefrontal cortex.

Stimulation Time The treatment duration of accelerated rTMS bears similarity to that of traditional rTMS, yet is fine-tuned within a nuanced range considering disease severity, individual variances, and stimulation parameters, aiming to optimize therapeutic outcomes.

Number of Treatment Sequences The strategy of compressing the treatment course and increasing the frequency in depression treatment inspires new avenues for OCD treatment. OCD is chronic and protracted, with traditional treatments often characterized by long durations and slow onsets. Thus, the treatment course could be moderately compressed (e.g., from 8–12 weeks to 4–6 weeks), and the weekly stimulation frequency increased (from 3–5 times to 5–7 times) [23]. Short-term intensive stimulation maintains the neural regulatory state and accelerates symptom improvement.

Clinical Recommendations

Accelerated rTMS may be considered for OCD patients requiring rapid symptom relief, particularly those with severe symptoms who have not responded adequately to traditional treatments. However, further research is needed to confirm its long-term efficacy and safety.

7.2.2 Deep TMS

dTMS is a new type of noninvasive physiotherapy technology based on the development of TMS, which has the advantages of deep stimulation site and wide range of action. Unlike standard transcranial magnetic stimulation, which uses a figure-8 coil, dTMS uses an H-coil, which stimulates a larger cortical volume more deeply.

7.2.2.1 Efficacy

For patients who remain symptomatic after first-line interventions such as SRIs and CBT treatments, dTMS becomes an adjunctive treatment option that can improve patients' OCD symptoms. Study has shown that OCD patients treated with dTMS experienced greater decreases in YBOCS scores and maintained significant decreases at the 1-month follow-up assessment [24]. One of the common co-morbidities of OCD is major depressive disorder (MDD). Study has shown that dTMS treatment improves depressive symptoms and obsessive-compulsive symptoms in patients with MDD co-morbidities OCD [25]. The best outcomes were obtained after 20 sessions of high-frequency dTMS with OCD patients, where, in addition to clinical improvement, patients showed amelioration of cognitive functions, specifically in cognitive control domains. A study found that after 20 sessions of dTMS for patients with OCD, in addition to the improvement of clinical symptoms, the patients' cognitive functions also improved [26].

7.2.2.2 Safety

The high-frequency dTMS using the H coil was well tolerated by patients. No severe adverse events such as seizures occurred, and the most frequent side effects included mild headache during or immediately after stimulation. In 2018, the U.S. Food and Drug Administration cleared the treatment of resistant OCD with high-frequency dTMS over the dorsomedial prefrontal (dmPFC) and anterior cingulate cortex (ACC) with the H7 coil. The H7 dTMS coil was specifically designed to directly target the mPFC and ACC.

7.2.2.3 Treatment Regimen

No consistent protocols have been developed for the stimulation target, frequency, and intensity of dTMS for OCD [27]. Current dTMS treatments for OCD predominantly focus on the dorsal mPFC and ACC. The high-frequency (HF; 20 Hz) dTMS over the mPFC-ACC alleviates OCD symptoms and may be used as a novel therapeutic intervention [28]. The H-coil was placed 4 cm anterior to the motor cortex of the foot, and 100% of the RMT of the leg was used as the stimulation intensity. A participant's RMT was determined before the first treatment and at the beginning of each week by ascertaining the coil position that elicited the minimal involuntary contractions of the feet (three of six attempts). An effective dTMS treatment protocol sets the frequency at 20 Hz, the intensity at 100% of the RMT, the pulse training time at 2 s, and the inter-training interval at 20 s, for a total of 50 training sessions and 2000 pulses per session. Treatment is given once a day for 6 weeks [24]. Another treatment regimen, in which each subject received 29 treatments over a 6-week period (5 treatments/week for weeks 1–5, 4 during the sixth week), also proved effective [25].

7.2.2.4 Clinical Recommendations

dTMS has been shown to be a safe and effective alternative treatment for OCD, particularly in patients with OCD who are nonresponsive or incomplete responders to medication or CBT, and has received regulatory approval. Currently, the effective course of treatment in dTMS for OCD is mostly set at 5–6 weeks, but there is still no agreement in this regard. Excessive duration of therapy may affect patient compliance. The majority of OCD patients benefitted from dTMS, and the onset of improvement usually occurs within 20 sessions [29]. Extending the treatment course beyond 29 sessions results in continued reduction of OCD symptoms, raising the prospect of value for extended treatment protocols in nonresponders.

7.2.3 Priming TMS

Priming TMS is a brief low-intensity high-frequency stimulation of the patient prior to low-frequency rTMS to enhance the neural response to rTMS. This approach, referred to as “priming stimulation,” involves providing low-frequency stimulation, for example, 1 Hz for 10–15 min.

7.2.3.1 Efficacy

The priming stimulation has been shown to markedly enhance the neural response to the low-frequency stimulation train. The pTMS was found to be more acceptable (i.e., with smaller dropout rate) than HF-rTMS, LF-rTMS, and sTMS. This intervention consists of inducing greater excitability suppression by priming a low-frequency protocol with a short period of higher-frequency stimulation—a mechanism described as homeostatic plasticity [30]. But priming stimulation has only been explored in neurophysiological experiments, and the effect of this approach in the treatment of OCD has not been studied.

7.2.3.2 Safety

The current study has not found that priming with 1 or 25 Hz rTMS results in an increase in serious adverse effects and that the procedure is well tolerated by all participants [31].

7.2.3.3 Treatment Regimen

Currently, pTMS for the treatment of OCD focuses on the dorsal mPFC and ACC. Currently, the stimulation target of the pTMS for OCD focuses on the dorsal mPFC and ACC, depending on the localization of the conventional transcranial magnetic stimulation performed next. Typically, effective pTMS sets the frequency to 1 Hz and the intensity to 100–160% of RMT. The stimulation time is set to 10–15 min. There is still no consensus on the optimal pTMS regimen for OCD treatment.

7.2.3.4 Clinical Recommendations

The addition of a subthreshold stimulation pretreatment before the r TMS regimen promises to be a new way to improve the efficacy of anti-compulsive therapy. Moreover, the pretreatment of rTMS is easier to administer and more tolerable to patients than more sequential and intense stimulation. It remains to be clarified whether the use of priming with low-frequency or high-frequency TMS initiation will improve the efficacy of rTMS in patients with OCD.

7.2.4 Synchronized TMS

Synchronized TMS is a treatment modality in which a rotating spherical neodymium magnet is positioned along the midline of the scalp during TMS stimulation, delivering synchronized stimulation according to an individual's alpha frequency.

7.2.4.1 Efficacy

It has been shown that synchronizing TMS pulses with ongoing u-oscillations in the electroencephalography (EEG) alpha-band (8–12 Hz) in human sensorimotor cortex results in a differential modulation of corticospinal excitability [32]. To realize the full potential of TMS to effectively modulate brain networks, each stimulus must be synchronized with the individual's instantaneous brain state.

7.2.4.2 Safety

There are no serious adverse events in subjects treated with synchronized EEG-TMS. A very small number of subjects reported mild transient discomfort at the site of stimulation and mild headache lasting until the next day after TMS. These side effects did not affect the subjects and were fully consistent with the study. No other side effects were reported.

7.2.4.3 Treatment Regimen

Real-time alpha-synchronized DLPFC stimulation: For brain oscillation-synchronized stimulation of the left DLPFC, the Electroencephalogram(EEG)-TMS set-up is used, with the capability of analyzing EEG signals in real-time and triggering TMS pulses depending on the instantaneous oscillatory phase of the recorded EEG signal [33]. To synchronize the TMS pulses with the instantaneous oscillatory alpha activity in left DLPFC, the alpha oscillations can be extracted at the sensor level using a Hjorth Laplace spatial filter centered on the electrode F5, which is the closest electrode to the target stimulation site and using four adjacent electrodes. Applying an autoregressive forward prediction method on the Hjorth-F5 signal led to an accurate prediction of the phase of instantaneous alpha oscillations, sufficient to trigger the 100 Hz TMS triple pulses at the predefined target phase [33]. In addition, EEG synchronized left prefrontal TMS for treatment-resistant depression is feasible [34], however, it has not been validated in OCD.

7.2.4.4 Clinical Recommendations

Alpha-synchronized rTMS of the left DLPFC is feasible, safe and has specific single-session neuromodulatory effects in patients with antidepressant-resistant MDD.

However, no study has yet clarified its effectiveness in OCD. That is why it is less used in clinical practice compared to rTMS. In future studies, physiological investigations are necessary to better understand which brain regions and which brain oscillations may serve as suitable targets for directional modulation in relation to OCD disorders, and clinical trials with repeated stimulation sessions are necessary to investigate the therapeutic potential and efficacy of brain oscillatory synchronized TMS in OCD as compared to current TMS therapies.

7.2.5 TBS

The recently introduced stimulation protocol, known as Theta Burst Stimulation (TBS), was first described by Huang et al. in the context of the human motor cortex [35]. TBS delivers bursts of pulses at a theta frequency, specifically triplets of 50 Hz pulses repeated at a rate of 5 Hz, to emulate the brain's natural rhythmic activity. Depending on the pattern of stimulation, TBS can have either excitatory or inhibitory effects on neural activity. Intermittent Theta Burst Stimulation (iTBS), characterized by intermittent bursts, induces an excitatory effect, whereas Continuous TBS (cTBS), which involves uninterrupted pulse delivery, results in an inhibitory effect [36]. Compared to conventional repetitive Transcranial Magnetic Stimulation (rTMS), TBS at 50 Hz can achieve more pronounced and enduring modulations of cortical excitability [37].

TBS represents an innovative approach within the realm of TMS. It is characterized by its use of a reduced stimulation intensity, operating at about 70% of the individual's motor threshold, in contrast to the 100% typically associated with the resting motor threshold. Additionally, TBS protocols are notably brief, lasting approximately 40 s, in stark comparison to the 15–20 min required for conventional low-frequency or high-frequency TMS treatments. The advantages of TBS become evident when considering patient experience. By enabling shorter treatment durations and lower intensity levels, TBS minimizes discomfort and enhances tolerability. This feature positions TBS as a more accessible and agreeable choice for patients undergoing noninvasive brain stimulation therapies, thereby increasing its clinical appeal and applicability.

A comprehensive review has assessed the potential of TBS in the treatment of OCD [38]. Findings from Randomized Controlled Trials (RCTs) [39–45] and various open-label studies [46–48] employing TBS suggest that, for individuals with OCD, TBS does not provide a clinically meaningful benefit over sham stimulation or traditional rTMS protocols. The active treatment response rates observed in these studies fall within the range of 16–28%. This lack of superior efficacy could be due to the fact that some patients may not receive an adequate number of treatment sessions in both TMS and TBS regimens. The FDA-approved guidelines recommend a minimum of 29 sessions; however, many clinical trials involve only 10–20 sessions.

This insight underscores the importance of maintenance therapy, where the frequency of TMS treatments can be adjusted over time according to individual patient responses. While the reviewed OCD trials did not explore or address the potential for maintenance therapy, it is plausible that nonresponders, partial responders, or those requiring sustained support could derive significant benefits from ongoing treatment. Moreover, a randomized controlled trial that compared TBS and rTMS in treating 45 patients with OCD over a 5-day period revealed a notable difference in response rates: 60% for TBS versus 31% for rTMS [45]. This finding indicates that, with continued refinement and optimization of treatment protocols, TBS has a promising outlook for delivering effective therapy, particularly when rapid symptom improvement is crucial. Supported by emerging models and advanced technologies, TBS retains considerable untapped potential for mitigating compulsive symptoms in OCD patients and may also offer benefits for associated comorbidities. A study involving 20 Hz theta burst stimulation (TBS-20 Hz) guided by neuronavigation, targeting the bilateral DLPFC and the right orbitofrontal cortex (ROFC), has been used to treat patients with OCD comorbid with depression and anxiety. Patients completed the Y-BOCS–Self-Report, the Beck Depression Inventory-II (BDI-II), and the Beck Anxiety Inventory (BAI) weekly to measure changes in OCD, depressive, and anxiety symptoms, respectively. The neuronavigated TBS-20 Hz was sequentially applied to the right DLPFC, left DLPFC, and then the ROFC. The results showed that 64% of patients achieved remission from OCD (Y-BOCS-SR $\leq$ 14) after an average treatment duration of 6.1 weeks. For MDD, 58% of patients reached remission (BDI $<$ 13) within an average of 4.1 weeks, while 62% of patients experienced remission from generalized anxiety disorder/panic disorder (GAD/PD) (BAI $<$ 8) after an average of 4.3 weeks. The use of neuronavigation-assisted TBS-20 Hz sequential stimulation of RDLPFC and LDLPFC, followed by ROFC, significantly alleviated symptoms of OCD, MDD, and GAD/PD [49]. Moreover, while pharmacotherapy and psychotherapy are available for treating refractory OCD, the high likelihood of treatment resistance necessitates the exploration of alternative approaches. TBS technology offers assistance in modulating the activity of dysfunctional areas in patients with refractory OCD, particularly targeting the cortico-striato-thalamo-cortical circuit, to restore a functional level comparable to what is generally accepted as normal [50].

The efficacy of TBS stems from its ability to modulate neural activity in a targeted manner. TMS, the technology underlying TBS, is a safe, non-painful, and noninvasive method that influences brain function by generating electromagnetic fields that penetrate the scalp and skull to reach the cortex. TBS specifically enhances cortical excitability by emulating the natural θ rhythm of the brain, thereby facilitating synaptic plasticity and neurotransmission [51]. When applied, TBS creates time-varying pulsed magnetic fields at the scalp level, which can penetrate cortical tissues to a depth of about 2 cm. The subsequent electrophysiological effects are influenced by multiple parameters of the magnetic field, such as the frequency and pattern of stimulation [52]. TBS treatment protocols are currently classified into several modes: iTBS, asymmetric intermittent Theta Burst Stimulation (aiTBS), cTBS, and bilateral TBS. Among these, the two predominant forms utilized in contemporary clinical studies are iTBS and cTBS, each designed to elicit distinct patterns of neural response and therapeutic outcomes.

7.2.5.1 iTBS

iTBS involves the administration of 2-s trains of TBS, delivered at intervals of 8 s, over a span of 192 s [totaling 600 pulses] [53].

Efficacy

The therapeutic efficacy of iTBS may be linked to its ability to significantly enhance lagged nonlinear coherence (phase synchronization) within the prefrontal regions of interest, including the ACC, superior frontal gyrus (SFG), and middle frontal gyrus (MFG), across the beta1, beta2, and theta frequency bands. Moreover, iTBS has been shown to induce notable changes in long-range connectivity between fronto-parietal areas, specifically in the theta, alpha1, and beta2 frequency bands, as evidenced by increased lagged linear coherence [41]. A pilot study employing a randomized crossover design utilized iTBS targeted at the left DLPFC to treat patients with OCD who had inadequate responses to pharmacotherapy and CBT. The findings indicated that iTBS delivered to the left DLPFC was both safe and effective, providing sustained relief from OCD symptoms for up to 3 months [41]. Targeting the L-DLPFC with iTBS has proven effective for individuals with treatment-resistant OCD. Additionally, a case report demonstrated that combining rTMS and iTBS, with targets including the bilateral DLPFC and the SMA, was equally effective in managing refractory OCD as traditional rTMS treatments [54].

Safety

iTBS, as a specialized form of TMS, can commonly lead to side effects such as headaches, localized scalp discomfort, paresthesia, facial nerve stimulation, dizziness, somnolence, and the induction of hypomanic episodes. Severe potential acute adverse reactions, though rare, could include seizures or cognitive disturbances. However, in the current clinical trials for OCD, no significant adverse reactions to iTBS have been noted. Specifically, in a study targeting the left dorsolateral prefrontal cortex (L-DLPFC) with iTBS for patients with inadequate responses to pharmacotherapy and CBT, all participants reported no adverse events during or following the treatment period [41]. Given the limited research on the dose–response relationship in iTBS, it remains uncertain whether higher intensity or more frequent iTBS sessions, administered at 80% of the RMT or active motor threshold (AMT), provide a superior risk-to-benefit profile compared to standard iTBS protocols. Theoretically, lower-intensity protocols might retain advantageous safety characteristics and warrant further investigation to fully understand their therapeutic potential.

Treatment Regimen

rTMS and TBS protocols are usually calibrated based on the minimal intensity required to induce a motor response in the motor cortex, referred to as the motor threshold. This threshold can be established through visual observation or by employing electromyography. In visual assessment, the motor threshold is commonly defined as the lowest stimulation intensity that results in visible movement of the target hand muscle in at least 5 out of 10 attempts. When using electromyography, the resting motor threshold is defined as the lowest intensity that elicits a motor-evoked potential (MEP) with a peak-to-peak amplitude of at least 50 μV in at least 5 out of 10 trials [55].

For iTBS, a widely used protocol involves delivering bursts of pulses at a frequency of 5 Hz, with each burst lasting 2 s, followed by an 8-s inter-burst interval. The entire sequence lasts approximately 3 min and comprises 600 pulses. In terms of target areas, recent clinical research indicates that focusing on the left frontolateral region of the dorsolateral prefrontal cortex (LF-R-DLPFC), the supplementary motor area, and the L-DLPFC may yield more effective outcomes for the treatment of OCD [8].

Clinical Recommendations

In general, the application of TMS for treating OCD is advised only after first-line and second-line treatments, as well as well-established augmentation strategies, have been fully explored and found inadequate, similar to the approach taken with other somatic interventions. The evidence base for TMS in OCD remains limited, largely because the variability in study designs, target brain regions, treatment durations, and stimulation parameters across double-blind sham-controlled trials has introduced significant heterogeneity, complicating the assessment of its true efficacy. Despite these challenges, recent meta-analyses and updated clinical guidelines have begun to reflect more favorable outcomes, indicating a growing recognition of TMS's potential in the management of OCD [56].

7.2.5.2 Accelerated iTBS

aiTBS is an expedited form of treatment that involves delivering multiple sessions within a single day over a brief treatment period, typically spanning just a few days.

Efficacy

To date, there has been no validation of the efficacy of aiTBS interventions specifically in patients with OCD. Existing aiTBS interventions have primarily focused on patients with MDD. In a randomized, double-blind, sham-controlled crossover trial, 45 MDD patients received aiTBS applied to the left DLPFC. In this accelerated neurostimulation paradigm, the placebo response was associated with reduced perfusion in brain regions linked to higher cognitive processes, leading to a decrease in suicidal ideation [57]. Moreover, aiTBS has been found to be more effective than sham stimulation in reducing depressive symptoms in patients with treatment-resistant Bipolar Disorder (BD) [58].

Safety

All existing studies have consistently reported that aiTBS is safe and well-tolerated, with no serious adverse events noted. The most commonly reported side effects include fatigue, headaches, and local discomfort at the stimulation site, all of which are similar in frequency and severity to those observed with conventional rTMS protocols. Notably, there have been no reports of seizures or mood switches in patients undergoing aiTBS treatment.

Treatment Regimen

aiTBS is defined as administering at least three iTBS sessions per day, for a minimum of 4 days per week, with the total stimulation/treatment period lasting 1 week [59].

The intensity of aiTBS is typically expressed as a percentage of the individual's MT. The MT is the minimum intensity of stimulation required to elicit a visible muscle contraction in a specific muscle group. The optimal stimulation intensity may vary depending on the disorder being treated and the individual patient characteristics. Generally, intensities ranging from 80% to 120% of the MT have been used in clinical studies. However, higher intensities may increase the risk of adverse effects, while lower intensities may be less effective.

aiTBS typically uses a frequency of 50 Hz, which is delivered in bursts of three pulses at 5 Hz. The intermittent nature of the stimulation allows for longer periods of rest between bursts, reducing the risk of neuronal fatigue and enhancing the therapeutic effect. The frequency of aiTBS may be adjusted based on the specific disorder and the patient's response. For example, lower frequencies may be used in patients with a history of seizures or those at risk of adverse effects.

The duration of aiTBS treatment sessions can vary depending on the disorder and the treatment protocol. Generally, sessions last between 10 and 30 min. The total number of treatment sessions may also vary, ranging from a few sessions to several weeks of daily or weekly treatments.

The number of treatment sequences refers to the number of times the stimulation is delivered in a single session. Multiple sequences may be used to increase the total amount of stimulation and enhance the therapeutic effect. However, the optimal number of sequences may vary depending on the disorder and the patient's tolerance. Generally, two to four sequences are used in clinical studies, with a rest period of a few minutes between sequences.

Clinical Recommendations

Considering the current paucity of research on aiTBS interventions for OCD, there is a critical need for more clinical evidence to support its use in this patient population. Given the accessibility and potential benefits of aiTBS, it is feasible to consider aiTBS as an adjunctive therapy and as a supplementary treatment option when standard pharmacological or psychological treatments have not been effective. This approach could provide valuable alternatives for patients with refractory OCD, enhancing the therapeutic arsenal available to clinicians.

7.2.5.3 Continuation TBS

cTBS is a form of rTMS that exerts faster and more enduring effects on synaptic plasticity compared to conventional rTMS. cTBS involves the uninterrupted administration of TBS for 40 s, which also totals 600 pulses [53].

Efficacy

Overall, cTBS has demonstrated notable efficacy in improving obsessive symptoms. Compared to iTBS, cTBS exhibits more significant short-term benefits in reducing these symptoms. In a randomized double-blind study involving 69 patients with OCD, both iTBS and cTBS were applied to the left OFC using 600 pulses at 110% of the resting motor threshold. The study found that both forms of TBS modulated OFC activation as expected. However, cTBS showed a more pronounced beneficial effect on the acute laboratory assessment of compulsive behaviors 90 min after stimulation, compared to iTBS. Importantly, these acute behavioral improvements persisted for up to 1 week following cTBS treatment [60].

Furthermore, cTBS administered using an accelerated protocol has shown more pronounced therapeutic effects in treating OCD. A study included 32 OCD patients who were randomly assigned to either active or sham cTBS groups. In the active cTBS group, an accelerated protocol was employed, involving bursts of three stimuli repeated at 5 Hz, with each burst delivered at 50 Hz and 80% of the motor threshold (MT). Participants underwent two sessions per day, each session consisting of 900 pulses, totaling 30 sessions over 3 weeks (10 sessions per week). The findings revealed that adjunctive accelerated cTBS led to significant improvements in psychopathology, disease severity, and depressive symptoms among OCD patients [44]. Compared to traditional 1-Hz rTMS, the accelerated high-dose Theta Burst Stimulation (ahTBS) protocol did not show a significant difference in clinical effectiveness after 5 days of treatment. Both groups experienced significant reductions in Y-BOCS scores, and there were no statistically significant differences in secondary outcomes, such as depressive symptoms, anxiety symptoms, and response rates, indicating the non-inferiority of ahTBS. However, a trend toward a higher response rate was observed in the ahTBS group. Neuropsychological testing showed that neither treatment led to adverse cognitive side effects. These results suggest that accelerated high-dose TBS is equally safe for treating OCD and demonstrates comparable short-term efficacy to traditional 1-Hz rTMS [45]. In a study involving OCD participants, an intensified novel cTBS protocol targeting the orbitofrontal cortex (OFC) was administered. Patients received 10 sessions of cTBS, with two sessions per day (a total of 1200 pulses per day; intensive protocol), 5 days per week. Significant group-by-time effects were observed for obsessions, compulsions, HAMA, HAMD, and CGI scores at baseline, immediately after the last treatment, and 2 weeks posttreatment. However, when controlling for confounding variables, only the HAMA scores and CGI effects remained statistically significant. We conclude that intensified OFC cTBS in OCD is well-tolerated and results in clinically meaningful improvements in anxiety symptoms and overall severity. This improvement in anxiety symptoms may be due to the modulation of state-dependent dysregulation in OCD [40].

In the context of treatment-resistant OCD, the efficacy of cTBS appears to be somewhat limited. Meta-analyses have indicated that, for cTBS, there were no significant differences in Y-BOCS scores or clinical response rates between the active stimulation group and the sham group. A reasonable explanation for this finding is that the three TBS studies utilized a stimulation intensity of 70% or 80% of the

RMT, delivering 600 pulses per session, which may be insufficient to achieve therapeutic benefits in patients with treatment-resistant OCD [61]. Furthermore, a foundational study investigated whether modulating activity in the right OFC using cTBS could impact symptoms of OCD. In this study, 28 participants with treatment-resistant OCD were randomized in a double-blind manner to receive either active or sham cTBS targeting the OFC, administered twice daily for 5 days a week over a 2-week period. The Y-BOCS was used to assess clinical response to the treatment. Following the two-week treatment period, there were no statistically significant differences between the groups in Y-BOCS scores or other secondary outcomes, such as anxiety symptoms and response rates. While the active group demonstrated a significant improvement in depressive symptoms, this difference was not sustained and became nonsignificant by the 6-week follow-up. These results suggest that although cTBS targeting the right OFC is safe for treating symptoms of treatment-resistant OCD, a two-week treatment duration may be insufficient to achieve a clinically significant effect [43]. Researchers evaluated the therapeutic efficacy and tolerability of cTBS applied to the SMA in patients with treatment-resistant OCD. Thirty outpatient participants with treatment-resistant OCD were randomly assigned to receive either active cTBS or sham cTBS for 6 weeks (five times per week). Each treatment session consisted of 600 stimuli delivered at 70% of the resting motor threshold. Patients were assessed at baseline, at the end of treatment (Week 6), and at follow-up (Week 12). A response to treatment was defined as a reduction of at least 25% on the Y-BOCS. The results showed no statistically significant differences in treatment effects between the active cTBS group and the sham cTBS group. There were no differences in response rates at Week 6 and Week 12. Both groups exhibited similar improvements in depressive and anxiety symptoms. Overall, cTBS was found to be safe but did not significantly improve symptoms in patients with treatment-resistant OCD. Further research is needed to determine the optimal parameters for cTBS in treating OCD [39].

Safety

As a variant of TBS, cTBS typically presents with side effects that are similar to those associated with TBS and other forms of rTMS. Common side effects may include mild to moderate headaches, discomfort or pain at the stimulation site, muscle twitching or tremors, transient mood changes such as anxiety or emotional fluctuations, temporary hearing loss or tinnitus, and, in extremely rare cases, seizure induction, especially in individuals with a history of epilepsy. These side effects are generally temporary and subside soon after the end of the treatment session. In very rare instances, more serious side effects such as severe mood disturbances or cognitive impairments could potentially occur, although these are highly uncommon. To date, studies on cTBS in patients with OCD have not reported any such severe adverse reactions. The side effects observed in these studies have been limited to the milder, more common types mentioned above, indicating that cTBS is generally well-tolerated in this patient population.

Treatment Regimen

Current cTBS treatments for OCD predominantly focus on the DLPFC and the SMA. The DLPFC is among the most frequently targeted regions in OCD therapy due to its involvement in cognitive control and emotional regulation. Stimulating the DLPFC can enhance patients' ability to manage obsessive thoughts and reduce compulsive behaviors. The SMA, which plays a key role in motor control and behavioral planning, is another important target. Stimulation of the SMA can help mitigate motor-related compulsions, such as repetitive movements or rituals, which are common in OCD. Beyond the DLPFC and SMA, some research has also investigated other brain areas as potential targets for stimulation, including the OFC and the ACC. However, the effectiveness of these alternative targets remains to be fully validated through additional research.

The intensity of cTBS is typically expressed as a percentage of the stimulator's output. Currently, there is no consensus on the optimal stimulation intensity for treating OCD with cTBS. Generally, the stimulation intensity ranges between 80% and 120% of the RMT. Higher stimulation intensities may enhance treatment efficacy but can also increase the risk of adverse reactions. Some studies have identified 85% RMT as an effective stimulation intensity, gradually escalating to 115% RMT during the intervention period [60]. The stimulation frequency of continuous Theta Burst Stimulation (cTBS) is typically set at 50 Hz. However, some studies have also explored the impact of different stimulation frequencies on the treatment of OCD. The duration of cTBS stimulation is usually 40 s. During this period, the stimulator delivers magnetic pulses at a specified frequency and intensity. Shorter stimulation durations may reduce treatment efficacy, while longer durations could increase the risk of adverse reactions. Therefore, when determining the stimulation duration, it is important to balance the patient's tolerance with the desired treatment outcome. The number of treatment sessions for using cTBS to treat OCD typically ranges from 10 to 20 sessions. Treatments are generally administered 3 to 5 times per week over a period of 2 to 4 weeks.

Some enhanced protocols for cTBS intensify the intervention frequency by using a stimulation intensity of 80% MT, delivered continuously in triplets at a 50 Hz frequency, and repeated at a 5 Hz frequency. Treatments are administered twice daily, with each session consisting of 900 pulses, and at least a 5-hour interval between sessions. The total number of treatments is 30, with sessions spaced over 3 weeks. This treatment regimen, which involves more than 1000 pulses per day for over 2 weeks, is referred to as an "accelerated protocol"[44]. Another modified accelerated high-dose theta burst stimulation protocol has a total intervention duration of 5 consecutive days. Each session of the modified continuous TBS consists of 1800 pulses, with each set of pulses comprising 600 theta bursts. Each theta burst is made up of 3 pulses at 50 Hz, and these theta bursts are repeated at a frequency of 5 Hz. Treatment is administered 10 times per day, with each session spaced 50 min apart. The stimulation intensity is set at 80% of the resting motor threshold (RMT) [62].

Clinical Recommendations

Overall, compared to other TBS paradigms, cTBS has been studied more extensively in the context of OCD. However, due to the heterogeneity of parameters and targets used in these studies, it is not yet conclusive whether cTBS effectively improves OCD symptoms. Nonetheless, cTBS and its intensified protocols still hold significant clinical potential. More precise neuro-navigation and more specific treatment targets remain key focuses for future research.

7.2.5.4 Bilateral TBS

Bilateral TBS involves the simultaneous application of TBS to both sides of the brain. This method of stimulation can more effectively induce plasticity changes in neurons. It works by influencing the balance between long-term potentiation and long-term depression.

Efficacy

The SMA is a commonly used target in bilateral TBS treatment for patients with OCD. A study administered 20 consecutive sessions of bilateral cTBS to the SMA in 29 patients with OCD and 29 healthy controls. The results showed that 11 out of the 29 patients (37.93%) responded to the bilateral cTBS, with significant clinical symptom relief observed in the OCD patients following the bilateral SMA cTBS treatment. Additionally, the functional connectivity between the Cerebelum_Crus2_L and Frontal_Inf_Tri_L in OCD patients indicated a favorable prognosis for cTBS efficacy. No serious adverse events occurred in any of the patients. Intervention with bilateral SMA cTBS can significantly improve symptoms in moderate to severe medication-refractory OCD patients. Pretreatment functional connectivity may serve as a valuable potential predictor of treatment outcomes [46]. In a single-blind, sham-controlled design of cTBS applied to the bilateral SMA for patients with OCD, 54 OCD patients were randomly assigned to receive either active or sham cTBS treatment over the bilateral SMA for 4 weeks (5 times per week, totaling 20 sessions). Patients were evaluated at baseline (week 0), end of treatment (week 4), and follow-up (week 8). The results showed no significant difference in the treatment response rate between the fourth and eighth weeks. At week 4, there was a significant difference in improvements in depression and anxiety between the two groups. There were no significant differences in treatment satisfaction and dropout rates between the groups. Bilateral SMA cTBS treatment was found to be safe and tolerable, significantly improving depression and anxiety in OCD patients, but it was not sufficient to improve OCD symptoms [8]. For some OCD patients who respond poorly to medication, open-label case reports have indicated that bilateral cTBS targeting the MA may alleviate obsessive symptoms in patients with OCD [47].

Furthermore, combining iTBS and cTBS to target the DLPFC may offer an effective treatment option for perinatal OCD. This approach can help address practical clinical issues such as medication safety for breastfeeding mothers and the specific needs of this population. A clinical case report of a dual-stimulation protocol using both iTBS and cTBS for postpartum OCD provides evidence that this combined treatment may effectively alleviate obsessive symptoms in the short term [63]. Additionally, a daily bilateral TBS regimen—administering cTBS to the right DLPFC and iTBS to the left DLPFC—has been validated for its efficacy in treating MDD over a 6-week period. This protocol may also benefit patients with comorbid OCD and depression [64]. The combination of iTBS and cTBS offers a promising therapeutic strategy, potentially addressing both conditions simultaneously while minimizing concerns related to pharmacological interventions in the perinatal period.

Safety

To date, the safety and tolerability of bilateral TBS appear comparable to those of unilateral stimulation, with no serious adverse events reported among all patients in current studies [46]. A singular case report further supports these findings, noting that the patient experienced no discomfort and had a positive overall experience with the treatment. Importantly, this includes postpartum women, who have shown an ability to tolerate the intervention well. The straightforward nature of the treatment setup, combined with the ease of its implementation, significantly enhances its practicality in clinical practice. Nevertheless, the requirement for specialized skills and knowledge to administer the treatment could pose a limitation, potentially restricting access in settings with limited resources [63].

Treatment Regimen

The fundamental parameters of Bilateral TBS do not differ significantly from other TBS paradigms. However, it is crucial to ensure the symmetry of stimulation sites bilaterally, which is essential for maintaining balanced and coordinated brain function. Additionally, since Bilateral TBS involves simultaneous stimulation of both cerebral hemispheres, it is imperative to maintain balanced stimulation intensities on both sides. Any imbalance in stimulation intensity could lead to asymmetrical modulation of brain function, potentially compromising treatment efficacy. To determine the appropriate stimulation intensity, the RMT can be measured for each patient to establish individualized stimulation levels. Subsequently, these levels should be adjusted to ensure that the stimulation intensity is essentially symmetrical between the two hemispheres. This adjustment is critical to achieving the desired therapeutic effect while minimizing potential side effects. Furthermore, the synchrony of stimulation timing is paramount in Bilateral TBS to ensure that both cerebral hemispheres are stimulated concurrently. Asynchronous stimulation could disrupt the coordination of brain functions and reduce the effectiveness of the treatment. To achieve this synchrony, synchronized triggering devices or specialized stimulation software can be employed to ensure that both stimulators start and end their pulses at precisely the same time. Given that bilateral TBS simultaneously targets both cerebral hemispheres, it may carry a relatively higher safety risk compared to unilateral stimulation. Therefore, rigorous safety monitoring is required during bilateral TBS sessions. Safety monitoring should include the observation of adverse reactions such as headaches, dizziness, and nausea, as well as the assessment of neurophysiological indicators like electroencephalograms and electromyograms. Should any adverse reactions or abnormal neurophysiological indicators be detected, the stimulation parameters should be promptly adjusted, or the stimulation should be halted if necessary.

Clinical Recommendations

Overall, the accelerated TBS paradigm involving sequential bilateral iTBS and cTBS has demonstrated comparable efficacy and response rates to conventional, with the notable advantage of achieving these outcomes in a significantly shorter treatment duration. Compared to more invasive neuromodulatory interventions such as Vagus Nerve Stimulation (VNS) and DBS, TBS offers a markedly less invasive alternative, making it a preferable choice from both a clinical safety and patient acceptability standpoint. However, the body of research on TBS for treating OCD remains limited. Existing studies often suffer from small sample sizes, heterogeneous stimulation parameters, and diverse cortical targets, which have not been standardized. Moreover, many studies do not utilize image-guided TBS (neuronavigation), which could enhance the precision and effectiveness of the treatment by ensuring accurate targeting of specific brain regions. Given these limitations, there is an urgent need for well-designed, large-scale, and methodologically rigorous studies to investigate the potential of TBS as an efficacious and cost-effective therapeutic intervention for OCD. Such research should aim to standardize TBS protocols, optimize stimulation parameters, and explore the use of advanced neuroimaging techniques to guide target selection and monitor treatment effects. Furthermore, a deeper understanding of the neural mechanisms underlying TBS's effects on brain network dynamics could provide valuable insights into how focal perturbations in neural activity influence remote brain regions and their functional connectivity. This knowledge could lead to more personalized and effective TBS treatments, potentially enhancing clinical outcomes for patients with OCD.

7.2.6 MST

Magnetic seizure therapy (MST) is a novel noninvasive neuromodulation technique. Unlike traditional electroconvulsive therapy (ECT), MST uses magnetic pulses instead of electric current to induce therapeutic seizures, thereby regulating the release of neurotransmitters and communication between neurons in the brain. ECT has demonstrated some success in case reports and series [65], but its use is limited by concerns over neurocognitive side effects and societal stigma. Similar to ECT, which has shown response in a significant proportion of treatment-resistant cases, MST has the potential to target relevant brain regions [66] in the pathophysiology of OCD. Computational analyses and electric field modeling [67] show that MST with frontal coil placement can focus on midline cortical areas like the cingulate cortex which is which are a part of the CSTC circuit [68] implicated in OCD.

A pilot study [69] shows that the efficacy of MST in treating obsessive-compulsive symptoms of treatment-resistant OCD patients is limited, with only one out of ten participants achieving a clinically meaningful reduction in Y-BOCS scores and no significant improvement in the overall group, suggesting that the current MST technique may not be an effective treatment option for obsessive-compulsive symptoms in this population. It is important to note that, to date, no randomized controlled trials specifically evaluating the efficacy of MST in OCD have been published. This lack of definitive clinical trial data represents a significant gap in our knowledge and highlights the need for further research in this area.

In conclusion, the current study did not demonstrate clinically meaningful effects of this MST technique in patients with OCD.

7.2.6.1 Safety

MST exhibits potential safety advantages in comparison to ECT. It is associated with less cognitive side effects, as evidenced by faster reorientation of patients' posttreatment and possible enhancements in memory and attention functions. In Tang's pilot study [69], a notably low incidence of severe adverse events was reported, further suggesting its relatively favorable safety profile in that context. However, the current understanding of MST's safety remains restricted due to several methodological drawbacks in previous research. These limitations encompass small sample sizes, which curtail the statistical power and generalizability of the findings, a lack of blinding and randomization procedures that may introduce biases, and substantial variability in study methods such as differences in MST stimulation parameters, patient selection criteria, and outcome assessment measures. This situation underlines the necessity for further meticulously designed investigations to comprehensively evaluate MST's safety.

Notably, studies conducted on major depressive disorder [70, 71], bipolar depression [70], and schizophrenia [72] have not only demonstrated its clinical effectiveness but also indicated that it has little to no neurocognitive adverse effects, providing additional support for its potential as a viable treatment option with a favorable safety profile.

7.2.6.2 Treatment Regimen

Due to the current limited research, an accurate treatment plan cannot be provided. Based on a review of the existing studies, we conclude that the stimulation target mainly focuses on the frontal cortex (by placing the MagVenture twin coil centered at the F3 and F4 sites in the international 10–20 system) [69] and also involves vertex stimulation (with the twin coils placed more posteriorly on the left and right hemispheres, centered at the vertex [70]). The stimulation intensity is delivered as high-intensity magnetic field pulses at 100% machine output, and the seizure threshold is reached by increasing the stimulation duration. Regarding the stimulation frequency, different frequencies were evaluated, including high-frequency (100 Hz) frontal placement, medium-frequency (50–60 Hz) frontal placement, low-frequency (25 Hz) frontal placement, and high-frequency vertex placement. In terms of stimulation time, the initial stimulation duration and incremental duration vary with different frequencies. For example, the initial stimulation duration for 25 Hz is 4 s with an incremental duration of 4 s; for 50 and 60 Hz, the initial duration is 2 s with an incremental duration of 2 s; and for 100 Hz, the initial duration is also 2 s with an incremental duration of 2 s. Moreover, the maximum stimulation duration differs according to the frequency, being 20 s for 25 Hz, 20 s for 50 Hz, 16 s for 60 Hz, and 10 s for 100 Hz. Concerning the number of treatment sequences, the actual number of treatments varies among individuals.

However, the above parameters are for reference only, and more experiments are still needed to verify the stimulation target, intensity, frequency, time, and number of treatment sequences for treating OCD before an effective treatment regimen can be determined.

7.2.6.3 Clinical Recommendations

Currently, the efficacy of MST in OCD is also not well established. The majority of studies on MST predominantly consist of case studies, small case series, or small observational investigations [73]. Consequently, there exists a significant risk of publication bias. Moreover, no RCTs are available, which further limits the robustness of the evidence base regarding the efficacy and safety of MST.

Available practice guidelines for treatment of OCD currently do not recommend ECT [68], which likely reflects the low quality and quantity of the available evidence. MST starts with a more focal area of stimulation compared with ECT but eventually spreads to the rest of the brain as a generalized seizure like in ECT. A potential advantage of MST is that the e-field is concentrated in the superficial cortex and the stimulation of other brain areas is at a lesser intensity. Thus, the therapeutic effects of MST are thought to be equal to ECT, with fewer cognitive side effects [74]. However, the use of therapeutic seizures may not be an effective mechanism of treatment in OCD, through either ECT or MST.

Given the current state of knowledge, more research is needed. Randomized controlled trials of MST in OCD should be conducted to establish its efficacy and safety definitively. Clinicians should consider MST as a potential treatment option for treatment-resistant OCD patients, especially those who have not responded well to traditional first-line treatments and augmentation strategies. However, until more conclusive evidence is available, it should be used with caution and in the context of a comprehensive treatment plan that includes other evidence-based therapies and careful patient selection and monitoring.

In conclusion, MST holds promise as a treatment for OCD, but further investigation is crucial to determine its optimal role in the management of this debilitating disorder (Table 7.1).

Table 7.1 TMS, TBS, and MST in OCD

TMS	Clinical recommendations				Levels + references
	Target	Frequency	Length (min)	Duration (sessions)	
aTMS	Right OFC	LF (1 Hz)	20 (1200 pulses/session)	10 (1 week, 5 days/ weeks, 2 sessions/day)	Unclear [75]
	dACC	LF (1 Hz)	34 (1200 pulses/session)	20 (2 weeks, 5 days/ weeks, 2 sessions/day)	Unclear [76]
	mPFC/ACC	HF (20 Hz, dTMS)	18 (2000 pulses/session)	30 (3 weeks, 5 days/ weeks, 2 sessions/day)	Unclear [77]
	SMA	HF (50 Hz, cTBS)	3 (900 pulses/session)	30 (3 weeks, 5 days/ weeks, 2 sessions/day)	Unclear [44]
	Left OFC	HF (50 Hz, cTBS)	0.67 (600 pulses/session)	10 (1 weeks, 5 days/ weeks, 2 sessions/day)	Unclear [40]
cTBS	Bilateral DLPFC and the ROFC	20 Hz	3600–4800 inhibitory pulses over RDLPFC,1800-2700 inhibitory pulses over ROFC(2–3 min)	90–95%	Not clear [78]
	Left orbitofrontal cortex (LOFC)	50 Hz	One continuous train of 40 s	110%	Not clear [78]
	Bilateral supplementary motor area (SMA)	1 Hz	(1 Hz, 90% of RMT) 20 sessions/day	90%	Not clear [78]
ahTBS	Supplementary motor area (SMA)	50 Hz	Two sessions per day, each session consisting of 900 pulses, totaling 30 sessions over 3 weeks (10 sessions per week)	80%	Not clear [78]

(continued)

Table 7.1 (continued)

	Clinical recommendations				
TMS	Target	Frequency	Length (min)	Duration (sessions)	Levels + references
iTBS	Left dorsolateral prefrontal cortex (L-DLPFC)	50 Hz	20 trains of 10 bursts (short bursts of 3 stimuli at 50 Hz, repeated at 5 Hz) given with 8 s intervals, 600 stimuli, 200 s	80%	Not clear [78]
piTBS	Not clear	Not clear	Not clear	Not clear	Not clear
dTMS	mPFC-ACC	HF (20 Hz)	Training time at 2 s + inter-training interval at 20 s, repeat 50 times.	20	Adjuvant therapy [24, 25, 28]
pTMS	Consistent with the target of the upcoming rTMS	LF (1 Hz)	10–15	Consistent with the upcoming rTMS	Still no consensus in OCD
sTMS	Left DLPFC	Individual alpha frequency (range 8–13 Hz)	40 TMS pulses, a total of 75 rTMS pulse trains	30	Still no consensus in OCD
MST	Not known	Not known	Not known	Not known	An investigational treatment [78]

Neuronavigated Right Orbitofrontal 20 Hz Theta Burst Transcranial Magnetic Stimulation Augmentation for Obsessive-Compulsive Disorder with Comorbid Depression and Anxiety Disorders: An Open-Label Study
Effect of Experimental Manipulation of the Orbitofrontal Cortex on Short-Term Markers of Compulsive Behavior: A Theta Burst Stimulation Study
The functional connectivity predictor of therapeutic effect of continuous theta burst stimulation on obsessive-compulsive disorder: A preliminary study
Adjunctive neuronavigated accelerated continuous theta-burst stimulation in obsessive-compulsive disorder: a randomized sham-controlled study
Theta burst stimulation for the treatment of obsessive-compulsive disorder: a pilot study
Clinical studies have confirmed that dTMS alleviates obsessive-compulsive symptoms as a novel therapeutic intervention
The CANMAT Guidelines mention MST as an investigational treatment due to limited available evidence
Abbreviations: *HF* high frequency, *LF* low frequency, *OFC* orbitofrontal cortex, *ACC* anterior cingulate cortex, *dACC* dorsal ACC, *mPFC* medial prefrontal cortex, *SMA* supplementary motor area

7.3 tDCS

7.3.1 Conventional tDCS

7.3.1.1 Efficacy

Transcranial direct current stimulation (tDCS) is regarded as a noninvasive technique for brain stimulation, involving low-intensity continuous current delivered through scalp electrodes to alter brain function [79]. This brain stimulation technique is able to modulate neural excitability without triggering action potentials, as the current is always subthreshold, and can facilitate or inhibit neural activities according to electrode polarity [80]. tDCS has demonstrated promise in easing symptoms of different neuropsychiatric conditions. While most research on tDCS in psychiatric disorders has been centered on depression and schizophrenia, there have been only a small number of studies on OCD and related disorders. However, tDCS has been proposed as a treatment for refractory OCD [81].

Volpato et al. probably described the first application of tDCS in an OCD patient [82]. This was a single case study with active and sham tDCS cross-over. The stimulation montage involved cathodal stimulation of left DLPFC, with a primary aim of comparing the effect of tDCS and rTMS on resting-state brain activity. The study reported no change observed in OC symptoms, whereas improvement in depression was noted after true tDCS. Following this, Gowda et al. conducted the first RCT study on tDCS for the intervention of OCD [83]. A total of 25 patients with refractory OCD were recruited in this study, with 12 cases in the real stimulation group and 13 cases in the sham stimulation group. The anode was placed on the left preSMA, and the cathode was placed on the right supraorbital region. Each intervention lasted for 20 min, twice a day, for a total of 5 days. The results showed that there were 4 subjects in the real stimulation group with a reduction rate of the Y-BOCS score $\geq$35%, while there were 0 cases in the sham stimulation group. These results suggest that tDCS intervention may be effective for refractory OCD.

Currently, most of the studies on tDCS for the intervention of OCD are case studies and open-label studies. The influences of individual differences and placebo effects on the research results cannot be excluded. Therefore, large-sample, multicenter RCTs are required to further confirm the efficacy and explore the optimal parameters for treatment.

7.3.1.2 Safety

tDCS is a noninvasive neuromodulation technique which delivers low-amplitude direct currents (1–2 mA) to the brain between two large surface electrodes (anode and cathode) positioned on distinct areas of the scalp with a rubber headband [79], which, depending on the polarity, intensity and duration of the current flow are capable of modulating cortical excitability by depolarizing or hyperpolarizing neuronal resting membrane potentials of different brain areas involved in cortico-subcortical loops [84].

The currents penetrate the skull and enter the brain from the anode, travel through the tissues and exit via the cathode. The anodal electrode generally increases cortical excitability and the cathodal electrode decreases. As far as side effects are concerned, tDCS was generally safe and well tolerated. There were no major clinical or cognitive side effects. The only side effects in a minority of patients were mild headache, facial pain, local tingling or itching and redness of the skin at the site of stimulation spontaneously resolved. The technique is relatively safe with side effects limited to local tingling and skin irritation [73, 85–87].

7.3.1.3 Treatment Regimen

Common stimulation targets include SMA, OFC, and pre-SMA. In a RCT exploring the use of tDCS as an adjunctive treatment, the cathode was placed over SMA, while the anode was positioned over the left deltoid in patients with treatment-resistant OCD [88]; in a RCT by Bation et al., the cathode was placed over the OFC and the anode over the right cerebellum [89] for treatment-resistant OCD; and as mentioned in the protocol by Green et al., participants received 2 mA anodal stimulation over the pre-SMA with the cathode positioned over the OFC for general OCD patients [90]. Typically, tDCS employs a low-amplitude electric current. Most studies utilize intensities spanning from 1 to 3 mA, while in numerous other studies, an intensity of around 2 mA has been adopted. In several studies, tDCS sessions were administered daily or on alternate days. For instance, in the case series by Basu et al., patients received tDCS for 20 min, for six sessions (one session per day) [91]; in the study by Ghina Harika-Germaneau et al., 21 treatment-resistant OCD outpatients received 10 sessions of tDCS with each treatment session consisting of 2 mA stimuli for 30 min [92]; and in the randomized sham-controlled trial, 43 patients with treatment-resistant OCD underwent 30 min active or sham tDCS sessions on 20 consecutive weekdays (4 weeks). The duration of tDCS sessions varies across studies, generally ranging from 20 to 30 min. Also, Different studies have employed different numbers of treatment sequences. In the case series by Basu et al., patients received six sessions of tDCS [91]; in the study by Harika-Germaneau et al., patients received 10 sessions of tDCS [92]; and in the randomized sham-controlled trial by Silva et al., patients underwent 20 sessions of tDCS over 4 weeks [88].

In conclusion, the treatment regimen for tDCS in OCD depends on various factors such as the patient's characteristics, the severity of OCD symptoms, and the response to previous treatments. However, common elements include targeting areas such as the SMA, OFC, and pre-SMA, using stimulation intensities of around 1–3 milliamperes, administering sessions with frequencies ranging from daily to alternate days, and session durations of 20–30 min. The number of treatment sequences also varies, but typically ranges from a few sessions to several weeks of treatment. Further research is needed to optimize and standardize the treatment regimen for tDCS in OCD to enhance its efficacy and reliability.

7.3.1.4 Clinical Recommendations

In 2020, the latest clinical guidelines on the application of tDCS in neurological and psychiatric disorders were published by Oxford University Press on behalf of the International Society for Neuropsychopharmacology [93].

Due to the limited research on OCD, this clinical guideline included some non-randomized controlled trials (except for two Class III crossover randomized controlled trials in OCD, the rest were all Class IV studies), and all the subjects were patients resistant to drug treatment, using tDCS as an adjunctive treatment. As for the recommendation and relevant basis based on the research, the recommendation is that anodal pre-SMA tDCS may be effective in improving OCD symptoms (Level C). The basis is that one Class II study and multiple Class IV studies showed that pre-SMA/SMA tDCS has certain potential, but the impact of polarity is unclear. For example, in one Class II study, the Y-BOCS score of patients was significantly improved after anodal pre-SMA tDCS treatment, and in the open-label extension stage, the scores of patients who did not reach the remission standard were further improved; in a case series study, by stimulating the pre-SMA and SMA with a specific anode, the Y-BOCS score, depression and anxiety symptoms of patients were improved, and the effect could last for 1–2 months. The Functional Magnetic Resonance Imaging (fMRI) of one patient showed an increase in the activity of the left SMA. However, there are also cases with different results. For example, in two cases of SMA cathodal tDCS treatment, one case had a mild and delayed response, and the other case improved after 20 sessions of treatment, and the Y-BOCS score increased by 45% 6 months later; in one case, the Y-BOCS score of the patient worsened after receiving anodal pre-SMA tDCS treatment, and improved after changing to cathodal tDCS; there are also some studies using different stimulation sites or methods with different results. Regarding the limitations and future directions of the research, the current research has a small sample size and a low evidence level.

tDCS has shown some efficacy in the treatment of obsessive-compulsive disorder, but there is heterogeneity in studies, and further research is needed to determine the optimal stimulation parameters and clinical application. In the future, more high-quality RCTs should be carried out to optimize the parameters of tDCS to improve its efficacy and safety in clinical treatment.

7.3.2 HD-tDCS

7.3.2.1 Efficacy

tDCS induces long-lasting NMDA receptor-dependent cortical plasticity via persistent subthreshold polarization of neuronal membranes. Conventional bipolar tDCS is applied with two large (35 cm^2) rectangular electrodes, however, this type of stimulation may lead to dispersed current passing through the brain cortex, and the area exhibiting the most significant current density might not correspond directly beneath the electrode. This can lead to imprecise targeting of the stimulated brain area, reducing the experiment's repeatability and reliability.

Therefore, a newly designed 4 × 1 High-definition transcranial direct current stimulation (HD-tDCS) protocol was proposed for more focal stimulation according to the results of computational modeling. HD-tDCS utilizes small disc electrodes deployed in 4 × 1 ring configuration whereby the physiological effects of the induced electric field are thought to be grossly constrained to the cortical area circumscribed by the ring. It provides a more precise targeting, with better sustainability, and longer-lasting effects [94].

So far, studies using HD-tDCS for the treatment of OCD mostly consists of case report [95, 96], and many studies have included patients with comorbid conditions. Only an open-label study on HD-tDCS for OCD involved 24 patients who received 1.5 mA HD-tDCS stimulation for 20 min each day, totalling 10 sessions. The result of this study demonstrated significant reductions in Y-BOCS, BDI-II, and BAI scores after the stimulation [86]. Recently, Wang et al. conducted a randomized controlled trial on the treatment of OCD with HD-tDCS [97]. The results indicated that HD-tDCS was acceptable and safe. However, compared with the sham stimulation group, it did not demonstrate significant clinical effectiveness in the treatment of moderate to severe OCD. HD-tDCS has shown certain potential in the treatment of OCD. However, there are still some uncertainties regarding its efficacy at present. More RCTs are still needed to clarify the relevant efficacy.

7.3.2.2 Safety

Most clinical trials exploring tDCS effect for OCD [98] have reported mild adverse effects such as redness, itching, and tingling at the stimulation site. However, no serious adverse effects have been reported, and there were no observed mood changes, such as manic episodes, among the participants. Wang et al.'s RCT study on HD-tDCS in the treatment of OCD showed similar results. A total of 47 subjects were included in the experiment [97], and no participants withdrew from the trial due to adverse reactions or reported adverse events. In those trials, adverse effects reported in the active and sham stimulation groups often show no statistically significant differences. This, to some extent, indicates that tDCS treatment for OCD is generally well-tolerated and safe, offering prospects and feasibility for future clinical use. There is limited research on the use of HD-tDCS for OCD, and no serious adverse effects have been reported so far, suggesting a certain level of safety regarding this stimulation device. However, further research is needed to confirm this.

7.3.2.3 Treatment Regimen

For stimulation targets, the Pre-SMA has been chosen in studies like that by Narayanaswamy et al. where 14 patients with OCD having persistent symptoms despite adequate and stable treatment with SSRIs were administered HD-tDCS with anodal 2 mA over the right Pre-SMA [83]. Another study by Cinosi et al. targeted the OFC and the SMA [99]. Commonly, a stimulation intensity of 2 mA is used as in the above-mentioned studies, but different intensities might be explored as in the pilot studies. Santos et al. tested 3 × 2 min HD-tDCS3x1 at 1, 2 and 3 mA; with anode center, cathode center, or sham stimulation, in random order across days [100]. The stimulation frequency for most studies was 20 min twice a day for 5 consecutive days. In some case series, patients received two 20-min sessions with an intersession gap of 20 min [96]. Generally, sessions last around 20 min, but the total treatment time can vary depending on the number of sessions and the duration

Table 7.2 tDCS in obsessive-compulsive disorder

tDCS	Clinical recommendations					Levels + references
	Anode position	Cathode position	Intensity (mA)	Length (min)	Duration (sessions)	
Conventional tDCS	Pre-SMA	SO	≥0.571	≥20	10–22	Possibly effective (level C) [78]
HD-tDCS	Not known	Not known	Not known	Not known	Not known	Not known

The CINP Guidelines mention Conventional tDCS as possibly effective (Level C) in improving OCD symptoms
Abbreviations: *pre-SMA* pre-supplementary motor area, *SO* supraorbital

of the treatment sequence. The number of treatment sequences can vary. In some cases, a single course of treatment may be sufficient, while in others, multiple courses may be required [96].

In conclusion, the treatment regimen for HD-tDCS in OCD is still being explored and refined. Different studies have used various stimulation targets, intensities, frequencies, times, and numbers of treatment sequences. Further research is needed to determine the most effective and standardized treatment regimen for HD-tDCS in OCD.

7.3.2.4 Clinical Recommendations

At present, the stimulation parameters of HD-tDCS in the treatment of OCD have not been fully determined. Available practice guidelines for treatment of OCD currently do not recommend HD-tDCS [68]. Future studies are needed to further optimize the stimulation parameters, such as current intensity, stimulation duration, and stimulation number, to improve the therapeutic effect.

In conclusion, HD-tDCS, as a noninvasive brain stimulation technique, has some potential in the treatment of OCD. However, the current research on HD-tDCS in the treatment of OCD is still in its preliminary stage, and further studies are needed to optimize stimulation parameters, identify the best stimulation targets, explore combined treatment methods, and evaluate long-term efficacy and safety (Table 7.2).

7.4 TMS vs. tDCS

7.4.1 Efficacy

Currently, no studies directly compare the efficacy of TMS and tDCS in treating OCD. However, the evidence supporting TMS, particularly rTMS, is relatively robust. In contrast, while tDCS has shown some promise in treating OCD, the available research is limited by a smaller number of studies, inconsistent experimental protocols, and findings of limited significance [101]. To better evaluate the therapeutic potential of tDCS, more high-quality, randomized, sham-controlled trials are needed in the future.

7.4.2 Safety

The adverse effects of TMS and tDCS are as described above. Relatively more adverse effects occur with TMS than tDCS, but both are mild and short-lived. Overall, both TMS and tDCS are safe and well tolerated, with few serious adverse events.

7.5 Conclusion

Neuromodulation represents a promising therapeutic approach for managing OCD, including treatment-resistant cases, offering a generally favorable safety and tolerability profile. Among noninvasive techniques, rTMS targeting the DLPFC and pre-SMA has demonstrated robust efficacy, supported by strong clinical evidence. While dTMS shows some potential, additional high-quality studies are necessary to establish its effectiveness more conclusively. Evidence for iTBS/cTBS in OCD treatment remains limited, reflecting a need for further research in this area. tDCS is a notably safe and potentially effective intervention, particularly when targeting the SMA, pre-SMA, and OFC with cathodal placement. However, the small sample sizes and scarcity of randomized controlled trials limit the generalizability of current findings. For invasive neuromodulation, DBS offers considerable promise for refractory OCD, yet the optimization of protocols and broader consensus on its use require further investigation. In summary, while neuromodulation holds significant potential for OCD treatment, ongoing research is essential to refine techniques, validate efficacy, and expand clinical applications.

Acknowledgments The authors thank all the patient volunteers and their families who participated in the literature for their support of the scientific research.

Disclosure/Conflicts of Interest The authors of this chapter do not have any commercial associations that might pose a conflict of interest in connection with this manuscript.

Financial Support We gratefully acknowledge funding provided by the following National Natural Science Foundation of China, China (Nos.: 81801347; 81971597), Natural Science Foundation of Guangdong Province, China (No.: 2021A1515011034), Planned Science and Technology Project of Guangdong Province, China (No.: 2017B020227011), Jinan University Central University project (No.: 11624320).

References

1. Foa EB, Kozak MJ, Goodman WK, Hollander E, Jenike MA, Rasmussen SA. DSM-IV field trial: obsessive-compulsive disorder. Am J Psychiatry. 1995;152:90–6.

2. Shavitt RG, de Mathis MA, Oki F, Ferrao YA, Fontenelle LF, Torres AR, Diniz JB, Costa DL, do Rosário MC, Hoexter MQ, Miguel EC, Simpson HB. Phenomenology of OCD: lessons from a large multicenter study and implications for ICD-11. J Psychiatr Res. 2014;57:141–8.
3. Ruscio AM, Stein DJ, Chiu WT, Kessler RC. The epidemiology of obsessive-compulsive disorder in the National Comorbidity Survey Replication. Mol Psychiatry. 2010;15:53–63.
4. Fontenelle LF, Mendlowicz MV, Versiani M. The descriptive epidemiology of obsessive-compulsive disorder. Prog Neuro-Psychopharmacol Biol Psychiatry. 2006;30:327–37.
5. Macy AS, Theo JN, Kaufmann SCV, Ghazzaoui RB, Pawlowski PA, Fakhry HI, Cassmassi BJ, IsHak WW. Quality of life in obsessive compulsive disorder. CNS Spectr. 2013;18:21–33.
6. Meier SM, Mattheisen M, Mors O, Schendel DE, Mortensen PB, Plessen KJ. Mortality among persons with obsessive-compulsive disorder in Denmark. JAMA Psychiatry. 2016;73:268–74.
7. Hirschtritt ME, Bloch MH, Mathews CA. Obsessive-compulsive disorder: advances in diagnosis and treatment. JAMA. 2017;317:1358–67.
8. Fitzsimmons S, van der Werf YD, van Campen AD, Arns M, Sack AT, Hoogendoorn AW, other members of the TETRO Consortium, van den Heuvel OA. Repetitive transcranial magnetic stimulation for obsessive-compulsive disorder: a systematic review and pairwise/network meta-analysis. J Affect Disord. 2022;302:302–12.
9. Wang J, Hua G, Wang S, Guo G, Quan D, Yao S, Zheng H. Glutamatergic neurotransmission is affected by low-frequency repetitive transcranial magnetic stimulation over the supplemental motor cortex of patients with obsessive-compulsive disorder. J Affect Disord. 2023;325:762–9.
10. Elbeh KAM, Elserogy YMB, Khalifa HE, Ahmed MA, Hafez MH, Khedr EM. Repetitive transcranial magnetic stimulation in the treatment of obsessive-compulsive disorders: double blind randomized clinical trial. Psychiatry Res. 2016;238:264–9.
11. Mansur CG, Myczkowki ML, de Barros Cabral S, Sartorelli Mdo C, Bellini BB, Dias AM, Bernik MA, Marcolin MA. Placebo effect after prefrontal magnetic stimulation in the treatment of resistant obsessive-compulsive disorder: a randomized controlled trial. Int J Neuropsychopharmacol. 2011;14:1389–97.
12. Rossi S, Hallett M, Rossini PM, Pascual-Leone A, Safety of TMS Consensus Group. Safety, ethical considerations, and application guidelines for the use of transcranial magnetic stimulation in clinical practice and research. Clin Neurophysiol. 2009;120:2008–39.
13. Acevedo N, Bosanac P, Pikoos T, Rossell S, Castle D. Therapeutic neurostimulation in obsessive-compulsive and related disorders: a systematic review. Brain Sci. 2021;11:948.
14. Perera MPN, Mallawaarachchi S, Miljevic A, Bailey NW, Herring SE, Fitzgerald PB. Repetitive transcranial magnetic stimulation for obsessive-compulsive disorder: a meta-analysis of randomized, sham-controlled trials. Biol Psychiatry Cogn Neurosci Neuroimaging. 2021;6:947–60.
15. Sachdev PS, McBride R, Loo CK, Mitchell PB, Malhi GS, Croker VM. Right versus left prefrontal transcranial magnetic stimulation for obsessive-compulsive disorder: a preliminary investigation. J Clin Psychiatry. 2001;62:981–4.
16. Liang K, Li H, Bu X, Li X, Cao L, Liu J, Gao Y, Li B, Qiu C, Bao W, Zhang S, Hu X, Xing H, Gong Q, Huang X. Efficacy and tolerability of repetitive transcranial magnetic stimulation for the treatment of obsessive-compulsive disorder in adults: a systematic review and network meta-analysis. Transl Psychiatry. 2021;11:332.
17. Mantovani A, Lisanby SH, Pieraccini F, Ulivelli M, Castrogiovanni P, Rossi S. Repetitive transcranial magnetic stimulation (rTMS) in the treatment of obsessive-compulsive disorder (OCD) and Tourette's syndrome (TS). Int J Neuropsychopharmacol. 2006;9:95–100.
18. Adu MK, Eboreime E, Sapara AO, Greenshaw AJ, Chue P, Agyapong VIO. The use of repetitive transcranial magnetic stimulation for treatment of obsessive-compulsive disorder: a scoping review. Ment Illn. 2021;13:1–13.
19. Khedr EM, Elbeh K, Saber M, Abdelrady Z, Abdelwarith A. A double blind randomized clinical trial of the effectiveness of low frequency rTMS over right DLPFC or OFC for treatment of obsessive-compulsive disorder. J Psychiatr Res. 2022;156:122–31.

20. Ruffini C, Locatelli M, Lucca A, Benedetti F, Insacco C, Smeraldi E. Augmentation effect of repetitive transcranial magnetic stimulation over the orbitofrontal cortex in drug-resistant obsessive-compulsive disorder patients: a controlled investigation. Prim Care Companion J Clin Psychiatry. 2009;11:226–30.
21. George MS. Whither TMS: a one-trick pony or the beginning of a neuroscientific revolution? Am J Psychiatry. 2019;176:904–10.
22. Dougherty DD, Brennan BP, Stewart SE, Wilhelm S, Widge AS, Rauch SL. Neuroscientifically informed formulation and treatment planning for patients with obsessive-compulsive disorder: a review. JAMA Psychiatry. 2018;75:1081–7.
23. Fitzgerald PB, Hoy KE, Elliot D, Susan McQueen RN, Wambeek LE, Daskalakis ZJ. Accelerated repetitive transcranial magnetic stimulation in the treatment of depression. Neuropsychopharmacology. 2018;43:1565–72.
24. Carmi L, Tendler A, Bystritsky A, Hollander E, Blumberger DM, Daskalakis J, Ward H, Lapidus K, Goodman W, Casuto L, Feifel D, Barnea-Ygael N, Roth Y, Zangen A, Zohar J. Efficacy and safety of deep transcranial magnetic stimulation for obsessive-compulsive disorder: a prospective multicenter randomized double-blind placebo-controlled trial. Am J Psychiatry. 2019;176:931–8.
25. Smárason O, Boedeker PJ, Guzick AG, Tendler A, Sheth SA, Goodman WK, Storch EA. Depressive symptoms during deep transcranial magnetic stimulation or sham treatment for obsessive-compulsive disorder. J Affect Disord. 2024;344:466–72.
26. Laskov O, Klírová M. Effects of deep transcranial magnetic stimulation (dTMS) on cognition. Neurosci Lett. 2021;755:135906.
27. Lusicic A, Schruers KRJ, Pallanti S, Castle DJ. Transcranial magnetic stimulation in the treatment of obsessive–compulsive disorder: current perspectives. Neuropsychiatr Dis Treat. 2018;14:1721–36.
28. Carmi L, Alyagon U, Barnea-Ygael N, Zohar J, Dar R, Zangen A. Clinical and electrophysiological outcomes of deep TMS over the medial prefrontal and anterior cingulate cortices in OCD patients. Brain Stimul. 2018;11:158–65.
29. Roth Y, Tendler A, Arikan MK, Vidrine R, Kent D, Muir O, MacMillan C, Casuto L, Grammer G, Sauve W, Tolin K, Harvey S, Borst M, Rifkin R, Sheth M, Cornejo B, Rodriguez R, Shakir S, Porter T, Kim D, Peterson B, Swofford J, Roe B, Sinclair R, Harmelech T, Zangen A. Real-world efficacy of deep TMS for obsessive-compulsive disorder: post-marketing data collected from twenty-two clinical sites. J Psychiatr Res. 2021;137:667–72.
30. Brunoni AR, Chaimani A, Moffa AH, Razza LB, Gattaz WF, Daskalakis ZJ, Carvalho AF. Repetitive transcranial magnetic stimulation for the acute treatment of major depressive episodes. JAMA Psychiatry. 2017;74:143–52.
31. Chung CLH, Mak MKY, Hallett M. Transcranial magnetic stimulation promotes gait training in Parkinson disease. Ann Neurol. 2020;88:933–45.
32. Zrenner C, Desideri D, Belardinelli P, Ziemann U. Real-time EEG-defined excitability states determine efficacy of TMS-induced plasticity in human motor cortex. Brain Stimul. 2018;11:374–89.
33. Zrenner B, Zrenner C, Gordon PC, Belardinelli P, McDermott EJ, Soekadar SR, Fallgatter AJ, Ziemann U, Müller-Dahlhaus F. Brain oscillation-synchronized stimulation of the left dorsolateral prefrontal cortex in depression using real-time EEG-triggered TMS. Brain Stimul. 2020;13:197–205.
34. George MS, Huffman S, Doose J, Sun X, Dancy M, Faller J, Li X, Yuan H, Goldman RI, Sajda P, Brown TR. EEG synchronized left prefrontal transcranial magnetic stimulation (TMS) for treatment resistant depression is feasible and produces an entrainment dependent clinical response: a randomized controlled double blind clinical trial. Brain Stimul. 2023;16:1753–63.
35. Huang YZ, Edwards MJ, Rounis E, Bhatia KP, Rothwell JC. Theta burst stimulation of the human motor cortex. Neuron. 2005;45:201–6.

36. Suppa A, Huang YZ, Funke K, Ridding MC, Cheeran B, Di Lazzaro V, Ziemann U, Rothwell JC. Ten years of theta burst stimulation in humans: established knowledge, unknowns and prospects. Brain Stimul. 2016;9:323–35.
37. Rounis E, Huang YZ. Theta burst stimulation in humans: a need for better understanding effects of brain stimulation in health and disease. Exp Brain Res. 2020;238:1707–14.
38. Acevedo N, Castle D, Rossell S. The promise and challenges of transcranial magnetic stimulation and deep brain stimulation as therapeutic options for obsessive-compulsive disorder. Expert Rev Neurother. 2024;24:145–58.
39. Harika-Germaneau G, Rachid F, Chatard A, Lafay-Chebassier C, Solinas M, Thirioux B, Millet B, Langbour N, Jaafari N. Continuous theta burst stimulation over the supplementary motor area in refractory obsessive-compulsive disorder treatment: a randomized sham-controlled trial. Brain Stimul. 2019;12:1565–71.
40. Dutta P, Dhyani M, Garg S, Tikka SK, Khattri S, Mehta S, Mishra J. Efficacy of intensive orbitofrontal continuous theta burst stimulation (iOFcTBS) in obsessive compulsive disorder: a randomized placebo controlled study. Psychiatry Res. 2021;298:113784. https://doi.org/10.1016/j.psychres.2021.113784. Epub 2021 Feb 4. PMID: 33582525.
41. Naro A, Billeri L, Cannavo A, De Luca R, Portaro S, Bramanti P, Calabro RS. Theta burst stimulation for the treatment of obsessive-compulsive disorder: a pilot study. J Neural Transm (Vienna). 2019;126:1667–77.
42. Guo Q, Wang K, Han H, Li P, Cheng J, Zhu J, Wang Z, Fan Q. Continuous theta burst stimulation over the bilateral supplementary motor area in obsessive-compulsive disorder treatment: a clinical randomized single-blind sham-controlled trial. Eur Psychiatry. 2022;65:e64.
43. Liu W, Shao H, Liao J, Yang D, Ma M, Yang J. Continuous theta-burst stimulation over the right orbitofrontal cortex in treatment-resistant obsessive-compulsive disorder treatment: a randomized sham-controlled trial. Int J Gen Med. 2021;14:3109–18.
44. Mukherjee A, Kumre PK, Goyal N, Khanra S. Adjunctive neuronavigated accelerated continuous theta-burst stimulation in obsessive-compulsive disorder: a randomized sham-controlled study. CNS Spectr. 2023;28:495–504. https://doi.org/10.1017/S1092852922000980. Epub 2022 Sept 5. PMID: 36059099.
45. Jiang J, Wan K, Liu Y, Tang Y, Tang W, Liu J, Ma J, Xue C, Chen L, Qian H, Liu D, Shen X, Fan R, Wang Y, Wang K, Ji G, Zhu C. A controlled clinical study of accelerated high-dose theta burst stimulation in patients with obsessive-compulsive disorder. Neural Plast. 2023;2023:2741287.
46. Li P, Cheng J, Fan Q, Lin L, Zhou S, Gao J, Tang Y, Yuan T, Wang Z. The functional connectivity predictor of therapeutic effect of continuous theta burst stimulation on obsessive-compulsive disorder: a preliminary study. J Affect Disord. 2022;311:231–8.
47. Noda Y, Fujii K, Tokura F, Nakajima S, Kitahata R. A case series of continuous theta burst stimulation treatment for the supplementary motor area twice a day in patients with obsessive-compulsive disorder: a real world TMS registry study in Japan. J Pers Med. 2023;13:875.
48. Syed FA, Naik SS, Arumugham SS, Mehta UM, Thirthalli J, Reddy YCJ. Adjuvant intermittent theta burst stimulation over dorsomedial prefrontal cortex in treatment-resistant obsessive-compulsive disorder type: letter to the editor. Brain Stimul. 2021;14:74–6.
49. Stubbeman WF, Yang J, Converse J, Gencosmanoglu M, Morales Ortega D, Morris J, Sobocinski A, Li V, Gunawardane G, Edelen Y, Khairkhah R, Perez J. Neuronavigated right orbitofrontal 20 Hz theta burst transcranial magnetic stimulation augmentation for obsessive-compulsive disorder with comorbid depression and anxiety disorders: an open-label study. Brain Sci. 2024;14:483.
50. Rapinesi C, Kotzalidis GD, Ferracuti S, Sani G, Girardi P, Del Casale A. Brain stimulation in obsessive-compulsive disorder (OCD): a systematic review. Curr Neuropharmacol. 2019;17:787–807.
51. Di Lazzaro V, Pilato F, Dileone M, Profice P, Oliviero A, Mazzone P, Insola A, Ranieri F, Meglio M, Tonali PA, Rothwell JC. The physiological basis of the effects of intermittent theta burst stimulation of the human motor cortex. J Physiol. 2008;586:3871–9.

52. Lisanby SH, Gutman D, Luber B, Schroeder C, Sackeim HA. Sham TMS: intracerebral measurement of the induced electrical field and the induction of motor-evoked potentials. Biol Psychiatry. 2001;49:460–3.
53. Neuteboom D, Zantvoord JB, Goya-Maldonado R, Wilkening J, Dols A, van Exel E, Lok A, de Haan L, Scheepstra KWF. Accelerated intermittent theta burst stimulation in major depressive disorder: a systematic review. Psychiatry Res. 2023;327:115429.
54. Chou PH, Sack AT, Su KP. Targeting three brain regions (bilateral SMA, left and right DLPFC) sequentially in one session using combined repetitive transcranial magnetic stimulation and intermittent theta-burst stimulation in treatment-refractory obsessive-compulsive disorder: a case report. Clin Psychopharmacol Neurosci. 2022;20:773–6.
55. Cocchi L, Zalesky A, Nott Z, Whybird G, Fitzgerald PB, Breakspear M. Transcranial magnetic stimulation in obsessive-compulsive disorder: a focus on network mechanisms and state dependence. Neuroimage Clin. 2018;19:661–74.
56. Dell'Osso B, Cremaschi L, Oldani L, Altamura AC. New directions in the use of brain stimulation interventions in patients with obsessive-compulsive disorder. Curr Med Chem. 2018;25:5712–21.
57. Baeken C, Wu GR, van Heeringen K. Placebo aiTBS attenuates suicidal ideation and frontopolar cortical perfusion in major depression. Transl Psychiatry. 2019;9:38.
58. Sheline YI, Makhoul W, Batzdorf AS, Nitchie FJ, Lynch KG, Cash R, Balderston NL. Accelerated intermittent theta-burst stimulation and treatment-refractory bipolar depression: a randomized clinical trial. JAMA Psychiatry. 2024;81:936–41.
59. Lefaucheur JP, Aleman A, Baeken C, Benninger DH, Brunelin J, Di Lazzaro V, Filipovic SR, Grefkes C, Hasan A, Hummel FC, Jaaskelainen SK, Langguth B, Leocani L, Londero A, Nardone R, Nguyen JP, Nyffeler T, Oliveira-Maia AJ, Oliviero A, Padberg F, Palm U, Paulus W, Poulet E, Quartarone A, Rachid F, Rektorova I, Rossi S, Sahlsten H, Schecklmann M, Szekely D, Ziemann U. Evidence-based guidelines on the therapeutic use of repetitive transcranial magnetic stimulation (rTMS): an update (2014–2018). Clin Neurophysiol. 2020;131:474–528.
60. Price RB, Gillan CM, Hanlon C, Ferrarelli F, Kim T, Karim HT, Renard M, Kaskie R, Degutis M, Wears A, Vienneau EP, Peterchev AV, Brown V, Siegle GJ, Wallace ML, Ahmari SE. Effect of experimental manipulation of the orbitofrontal cortex on short-term markers of compulsive behavior: a theta burst stimulation study. Am J Psychiatry. 2021;178:459–68.
61. Zhou S, Fang Y. Efficacy of non-invasive brain stimulation for refractory obsessive-compulsive disorder: a meta-analysis of randomized controlled trials. Brain Sci. 2022;12:943.
62. Cole EJ, Stimpson KH, Bentzley BS, Gulser M, Cherian K, Tischler C, Nejad R, Pankow H, Choi E, Aaron H, Espil FM, Pannu J, Xiao X, Duvio D, Solvason HB, Hawkins J, Guerra A, Jo B, Raj KS, Phillips AL, Barmak F, Bishop JH, Coetzee JP, DeBattista C, Keller J, Schatzberg AF, Sudheimer KD, Williams NR. Stanford accelerated intelligent neuromodulation therapy for treatment-resistant depression. Am J Psychiatry. 2020;177:716–26.
63. Zhang Y, Suo X, Pan L, Wang X, Xu W, Xu J, Gao J. Dual stimulation with iTBS-cTBS treatment in postpartum obsessive-compulsive disorder: a case report. Int J Neurosci. 2024:1–4. https://doi.org/10.1080/00207454.2024.2396417.
64. Plewnia C, Brendel B, Schwippel T, Nieratschker V, Ethofer T, Kammer T, Padberg F, Martus P, Fallgatter AJ. Treatment of major depressive disorder with bilateral theta burst stimulation: study protocol for a randomized, double-blind, placebo-controlled multicenter trial (TBS-D). Eur Arch Psychiatry Clin Neurosci. 2021;271:1231–43.
65. Dos Santos-Ribeiro S, de Salles Andrade JB, Quintas JN, Baptista KB, Moreira-de-Oliveira ME, Yucel M, Fontenelle LF. A systematic review of the utility of electroconvulsive therapy in broadly defined obsessive-compulsive-related disorders. Prim Care Companion CNS Disord. 2018;20:18r02342.
66. Lee WH, Lisanby SH, Laine AF, Peterchev AV. Comparison of electric field strength and spatial distribution of electroconvulsive therapy and magnetic seizure therapy in a realistic human head model. Eur Psychiatry. 2016;36:55–64.

67. McGovern RA, Sheth SA. Role of the dorsal anterior cingulate cortex in obsessive-compulsive disorder: converging evidence from cognitive neuroscience and psychiatric neurosurgery. J Neurosurg. 2017;126:132–47.
68. Katzman MA, Bleau P, Blier P, Chokka P, Kjernisted K, Van Ameringen M, Canadian Anxiety Guidelines Initiative Group on behalf of the Anxiety Disorders Association of Canada/Association Canadienne des troubles anxieux and McGill University, Antony MM, Bouchard S, Brunet A, Flament M, Grigoriadis S, Mendlowitz S, O'Connor K, Rabheru K, Richter PM, Robichaud M, Walker JR. Canadian clinical practice guidelines for the management of anxiety, posttraumatic stress and obsessive-compulsive disorders. BMC Psychiatry. 2014;14(Suppl 1):S1.
69. Tang VM, Blumberger DM, Weissman CR, Dimitrova J, Throop A, McClintock SM, Voineskos D, Rajji TK, Downar J, Knyahnytska Y, Mulsant BH, Fitzgerald PB, Daskalakis ZJ. A pilot study of magnetic seizure therapy for treatment-resistant obsessive-compulsive disorder. Depress Anxiety. 2021;38:161–71.
70. Cretaz E, Brunoni AR, Lafer B. Magnetic seizure therapy for unipolar and bipolar depression: a systematic review. Neural Plast. 2015;2015:521398.
71. Daskalakis ZJ, McClintock SM, Hadas I, Kallioniemi E, Zomorrodi R, Throop A, Palmer L, Farzan F, Thorpe KE, Tamminga C, Blumberger DM. Confirmatory efficacy and safety trial of magnetic seizure therapy for depression (CREST-MST): protocol for identification of novel biomarkers via neurophysiology. Trials. 2021;22:906.
72. Wu H, Jiang J, Cao X, Wang J, Li C. Magnetic seizure therapy for people with schizophrenia. Cochrane Database Syst Rev. 2023;6:CD012697.
73. Fontenelle LF, Coutinho ES, Lins-Martins NM, Fitzgerald PB, Fujiwara H, Yucel M. Electroconvulsive therapy for obsessive-compulsive disorder: a systematic review. J Clin Psychiatry. 2015;76:949–57.
74. Deng ZD, Lisanby SH, Peterchev AV. Electric field strength and focality in electroconvulsive therapy and magnetic seizure therapy: a finite element simulation study. J Neural Eng. 2011;8:016007.
75. Nauczyciel C, Le Jeune F, Naudet F, Douabin S, Esquevin A, Vérin M, Dondaine T, Robert G, Drapier D, Millet B. Repetitive transcranial magnetic stimulation over the orbitofrontal cortex for obsessive-compulsive disorder: a double-blind, crossover study. Transl Psychiatry. 2014 ;4(9):e436. https://doi.org/10.1038/tp.2014.62. PMID: 25203167.
76. Meek BP, Fotros A, Abo Aoun M, Modirrousta M. Improvements in error-monitoring and symptoms following low-frequency rTMS of dorsal anterior cingulate cortex in obsessive compulsive disorder; a randomized, sham-controlled study. Brain Cogn. 2021;154:105809. https://doi.org/10.1016/j.bandc.2021.105809. Epub 2021 Oct 4. PMID: 34619574.
77. Ozer U, Yucens B, Tumkaya S. Efficacy of accelerated deep transcranial magnetic stimulation wíth double cone coíl in obsessive-compulsive disorder: a double-blind, placebo-controlled study. J Psychiatr Res. 2024;171:325–31. https://doi.org/10.1016/j.jpsychires.2024.02.005. Epub 2024 Feb 2. Erratum in: J Psychiatr Res. 2025 Jan;181:647. https://doi.org/10.1016/j.jpsychires.2024.12.017. PMID: 38342033.
78. Keramatian K, Chithra NK, Yatham LN. The CANMAT and ISBD guidelines for the treatment of bipolar disorder: summary and a 2023 update of evidence. Focus (Am Psychiatr Publ). 2023;21:344–53.
79. Wagner T, Valero-Cabre A, Pascual-Leone A. Noninvasive human brain stimulation. Annu Rev Biomed Eng. 2007;9:527–65.
80. Paulus W. Transcranial electrical stimulation (tES – tDCS; tRNS, tACS) methods. Neuropsychol Rehabil. 2011;21:602–17.
81. Senco NM, Huang Y, D'Urso G, Parra LC, Bikson M, Mantovani A, Shavitt RG, Hoexter MQ, Miguel EC, Brunoni AR. Transcranial direct current stimulation in obsessive-compulsive disorder: emerging clinical evidence and considerations for optimal montage of electrodes. Expert Rev Med Devices. 2015;12:381–91.

82. Volpato C, Piccione F, Cavinato M, Duzzi D, Schiff S, Foscolo L, Venneri A. Modulation of affective symptoms and resting state activity by brain stimulation in a treatment-resistant case of obsessive-compulsive disorder. Neurocase. 2013;19:360–70.
83. Gowda SM, Narayanaswamy JC, Hazari N, Bose A, Chhabra H, Balachander S, Bhaskarapillai B, Shivakumar V, Venkatasubramanian G, Reddy YCJ. Efficacy of pre-supplementary motor area transcranial direct current stimulation for treatment resistant obsessive compulsive disorder: a randomized, double blinded, sham controlled trial. Brain Stimul. 2019;12:922–9.
84. Keeser D, Meindl T, Bor J, Palm U, Pogarell O, Mulert C, Brunelin J, Moller HJ, Reiser M, Padberg F. Prefrontal transcranial direct current stimulation changes connectivity of resting-state networks during fMRI. J Neurosci. 2011;31:15284–93.
85. Bikson M, Grossman P, Thomas C, Zannou AL, Jiang J, Adnan T, Mourdoukoutas AP, Kronberg G, Truong D, Boggio P, Brunoni AR, Charvet L, Fregni F, Fritsch B, Gillick B, Hamilton RH, Hampstead BM, Jankord R, Kirton A, Knotkova H, Liebetanz D, Liu A, Loo C, Nitsche MA, Reis J, Richardson JD, Rotenberg A, Turkeltaub PE, Woods AJ. Safety of transcranial direct current stimulation: evidence based update 2016. Brain Stimul. 2016;9:641–61.
86. Cheng J, Li P, Tang Y, Zhang C, Lin L, Gao J, Wang Z. Transcranial direct current stimulation improve symptoms and modulates cortical inhibition in obsessive-compulsive disorder: a TMS-EEG study. J Affect Disord. 2022;298:558–64.
87. Sabe M, Hyde J, Cramer C, Eberhard A, Crippa A, Brunoni AR, Aleman A, Kaiser S, Baldwin DS, Garner M, Sentissi O, Fiedorowicz JG, Brandt V, Cortese S, Solmi M. Transcranial magnetic stimulation and transcranial direct current stimulation across mental disorders: a systematic review and dose-response meta-analysis. JAMA Netw Open. 2024;7:e2412616.
88. Silva R, Brunoni AR, Goerigk S, Batistuzzo MC, Costa D, Diniz JB, Padberg F, D'Urso G, Miguel EC, Shavitt RG. Efficacy and safety of transcranial direct current stimulation as an add-on treatment for obsessive-compulsive disorder: a randomized, sham-controlled trial. Neuropsychopharmacology. 2021;46:1028–34.
89. Bation R, Mondino M, Le Camus F, Saoud M, Brunelin J. Transcranial direct current stimulation in patients with obsessive compulsive disorder: a randomized controlled trial. Eur Psychiatry. 2019;62:38–44.
90. Green PE, Loftus A, Anderson RA. Protocol for transcranial direct current stimulation for obsessive-compulsive disorder. Brain Sci. 2020;10:1008.
91. Basu I, Panda SP, Yadav P. Transcranial direct current stimulation as an augmentation therapy in patients with obsessive-compulsive disorder: a case series. Ind Psychiatry J. 2024;33:S226–8.
92. Harika-Germaneau G, Heit D, Chatard A, Thirioux B, Langbour N, Jaafari N. Treating refractory obsessive-compulsive disorder with transcranial direct current stimulation: an open label study. Brain Behav. 2020;10:e01648.
93. Fregni F, El-Hagrassy MM, Pacheco-Barrios K, Carvalho S, Leite J, Simis M, Brunelin J, Nakamura-Palacios EM, Marangolo P, Venkatasubramanian G, San-Juan D, Caumo W, Bikson M, Brunoni AR, Neuromodulation Center Working Group. Evidence-based guidelines and secondary meta-analysis for the use of transcranial direct current stimulation in neurological and psychiatric disorders. Int J Neuropsychopharmacol. 2021;24:256–313.
94. Kuo HI, Bikson M, Datta A, Minhas P, Paulus W, Kuo MF, Nitsche MA. Comparing cortical plasticity induced by conventional and high-definition 4 × 1 ring tDCS: a neurophysiological study. Brain Stimul. 2013;6:644–8.
95. Nayok SB, Pathak H, Suhas S, Thimmashetty VH, Shivaprakash P, Madhu Sudhan RM, Sreeraj VS, Narayanaswamy JC, Venkatasubramanian G. Concurrent conventional & high-definition transcranial direct current stimulation for treatment of schizophrenia with co-morbid obsessive-compulsive disorder: a case report. Brain Stimul. 2021;14:1483–5.
96. Parlikar R, Sreeraj VS, Chhabra H, Thimmashetty VH, Parameshwaran S, Selvaraj S, Shivakumar V, Narayanaswamy JC, Rao NP, Venkatasubramanian G. Add-on HD-tDCS for

obsessive-compulsive disorder with comorbid bipolar affective disorder: a case series. Asian J Psychiatr. 2019;43:87–90.

97. Wang Y, Cheng J, Ruan H, Zhang J, Zheng Z, Lin L, Gao J, Yuan C, Han H, Fan Q, Wang Z. Assessing the effectiveness of high-definition transcranial direct current stimulation for treating obsessive-compulsive disorder: results from a randomized, double-blind, controlled trial. Compr Psychiatry. 2024;135:152520.
98. Rachid F. Transcranial direct current stimulation for the treatment of obsessive-compulsive disorder? A qualitative review of safety and efficacy. Psychiatry Res. 2019;271:259–64.
99. Fineberg NA, Cinosi E, Smith MVA, Busby AD, Wellsted D, Huneke NTM, Garg K, Aslan IH, Enara A, Garner M, Gordon R, Hall N, Meron D, Robbins TW, Wyatt S, Pellegrini L, Baldwin DS. Feasibility, acceptability and practicality of transcranial stimulation in obsessive compulsive symptoms (FEATSOCS): a randomised controlled crossover trial. Compr Psychiatry. 2023;122:152371.
100. Santos TEG, Favoretto DB, Toostani IG, Nascimento DC, Rimoli BP, Bergonzoni E, Lemos TW, Truong DQ, Delbem ACB, Makkiabadi B, Moraes R, Louzada F, Bikson M, Leite JF, Edwards DJ. Manipulation of human verticality using high-definition transcranial direct current stimulation. Front Neurol. 2018;9:825.
101. Green PE, Loftus AM, Anderson RA. Transcranial direct current stimulation for obsessive compulsive disorder: a systematic review and CONSORT evaluation. J Int Neuropsychol Soc. 2024:1–12. https://doi.org/10.1017/S1355617724000602.

Attention Deficit Hyperactivity Disorder

8

Mengyue Gu, Jingyang Gu, Song Wang, Yudong Shi, and HuanZhong Liu

Abstract

Attention Deficit Hyperactivity Disorder (ADHD) is a prevalent neurodevelopmental condition marked by inattentiveness, hyperactive behaviors, and impulsivity. While pharmacological treatments provide clinical benefits for some patients, many individuals experience suboptimal therapeutic outcomes or adverse side effects. Emerging non-pharmacological approaches—particularly transcranial magnetic stimulation (TMS) and transcranial direct current stimulation (tDCS)—demonstrate promising potential for alleviating ADHD symptoms through targeted neuromodulation. These techniques may offer viable alternatives for medication-resistant patients or those seeking to minimize pharmaceutical interventions. Nevertheless, considerable variability persists across studies due to divergent treatment protocols (e.g., stimulation parameters, target brain regions), participant demographics, and intervention durations. This chapter systematically examines these critical sources of heterogeneity.

Keywords

ADHD · TMS · tDCS

M. Gu · J. Gu · S. Wang · Y. Shi · H. Liu (✉)
Department of Psychiatry, Chaohu Hospital of Anhui Medical University, AnHui, China
e-mail: huanzhongliu@ahmu.edu.cn

W. Zheng, Y. Ning (eds.), *TMS and tDCS for Psychiatric Disorders*,
https://doi.org/10.1007/978-981-96-8504-2_8

Abbreviations

ADHD	Attention Deficit Hyperactivity Disorder
ASD	Autism Spectrum Disorder
CS	Conditioning Stimulus
cTBS	Continuous iTBS
DLPFC	Dorsolateral Prefrontal Cortex
HF-Rtms	High-Frequency rTMS
iTBS	Intermittent TBS
LF-rTMS	Low-Frequency rTMS
MT	Motor Threshold
PPC	Posterior Parietal Cortex
ppTMS	Paired-pulse TMS
RMT	Resting Motor Threshold
SMA	Supplementary Motor Area
spTMS	Single-pulse TMS
TDC	Typically Developing Children
tDCS	Transcranial Direct Current Stimulation
TMS	Transcranial Magnetic Stimulation
vmPFC	Ventromedial Prefrontal Cortex

8.1 ADHD

8.1.1 Definition and Introduction

Attention-deficit/hyperactivity disorder (ADHD) is a common neurodevelopmental disorder characterized by symptoms of inattention, hyperactivity, and impulsivity. Typically beginning in childhood, ADHD can persist into adulthood. The global prevalence of ADHD in children and adolescents is approximately 5.29% [1], while the prevalence in early adulthood is around 2.5% [2], gradually declining to about 1% with age [3]. Males are more frequently diagnosed than females. The causes of ADHD are multifactorial, including genetic predispositions, environmental influences (e.g., maternal smoking during pregnancy, preterm birth), and neurophysiological abnormalities, such as dysfunctions in the prefrontal cortex [4, 5]. ADHD is associated with a range of cognitive deficits, particularly in executive functions [6]. Compared to unaffected control groups, individuals with ADHD show impairments in behavioral inhibition, working memory, attention, planning, and organizational skills, which contribute to poor academic performance, interpersonal relationship issues, and emotional challenges [7].

8.1.2 Therapeutic Method

The first-line treatment for ADHD includes central nervous system stimulants such as methylphenidate and amphetamines. Nonstimulant medications like norepinephrine reuptake inhibitor atomoxetine and α-2 adrenergic agonists guanfacine and clonidine are also options for children over 7 and adults. Multiple RCTs and meta-analyses have demonstrated significant short-term improvements in ADHD symptoms [8, 9]. However, in natural cohort studies (without the strict control of RCTs), long-term medication adherence declines, reducing the expected therapeutic effects [10]. Additionally, side effects such as delayed growth and the potential for stimulant misuse limit the widespread application of these medications, prompting the search for alternative treatments.

In recent years, non-pharmacological interventions like transcranial magnetic stimulation (TMS) and transcranial direct current stimulation (tDCS) have shown potential in treating ADHD by modulating neuronal activity to improve attention and impulsivity, especially in patients who cannot tolerate or wish to reduce medication. However, due to differences in treatment parameters, targeted brain regions, study populations, and treatment duration, results have been inconsistent. This chapter will focus on analyzing and describing these variations.

8.2 TMS

8.2.1 Single-Pulse TMS (spTMS)

8.2.1.1 Efficacy

TMS is a noninvasive brain stimulation technique, which was first proposed by Barker et al. based on the principle of electromagnetic induction [17]. TMS can enhance or attenuate neuronal activity by applying magnetic fields to specific brain regions, exploring the control of attention by individual brain regions, and enhancing understanding of potential treatments for ADHD. TMS can be divided into single-pulse TMS, paired-pulse TMS, repetitive TMS, and theta pulse stimulation modes. Single-pulse transcranial magnetic stimulation (spTMS) is a technique with a high spatiotemporal resolution [18]. In the late 1980s, Amassian et al. first proposed that spTMS could be used as a reversible virtual injury that alters and interferes with neural activity, thereby exploring the function of specific brain regions and accurately detecting the state of brain activity [19]. The therapeutic efficacy of spTMS is still in the experimental research phase. Research has shown that applying spTMS with primary somatosensory cortices (SI) and the contralateral posterior parietal cortex (PPC), both play a sequential and unique role in the neural processes of cross-modal association and working memory [20]. In addition, modulation of brain regions associated with attention deficits, such as the prefrontal cortex and parietal cortex, using spTMS can help the connection of attention neural networks by improving neuroplasticity and promoting better attention function [21–23]. Existing studies suggest that spTMS can help better understand the occurrence of ADHD, but its potential as a standalone treatment is still under investigation and more clinical trials are needed to determine efficacy.

8.2.1.2 Safety

The application of spTMS in children has not caused serious adverse effects. As early as 20 years ago, researchers evaluated the safety of spTMS in children, and the results showed that the risk of spTMS for children was very small [24, 25]. Common side effects include headache, scalp pain, numbness/tingling, ringing in ears, and nausea/vomiting. These side effects are usually transient and mild. Besides, TMS may increase the risk of seizures, especially in children with a history of epilepsy or brain injury [26, 27]. Compared with rTMS, spTMS has a lower stimulation frequency and a lower incidence of adverse reactions. Since factors such as age, health status, and neurodevelopmental status may all affect the safety of TMS, further studies and long-term follow-up data are still necessary to fully evaluate the safety of its application in children.

8.2.1.3 Treatment Regimen

The application of spTMS in the treatment of ADHD is in the early stage of research, and its treatment plan is still being explored. The spTMS protocol consisted of single-pulse discharges typically separated by 4–8 s [21]. Common target areas include the motor cortex, prefrontal cortex, and other regions involved in various cognitive functions [28]. Current studies mostly calibrate the stimulation intensity based on the motor threshold (MT), usually using an intensity of 80% to 120% of the Resting motor threshold (RMT), which is considered safe and affordable for patients, and 50–200 pulses per session can be used [27]. The evidence for the treatment of ADHD by spTMS is still limited, the optimal stimulation parameters, target area, and treatment regimen are not yet clear, and more clinical trials are needed to validate further research.

8.2.1.4 Clinical Recommendations

At present, spTMS is not widely used in clinical practice, but its combination with drug therapy can provide a promising multimodal approach. The combination of the two may result in a potential synergistic effect. Medications usually target the dopamine and norepinephrine neurotransmitter systems to relieve symptoms such as inattention and impulsivity. However, TMS can modulate brain activity in specific brain regions such as the prefrontal cortex to improve neuroplasticity [29, 30]. On the other hand, neuroimaging techniques, such as fMRI, EEG, and MEG, can not only identify the specific brain areas and their neural networks related to ADHD but also ensure the target selection and efficacy evaluation of TMS by accurately locating and detecting the dysfunction areas, so as to provide a comprehensive and personalized treatment plan for clinical treatment [31, 32].

8.2.2 Paired-Pulse TMS (ppTMS)

8.2.2.1 Efficacy

The current application of ppTMS in the field of mental illness is mainly in the assessment of mental illness status, and it has not yet been formalized. Previous studies have shown that the neural inhibitory motor circuits in children with ADHD can be disrupted [33]. The measurement of short-interval intracortical inhibition (SICI) by ppTMS can be used to quantify the core feature of ADHD, which is unrestrained impulsive behavior [34]. Studies have shown that children with ADHD experience a 30–40% reduction in resting state SICI, which is negatively correlated with the severity of ADHD symptoms, particularly excessive physical activity [35].

8.2.2.2 Safety

ppTMS has been recognized as a specialized tool for evaluating cortical circuits, and neurophysiological studies typically use the ppTMS paradigm to detect cortical excitability and inhibition, which is widely used in TMS research for ADHD [36].

8.2.2.3 Treatment Regimen

ppTMS outputs paired stimulation pulses each time, which are referred to as initial conditioning stimulus (CS) and subsequent test stimulus (TS), and by setting the stimulation intensity, frequency, and interval time between two pulses to achieve the purpose of stimulation, if the interval between CS and TS is 1-5 ms, SICI will be triggered [37].

8.2.2.4 Clinical Recommendations

The study of cortical excitability and inhibition in child and adolescent psychiatry using ppTMS can help identify new biomarkers and provide information for treatment selection and the characterization of endophenotypes and phronotypes [36]. Currently, several established ppTMS schemes exist for assessing the severity of ADHD and evaluating the effectiveness of drug therapy, which can aid in the selection of ADHD medications and foster the development of future ADHD treatments [38]. It is worth mentioning that the disruption of the promotion and inhibition circuits observed in pediatric patients did not manifest in adult ADHD patients [39]. ppTMS with specific intensity, frequency, and pattern can help detect, regulate, and even restore brain activity [21]. Overall, ppTMS has a promising future in the treatment of ADHD.

8.2.3 rTMS

8.2.3.1 Low-Frequency rTMS (LF-rTMS)

Efficacy

Low-frequency rTMS can normalize the imbalance of excitatory and inhibitory processes in the motor cortex through its inhibitory effect on the motor cortex, thus improving the symptoms of hyperactivity in ADHD patients. Some studies have used low-frequency rTMS to treat children with ADHD, which can effectively improve the children's attention deficit symptoms at school and hyperactivity/impulsivity symptoms at home, with a good safety profile [40].

Safety

A single-blind pseudo-stimulation randomized controlled study on 25 children with ADHD aged 8–14 years, intervening in their first somatomotor area with 1 Hz rTMS at an intensity of 80% (resting motor threshold, RMT) RMT for 900 consecutive stimulations, and monitoring their 64-lead EEG with the TMS-induced N100 wave amplitude, which showed a significant decrease in both N100 and MEP wave amplitudes after the low-frequency rTMS intervention. rTMS was well tolerated, with only three children reporting mild transient headaches before, during, or after rTMS, and EEG recordings showed no signs of epileptic activity.

Treatment Regimen

Niederhofer et al. showed [12] that stimulation of motor areas with low-frequency rTMS (1 Hz), 1200 stimuli per day for 5d per week, resulted in a significant improvement in ADHD symptoms after 4 weeks of treatment; it was also effective in children with ADHD treated with methylphenidate and helped to reduce the dose of the original medication used. In order to evaluate the safety and efficacy of low-frequency rTMS in ADHD boys with ADHD who did not respond to conventional treatment. Gómez et al. conducted [11] an open trial, 10 patients with ADHD aged 7 to 12 years were treated with a butterfly coil, with stimulation of the left DLPFC, at a frequency of 1 Hz, with 90% RMT, for 5 consecutive days, with 1500 stimulation per day. The treatment was well tolerated and completed. Seven subjects reported transient mild headache or localized discomfort that lasted only a few minutes and disappeared without medication. Two subjects reported neck pain. Overall, treatment of the left DLPFC with rTMS at 1 Hz was well tolerated safe and reliable by the children. Moreover, the behavioral problems of the children as rated by parents and teachers improved significantly after the treatment.

Clinical Recommendations

rTMS, as a kind of physical therapy, basic principle is that electric current stimulates brain tissues to release dopamine and promote cortical excitability, which in turn affects brain blood flow and metabolism. rTMS adopts different stimulation sites, stimulation intensity, stimulation frequency, and treatment duration, and its clinical efficacy on ADHD patients varies. rTMS has a good clinical effect on children with ADHD, and low-frequency transcranial magnetic stimulation therapy has been shown to be effective in treating children with ADHD. As shown in Table 8.1.

Table 8.1 TMS in ADHD

TMS	Clinical recommendations				References
	Target	Frequency	Stimulus per day (times)/ stimulation interval (s)	Duration (sessions)	
spTMS	Not clear				
ppTMS	Not clear				
LF-rTMS	L-DLPFC	1 Hz	1500 times	5	[11]
	Motor area	1 Hz	1200 times	5	[12]
HF-rTMS	R-DLPFC	20 Hz	30 s	1	[13]
	Bilateral prefrontal	18 Hz	20 s	20	[14]
	rPFC	18 Hz	20 s	15	[15]
	rPFC	10 Hz	2000 times	10	[16]
iTBS	Not clear				
cTBS	Not clear				

Abbreviations: *spTMS* Single-pulse TMS, *ppTMS* Paired-pulse TMS, *LF-rTMS* Low frequency rTMS, *HF-rTMS* High frequency rTMS; *iTBS* Intermittent TBS; *cTBS* Continuous iTBS, *L-DLPFC* Left-Dorsal lateral prefrontal cortex, *R-DLPFC* Right-Dorsal lateral prefrontal cortex, *rPFC* Right prefrontal cortex

8.2.3.2 High-Frequency rTMS (HF-rTMS)

Efficacy

A meta-analysis of fMRI in ADHD patients suggests that ADHD patients have underactivation in the right prefrontal cortex [41], and that HF-rTMS can have an activating effect on stimulated brain regions [42]. HF-rTMS acting on cortical motor areas or dorsolateral prefrontal cortex induces modulation of endogenous dopamine release into the caudate nucleus, which increases cortical excitability and improves frontal brain functioning and arousal levels in frontal regions, thus improving attention deficits.

Safety

A randomized, pseudo-stimulation-controlled crossover study of 9 patients with adult ADHD showed that high-frequency rTMS was safe with no serious adverse effects [16].

Treatment Regimen

One study was conducted on adult patients with ADHD using rTMS stimulation site for the right dorsolateral prefrontal cortex, stimulated at a 100% motor threshold intensity of 20 Hz, with a stimulation interval of 30 seconds. The results of the study showed that after a single session of true rTMS to the right prefrontal cortex using a HF-rTMS, there was a significant improvement in attention, but no significant changes in mood and anxiety [13]. However, in a randomized double-blind placebo-controlled rTMS treatment of adult ADHD patients with deep TMS with bilateral prefrontal stimulation, 20 daily sessions of transcranial magnetic stimulation for 4 consecutive weeks, with each training session consisting of 55 sets of impulses at a frequency of 18 Hz, spaced at intervals of 20 seconds, and with an intensity of 120% of the measured motor thresholds, but no bilateral deep rTMS was found to have beneficial effects [14]. We evaluated targeting the rPFC with multiple repetitive transcranial magnetic stimulation (rTMS) to 43 adult patients with ADHD with daily high-frequency (18 Hz) stimulation for a period of 3 weeks, which resulted in a significant improvement in clinical symptoms and documented EEG changes composed of the alpha and low-gamma power that were highly correlated with clinical symptoms [15].

HF-rTMS is widely used not only in adult ADHD patients but also in children and adolescents with ADHD.

In a randomized, pseudo-stimulation-controlled crossover study of 9 patients with ADHD aged 15–20 years, rTMS stimulation was performed at the site of the right prefrontal cortex at a frequency of 10 Hz (100% motor threshold), with 2000 pulses per session, 5 times per week for 2 weeks. The results showed that TMS was safe with no serious adverse effects, significant improvements in overall clinical impression and ADHD-IV scores were observed in both the TMS stimulation and pseudostimulation groups, but no significant differences were found between the two groups, and participants in the group receiving effective transcranial magnetic stimulation began to show significant improvement in clinical symptoms early in the treatment period, noting that a large-sample controlled study is still needed [16].

Clinical Recommendations

HF-rTMS plays an active role in the treatment of adults with ADHD. In addition, when combined with drug therapy, the dose of drugs can be reduced. A study found that 6-week rDLPFC HF-rTMS combined with atomoxetine was more effective than rTMS alone or atomoxetine alone [43], and the results of a recent meta-analysis showed that rTMS not only significantly improves the core symptoms of ADHD patients, but also has a significant effect in cognitive function improvement [44]. However, the limitation of current clinical studies on rTMS for ADHD is the small sample size, and more clinical studies with larger clinical sample sizes are still needed to further observe the efficacy of rTMS (as shown in Table 8.1).

8.2.4 TBS

8.2.4.1 Intermittent TBS (iTBS)

Efficacy

No independent study of iTBS in ADHD patients was reported. There was only one case-control study that investigated the use of a low-intensity rTMS protocol to measure motor cortex (M1) plasticity in youth with autism spectrum disorder (ASD) compared with typically developing children (TDC). Nine youth with ASD (two of them had combined ADHD) and nine TDC participated. This study demonstrated early evidence for a potential physiological biomarker of cortical plasticity in youth with ASD using a rapid low-intensity rTMS protocol with a discriminate measure at 20 minutes following stimulation [45].

Safety

No moderate or severe adverse events were identified by structured diagnostic interviews and no seizures occurred. One subject reported a mild headache (rated: minimal) following iTBS, which spontaneously resolved.

Treatment Regimen

A modified iTBS protocol consisted of bursts of 3 pulses at 30 Hz repeating every 200 ms for 2 seconds (1 train) with trains repeating every 10 seconds apart for a total of 300 pulses.

Clinical Recommendations

Unable to make clinical recommendations due to lack of clinical evidence, and more clinical studies are needed to further observe the efficacy and safety of iTBS.

8.2.4.2 Continuous iTBS (cTBS)

Efficacy

Similarly, no independent study of cTBS in ADHD patients was reported. But there are studies have shown that rTMS to be effective in reducing tics [46, 47]. To determine whether 8 sessions of cTBS over supplementary motor area (SMA) given over 2 days may reduce tics and motor cortical network activity in Tourette syndrome/chronic tic disorders. Steve W. Wu et al. conducted a randomized (1: 1), double-blind, sham-controlled trial of functional MRI (fMRI)-navigated, 30 Hz cTBS at 90% of resting motor threshold (RMT) over SMA in 12 patients ages 10–22 years. Comorbid ADHD (n = 8) was permitted. Active, fMRI-navigated cTBS administered in 8 sessions over 2 days to the SMA induced significant inhibition in the motor network (SMA, bilateral M1). However, both groups on average experienced tic reduction at 7 days [48].

Safety

All 12 patients who received cTBS completed the study without serious adverse events. Three participants complained of mild adverse events (abdominal pain, headaches, dry eyes) which resolved without medical intervention.

Treatment Regimen

The cTBS magnetic pulsing pattern consists of 3 pulses per burst (30 Hz) with bursts repeating 5 times per second for a total of 600 pulses per train.

Clinical Recommendations

Unable to make clinical recommendations due to lack of clinical evidence, and more clinical studies are needed to further observe the efficacy of cTBS.

8.3 tDCS

8.3.1 Conventional tDCS

8.3.1.1 Efficacy

Transcranial direct current stimulation (tDCS) is a noninvasive neuromodulation technique that has shown potential efficacy in alleviating ADHD symptoms. Studies indicate that different stimulation sites can positively influence various executive functions in ADHD patients. For example, the left dorsolateral prefrontal cortex (DLPFC) is a common target, and several studies have found that anodal tDCS stimulation of this area can increase correct response rates in Go/No-Go tasks, enhance selective attention, and improve inhibitory control functions [49, 54]. Similarly, the right DLPFC has also been found effective in improving attention deficit symptoms in adult ADHD patients [50, 55].

Additionally, stimulation of the ventromedial prefrontal cortex (vmPFC) has shown potential in regulating emotions and impulse control. For instance, Vöckel et al. demonstrated that anodal tDCS applied to the vmPFC significantly improved effort maintenance in high-effort tasks among children with ADHD [51]. In other studies, vmPFC stimulation was found to reduce delay discounting behavior, encouraging children with ADHD to choose larger delayed rewards [56]. However, while tDCS has shown improvements in several tasks, some studies did not find significant effects, possibly due to individual differences, stimulation parameters, and treatment designs [57, 58].

8.3.1.2 Safety

Overall, tDCS is considered a safe treatment method, with most studies reporting mild side effects such as scalp tingling, skin-burning sensations, or transient headaches. For example, Schertz et al. recorded mild scalp tingling and itching sensations in children with ADHD, but these symptoms typically resolved shortly after treatment [58]. However, certain populations may experience more noticeable adverse reactions. For instance, Westwood et al. reported negative emotional responses, such as irritability, sleep difficulties, and appetite loss among adolescent ADHD patients, indicating that tDCS may cause emotional fluctuations in specific groups [59]. This highlights the need for close monitoring during treatment, and adjustments or discontinuation may be necessary depending on the patient's response.

8.3.1.3 Treatment Regimen

The efficacy of tDCS in the treatment of ADHD is highly dependent on the detailed design of the treatment protocol, including the stimulation target, intensity, frequency, duration, and total number of sessions. First, the stimulation target usually involves brain regions associated with executive functions, emotional regulation, and impulse control, such as the left or right DLPFC and vmPFC. Stimulation of the left DLPFC is commonly used to improve attention and inhibitory control functions [49, 54], while the right DLPFC is targeted for attention regulation in adult ADHD patients [50, 55]. Meanwhile, stimulation of the vmPFC mainly addresses emotional and decision-making adjustments [51, 56].

Table 8.2 tDCS in ADHD

tDCS	Clinical recommendations					References
	Anode position	Cathode position	Intensity (mA)	Length (min)	Duration (sessions)	
Conventional tDCS	F3	Fp2	1.5	15	3	[49]
	F4	F3	2	30	28	[50]
	Fpz	F4	2	20	1	[51]
HD-tDCS	Right inferior frontal gyrus	Right inferior frontal gyrus	0.25/0.5	20	5	[52]
	Fp2	Fpz、Afz、AF4、AF8	1	20	10	[53]

Abbreviations: *tDCS* Transcranial direct current stimulation, *HD-tDCS* High-Definition Transcranial Direct Current Stimulation

Regarding stimulation intensity, current levels are typically set between 1.0 and 2.0 mA. Lower intensities are often applied in studies involving children to minimize discomfort or adverse effects [58, 60], while higher intensities are frequently used in adults to achieve more significant therapeutic outcomes [61, 62]. Each tDCS session generally lasts between 15 and 30 minutes; shorter durations are employed to reduce discomfort, especially in pediatric patients, while longer sessions promote more significant neuroplastic changes in adults. The frequency and total number of sessions also vary according to the treatment goals. Short-term studies often involve daily sessions over a few days [57], whereas long-term protocols may include multiple sessions per week over four weeks or longer [50, 58], with total sessions reaching 20 to 30 [55].

8.3.1.4 Clinical Recommendations

tDCS can be a supplementary treatment option for ADHD but should be applied with caution, especially in children and adolescents. Clinicians should prioritize targeting the left or right DLPFC for improving attention and executive functions while considering vmPFC for emotional and impulse control. Combining tDCS with other non-pharmacological therapies, such as cognitive training, may enhance effectiveness. Treatment protocols should be tailored based on patient age, symptom type, and initial response, with careful monitoring of potential side effects, such as mood changes. Further optimization of parameters is needed to maximize efficacy and minimize risks (as shown in Table 8.2).

8.3.2 High-Definition Transcranial Direct Current Stimulation (HD-tDCS)

8.3.2.1 Efficacy

Compared to traditional tDCS, the major advantage of HD-tDCS lies in its high-density electrode configuration, which allows the current to be more focused on the target area, reducing the impact on peripheral irrelevant areas. Currently, there are few studies on the use of HD-tDCS in treating ADHD. The initial study showed that after three treatments on 15 children and adolescents with ADHD (targeting the right inferior frontal gyrus), HD-tDCS significantly increased the average amplitude levels of P300 and N200 in patients compared to sham stimulation, but no significant improvement was observed in working memory tasks, and there was no significant difference compared to traditional tDCS [63]. Breitling-Ziegler et al. further found that after 5 days of HD-tDCS stimulation on the right inferior frontal gyrus of 33 children and adolescents with ADHD, compared to sham stimulation, 0.25 mA stimulation resulted in an increase in error rate in the nogo task, while 0.5 mA showed a reduction in omission errors. The therapeutic effect of HD-tDCS appears to be related to the current intensity [52]. Additionally, another study involving 56 adult ADHD patients found that while HD-tDCS targeting the right dorsolateral prefrontal cortex significantly improved cognitive measures of attention maintenance, it did not significantly alleviate overall symptoms of hyperactivity patients [53].

8.3.2.2 Safety

The side effects of HD-tDCS are similar to traditional tDCS, typically including scalp irritation, mild headache, and the like. However, due to its more focused current, local stimulation sensation may be slightly stronger than traditional tDCS, and its side effects are mostly transient and mild.

8.3.2.3 Treatment Regimen

Existing studies indicate that current intensity affects treatment outcomes, with low current potentially leading to impaired inhibitory behavior in ADHD patients. In addition, the stimulation area is also related to treatment outcomes. However, due to the limited number of studies and lack of evidence, specific stimulation parameters and treatment durations require further exploration.

HD-tDCS is especially suitable for patients who require precise control of brain regions, particularly individuals who show poor responsiveness to traditional tDCS or other noninvasive neurostimulation methods.

8.3.2.4 Clinical Recommendations

HD-tDCS is especially suitable for patients who require precise control of brain regions, particularly individuals who show poor responsiveness to traditional tDCS or other noninvasive neurostimulation methods. As shown in Table 8.2.

8.4 TMS vs. tDCS

8.4.1 Efficacy

Noninvasive brain stimulation (NIBS) is a potential tool to promote neuroplasticity by altering the brain activity of specific cortical regions and their networks. Among the NIBS techniques, the most commonly used are transcranial magnetic stimulation (TMS) and transcranial direct current stimulation (tDCS) [64]. Low-frequency (<1 Hz) TMS usually have an inhibitory effect, high-frequency (>5 Hz) TMS increase cortical excitability [42]. In contrast, tDCS regulates brain excitability by applying a low-amplitude (0.2–0.5 mA) current through electrodes on the scalp. The short-term effects of tDCS are achieved by resting potential, while the long-term effects are attributed to NMDA-dependent mechanisms and synaptic plasticity [65]. It is generally believed that anode tDCS can excite the function of the cortical region, while cathode tDCS has an inhibitory effect.

In treating ADHD patients, tDCS and TMS have shown different therapeutic effects in alleviating symptoms. The therapeutic effect of TMS is mainly reflected in the relief of symptoms such as inattention, while tDCS mainly improves the executive function of patients. The effectiveness of tDCS is also different for different populations, and tDCS can treat inattention in adult patients [50]. Low-frequency rTMS on the left DLPFC and high-frequency rTMS on the right DLPFC improved inattention, hyperactivity, and impulsivity in the subjects [23]. Anode-left DLPFC tDCS had the most significant effect on executive control function, while cathode-left DLPFC tDCS improved suppression control [66]. tDCS stimulation of the left DLPFC may improve impulsive symptoms of ADHD, supporting the therapeutic potential of tDCS in adults with ADHD. In addition to this, tDCS may also improve language fluency in children with ADHD [60]. Therefore, the two treatment methods can achieve complementary effects in the treatment of ADHD. In summary, physical nerve stimulation can be used to treat ADHD, and the therapeutic effect is promising.

8.4.2 Safety

Both TMS and tDCS are considered to be well-tolerated and have an acceptable safety profile [67]. In terms of safety, TMS is a noninvasive, noninvasive brain stimulation technique. In the current study, the main adverse effects observed included mild headache and localized scalp discomfort, and no serious adverse effects were observed. Generally speaking [68], tDCS is also a safe technique because it works with a low-amplitude current, which mainly causes tingling, burning, and discomfort on the scalp of the subject. But it is also mild and transient, within the tolerable range [69]. In summary, the two brain stimulation techniques can be used to treat a variety of mental disorders, including ADHD, and are reliable in safety.

8.5 Conclusion

The number of studies evaluating the efficacy of noninvasive brain stimulation (especially TMS and tDCS) in the treatment of children with ADHD is still small, and the study design is highly heterogeneous. Although the efficacy of TMS and tDCS is mixed, they both hold promise as treatment options for people with ADHD. Considering that the optimal indications, short-and long-term efficacy, and the interaction between different brain stimulation techniques in the treatment of ADHD children are still unclear, rigorous clinical studies are needed to confirm.

Noninvasive brain stimulation technology has the advantages of good safety and small adverse reactions, has shown certain efficacy in the treatment of ADHD children, and has great application prospects. It is expected that these approaches can help enhance our understanding of the underlying pathological mechanisms and treatments of ADHD, ultimately achieving effective outcomes.

Acknowledgments None.

Disclosure/Conflicts of Interest The authors declare no conflicts of interest in conducting this study or preparing the manuscript.

Financial Support None.

References

1. Polanczyk GV, Willcutt EG, Salum GA, et al. ADHD prevalence estimates across three decades: an updated systematic review and meta-regression analysis. Int J Epidemiol. 2014;43:434–42. https://doi.org/10.1093/ije/dyt261.
2. Song P, Zha M, Yang Q, et al. The prevalence of adult attention-deficit hyperactivity disorder: a global systematic review and meta-analysis. J Glob Health. 2021;11:04009. https://doi.org/10.7189/jogh.11.04009.
3. Vos M, Hartman CA. The decreasing prevalence of ADHD across the adult lifespan confirmed. J Glob Health. 2022;12:03024. https://doi.org/10.7189/jogh.12.03024.
4. Faraone SV, Bellgrove MA, Brikell I, et al. Attention-deficit/hyperactivity disorder. Nat Rev Dis Primers. 2024;10:11. https://doi.org/10.1038/s41572-024-00495-0.
5. Posner J, Polanczyk GV, Sonuga-Barke E. Attention-deficit hyperactivity disorder. Lancet. 2020;395:450–62. https://doi.org/10.1016/s0140-6736(19)33004-1.
6. Schachar RJ. Fifty years of executive control research in attention-deficit/hyperactivity disorder: what we have learned and still need to know. Neurosci Biobehav Rev. 2023;155:105461. https://doi.org/10.1016/j.neubiorev.2023.105461.
7. Onandia-Hinchado I, Pardo-Palenzuela N, Diaz-Orueta U. Cognitive characterization of adult attention deficit hyperactivity disorder by domains: a systematic review. J Neural Transm (Vienna). 2021;128:893–937. https://doi.org/10.1007/s00702-021-02302-6.
8. O'Connor L, Carbone S, Gobbo A, et al. Pediatric attention deficit hyperactivity disorder (ADHD): 2022 updates on pharmacological management. Expert Rev Clin Pharmacol. 2023;16:799–812. https://doi.org/10.1080/17512433.2023.2249414.
9. Radonjić NV, Bellato A, Khoury NM, et al. Nonstimulant medications for attention-deficit/hyperactivity disorder (ADHD) in adults: systematic review and meta-analysis. CNS Drugs. 2023;37:381–97. https://doi.org/10.1007/s40263-023-01005-8.
10. Charach A, Fernandez R. Enhancing ADHD medication adherence: challenges and opportunities. Curr Psychiatry Rep. 2013;15:371. https://doi.org/10.1007/s11920-013-0371-6.
11. Gómez L, Vidal B, Morales L, et al. Low frequency repetitive transcranial magnetic stimulation in children with attention deficit/hyperactivity disorder. Preliminary results. Brain Stimulation. 2014;7:760–2. https://doi.org/10.1016/j.brs.2014.06.001.
12. Niederhofer H. Effectiveness of the repetitive Transcranical magnetic stimulation (rTMS) of 1 Hz for attention-deficit hyperactivity disorder (ADHD). Psychiatr Danub. 2008;20:91–2.
13. Bloch Y, Harel EV, Aviram S, et al. Positive effects of repetitive transcranial magnetic stimulation on attention in ADHD subjects: a randomized controlled pilot study. World J Biol Psychiatry. 2010;11:755–8. https://doi.org/10.3109/15622975.2010.484466.

14. Paz Y, Friedwald K, Levkovitz Y, et al. Randomised sham-controlled study of high-frequency bilateral deep transcranial magnetic stimulation (dTMS) to treat adult attention hyperactive disorder (ADHD): negative results. World J Biol Psychiatry. 2018;19:561–6. https://doi.org/10.1080/15622975.2017.1282170.
15. Alyagon U, Shahar H, Hadar A, et al. Alleviation of ADHD symptoms by non-invasive right prefrontal stimulation is correlated with EEG activity. NeuroImage: Clin. 2020;26:102206. https://doi.org/10.1016/j.nicl.2020.102206.
16. Weaver L, Rostain AL, Mace W, et al. Transcranial magnetic stimulation (TMS) in the treatment of attention-deficit/hyperactivity disorder in adolescents and young adults: a pilot study. J ECT. 2012;28:98–103. https://doi.org/10.1097/YCT.0b013e31824532c8.
17. Barker AT, Jalinous R, Freeston IL. Non-invasive magnetic stimulation of human motor cortex. Lancet (London, England). 1985;1:1106–7. https://doi.org/10.1016/s0140-6736(85)92413-4.
18. Dugué L, VanRullen R. Transcranial magnetic stimulation reveals intrinsic perceptual and attentional rhythms. Front Neurosci. 2017;11:154. https://doi.org/10.3389/fnins.2017.00154.
19. Amassian VE, Cracco RQ, Maccabee PJ, et al. Suppression of visual perception by magnetic coil stimulation of human occipital cortex. Electroencephalogr Clin Neurophysiol. 1989;74:458–62. https://doi.org/10.1016/0168-5597(89)90036-1.
20. Ku Y, Zhao D, Hao N, et al. Sequential roles of primary somatosensory cortex and posterior parietal cortex in tactile-visual cross-modal working memory: a single-pulse transcranial magnetic stimulation (spTMS) study. Brain Stimul. 2015;8:88–91. https://doi.org/10.1016/j.brs.2014.08.009.
21. Jannati A, Oberman LM, Rotenberg A, et al. Assessing the mechanisms of brain plasticity by transcranial magnetic stimulation. Neuropsychopharmacology. 2023;48:191–208. https://doi.org/10.1038/s41386-022-01453-8.
22. Rubia K, Westwood S, Aggensteiner P-M, et al. Neurotherapeutics for attention deficit/hyperactivity disorder (ADHD): a. Review. 2021;10:2156.
23. Yen C, Valentine EP, Chiang M-C. The use of transcranial magnetic stimulation in attention optimization research: a review from basic theory to findings in attention-deficit/hyperactivity disorder and depression. Life (Basel). 2024;14:329. https://doi.org/10.3390/life14030329.
24. Gilbert DL, Garvey MA, Bansal AS, et al. Should transcranial magnetic stimulation research in children be considered minimal risk? Clin Neurophysiol. 2004;115:1730–9. https://doi.org/10.1016/j.clinph.2003.10.037.
25. Rossi S, Hallett M, Rossini PM, et al. Safety, ethical considerations, and application guidelines for the use of transcranial magnetic stimulation in clinical practice and research. Clin Neurophysiol. 2009;120:2008–39. https://doi.org/10.1016/j.clinph.2009.08.016.
26. Krishnan C, Santos L, Peterson MD, et al. Safety of noninvasive brain stimulation in children and adolescents. Brain Stimul. 2015;8:76–87. https://doi.org/10.1016/j.brs.2014.10.012.
27. Zewdie E, Ciechanski P, Kuo HC, et al. Safety and tolerability of transcranial magnetic and direct current stimulation in children: prospective single center evidence from 3.5 million stimulations. Brain Stimul. 2020;13:565–75. https://doi.org/10.1016/j.brs.2019.12.025.
28. Siebner HR, Funke K, Aberra AS, et al. Transcranial magnetic stimulation of the brain: what is stimulated? – a consensus and critical position paper. Clin Neurophysiol. 2022;140:59–97. https://doi.org/10.1016/j.clinph.2022.04.022.
29. Kessi M, Duan H, Xiong J, et al. Attention-deficit/hyperactive disorder updates. Front Mol Neurosci. 2022;15:925049. https://doi.org/10.3389/fnmol.2022.925049.
30. Rosson S, de Filippis R, Croatto G, et al. Brain stimulation and other biological non-pharmacological interventions in mental disorders: an umbrella review. Neurosci Biobehav Rev. 2022;139:104743. https://doi.org/10.1016/j.neubiorev.2022.104743.
31. Kowalczyk OS, Mehta MA, O'Daly OG, et al. Task-based functional connectivity in attention-deficit/hyperactivity disorder: a systematic review. Biol Psychiatry Global Open Sci. 2022;2:350–67. https://doi.org/10.1016/j.bpsgos.2021.10.006.
32. Liu G, Lu W, Qiu J, et al. Identifying individuals with attention-deficit/hyperactivity disorder based on multisite resting-state functional magnetic resonance imaging: a radiomics analysis. Hum Brain Mapp. 2023;44:3433–45. https://doi.org/10.1002/hbm.26290.

33. Buchmann J, Gierow W, Weber S, et al. Restoration of disturbed intracortical motor inhibition and facilitation in attention deficit hyperactivity disorder children by methylphenidate. Biol Psychiatry. 2007;62:963–9. https://doi.org/10.1016/j.biopsych.2007.05.010.
34. Harris AD, Gilbert DL, Horn PS, et al. Relationship between GABA levels and task-dependent cortical excitability in children with attention-deficit/hyperactivity disorder. Clin Neurophysiol. 2021;132:1163–72. https://doi.org/10.1016/j.clinph.2021.01.023.
35. Gilbert DL, Isaacs KM, Augusta M, et al. Motor cortex inhibition: a marker of ADHD behavior and motor development in children. Neurology. 2011;76:615–21. https://doi.org/10.1212/WNL.0b013e31820c2ebd.
36. Croarkin PE, Wall CA, Lee J. Applications of transcranial magnetic stimulation (TMS) in child and adolescent psychiatry. Int Rev Psychiatry. 2011;23:445–53. https://doi.org/10.3109/09540261.2011.623688.
37. Schneider MK, Retz W, Gougleris G, et al. Effects of long-acting methylphenidate in adults with attention deficit hyperactivity disorder: a study with paired-pulse transcranial magnetic stimulation. Neuropsychobiology. 2011;64:195–201. https://doi.org/10.1159/000326693.
38. Jannati A, Ryan MA, Kaye HL, et al. Biomarkers obtained by transcranial magnetic stimulation in neurodevelopmental disorders. J Clin Neurophysiol. 2022;39:135–48. https://doi.org/10.1097/WNP.0000000000000784.
39. Hoeppner J, Neumeyer M, Wandschneider R, et al. Intracortical motor inhibition and facilitation in adults with attention deficit/hyperactivity disorder. J Neural Transm (Vienna). 2008;115:1701–7. https://doi.org/10.1007/s00702-008-0091-y.
40. Helfrich C, Pierau SS, Freitag CM, et al. Monitoring cortical excitability during repetitive transcranial magnetic stimulation in children with ADHD: a single-blind, sham-controlled TMS-EEG study. PLoS One. 2012;7:e50073. https://doi.org/10.1371/journal.pone.0050073.
41. Norman LJ, Carlisi C, Lukito S, et al. Structural and functional brain abnormalities in attention-deficit/hyperactivity disorder and obsessive-compulsive disorder: a comparative meta-analysis. JAMA Psychiatry. 2016;73:815–25. https://doi.org/10.1001/jamapsychiatry.2016.0700.
42. Lefaucheur JP, Aleman A, Baeken C, et al. Evidence-based guidelines on the therapeutic use of repetitive transcranial magnetic stimulation (rTMS): an update (2014-2018). Clin Neurophysiol. 2020;131:474–528. https://doi.org/10.1016/j.clinph.2019.11.002.
43. Cao P, Xing J, Cao Y, et al. Clinical effects of repetitive transcranial magnetic stimulation combined with atomoxetine in the treatment of attention-deficit hyperactivity disorder. Neuropsychiatr Dis Treat. 2018;14:3231–40. https://doi.org/10.2147/ndt.S182527.
44. Chen YH, Liang SC, Sun CK, et al. A meta-analysis on the therapeutic efficacy of repetitive transcranial magnetic stimulation for cognitive functions in attention-deficit/hyperactivity disorders. BMC Psychiatry. 2023;23:756. https://doi.org/10.1186/s12888-023-05261-2.
45. Pedapati EV, Gilbert DL, Erickson CA, et al. Abnormal cortical plasticity in youth with autism Spectrum disorder: a transcranial magnetic stimulation case-control pilot study. J Child Adolesc Psychopharmacol. 2016;26:625–31. https://doi.org/10.1089/cap.2015.0183.
46. Kwon HJ, Lim WS, Lim MH, et al. 1-Hz low frequency repetitive transcranial magnetic stimulation in children with Tourette's syndrome. Neurosci Lett. 2011;492:1–4. https://doi.org/10.1016/j.neulet.2011.01.007.
47. Le K, Liu L, Sun M, et al. Transcranial magnetic stimulation at 1 hertz improves clinical symptoms in children with Tourette syndrome for at least 6 months. J Clin Neurosci. 2013;20:257–62. https://doi.org/10.1016/j.jocn.2012.01.049.
48. Wu SW, Maloney T, Gilbert DL, et al. Functional MRI-navigated repetitive transcranial magnetic stimulation over supplementary motor area in chronic tic disorders. Brain Stimul. 2014;7:212–8. https://doi.org/10.1016/j.brs.2013.10.005.
49. Soltaninejad Z, Nejati V, Ekhtiari H. Effect of anodal and Cathodal transcranial direct current stimulation on DLPFC on modulation of inhibitory control in ADHD. J Atten Disord. 2019;23:325–32. https://doi.org/10.1177/1087054715618792.

50. Leffa DT, Grevet EH, Bau CHD, et al. Transcranial direct current stimulation vs sham for the treatment of inattention in adults with attention-deficit/hyperactivity disorder: the TUNED randomized clinical trial. JAMA Psychiatry. 2022;79:847–56. https://doi.org/10.1001/jamapsychiatry.2022.2055.
51. Vöckel J, Kühnel A, Rossberg R, et al. Transcranial direct current stimulation enhances effort maintenance in ADHD. Brain Stimul. 2024;17:899–906. https://doi.org/10.1016/j.brs.2024.07.018.
52. Breitling-Ziegler C, Zaehle T, Wellnhofer C, et al. Effects of a five-day HD-tDCS application to the right IFG depend on current intensity: a study in children and adolescents with ADHD. Prog Brain Res. 2021;264:117–50. https://doi.org/10.1016/bs.pbr.2021.01.014.
53. Wang Y-C, Liu J, Wu Y-C, et al. A randomized, sham-controlled trial of high-definition transcranial direct current stimulation on the right orbital frontal cortex in children and adolescents with attention-deficit hyperactivity disorder. Front Psych. 2023;14:987093. https://doi.org/10.3389/fpsyt.2023.987093.
54. Guimarães RSQ, Bandeira ID, Barretto BL, et al. The effects of transcranial direct current stimulation on attention and inhibitory control of children and adolescents with attention-deficit/hyperactivity disorder (ADHD): study protocol for a randomized, sham-controlled, triple-blind, cross-over trial. Medicine (Baltimore). 2021;100:e24283. https://doi.org/10.1097/md.0000000000024283.
55. Mauche N, Ulke C, Huang J, et al. Treatment of adult attention-deficit hyperactivity disorder (ADHD) with transcranial direct current stimulation (tDCS): study protocol for a parallel, randomized, double-blinded, sham-controlled, multicenter trial (stim-ADHD). Eur Arch Psychiatry Clin Neurosci. 2024;274:71–82. https://doi.org/10.1007/s00406-023-01652-4.
56. Nejati V, Sarraj Khorrami A, Nitsche MA. Transcranial direct current stimulation improves reward processing in children with ADHD. J Atten Disord. 2021;25:1623–31. https://doi.org/10.1177/1087054720923094.
57. Guimarães RSQ, Bandeira ID, Barretto BL, et al. Efficacy and safety of transcranial direct current stimulation over the left dorsolateral prefrontal cortex in children and adolescents with attention-deficit/hyperactivity disorder: a randomized, triple-blinded, sham-controlled, cross-over trial. Front Psych. 2023;14:1217407. https://doi.org/10.3389/fpsyt.2023.1217407.
58. Schertz M, Karni-Visel Y, Genizi J, et al. Transcranial direct current stimulation (tDCS) in children with ADHD: a randomized, sham-controlled pilot study. J Psychiatr Res. 2022;155:302–12. https://doi.org/10.1016/j.jpsychires.2022.08.022.
59. Westwood SJ, Criaud M, Lam S-L, et al. Transcranial direct current stimulation (tDCS) combined with cognitive training in adolescent boys with ADHD: a double-blind, randomised, sham-controlled trial. Psychol Med. 2023;53:497–512. https://doi.org/10.1017/s0033291721001859.
60. Nejati V, Estaji R, Helisaz Z. Transcranial direct-current stimulation improves verbal fluency in children with attention deficit hyperactivity disorder (ADHD). Brain Sci. 2023;13:1257. https://doi.org/10.3390/brainsci13091257.
61. Allenby C, Falcone M, Bernardo L, et al. Transcranial direct current brain stimulation decreases impulsivity in ADHD. Brain Stimul. 2018;11:974–81. https://doi.org/10.1016/j.brs.2018.04.016.
62. Dubreuil-Vall L, Gomez-Bernal F, Villegas AC, et al. Transcranial direct current stimulation to the left dorsolateral prefrontal cortex improves cognitive control in patients with attention-deficit/hyperactivity disorder: a randomized behavioral and neurophysiological study. Biol Psychiatry Cogn Neurosci Neuroimag. 2021;6:439–48. https://doi.org/10.1016/j.bpsc.2020.11.006.
63. Breitling C, Zaehle T, Dannhauer M, et al. Comparison between conventional and HD-tDCS of the right inferior frontal gyrus in children and adolescents with ADHD. Clin Neurophysiol. 2020;131:1146–54. https://doi.org/10.1016/j.clinph.2019.12.412.

64. Polanía R, Nitsche MA, Ruff CC. Studying and modifying brain function with non-invasive brain stimulation. Nat Neurosci. 2018;21:174–87. https://doi.org/10.1038/s41593-017-0054-4.
65. Ganguly K, Poo M-M. Activity-dependent neural plasticity from bench to bedside. Neuron. 2013;80:729–41. https://doi.org/10.1016/j.neuron.2013.10.028.
66. Salehinejad MA, Ghanavati E, Glinski B, et al. A systematic review of randomized controlled trials on efficacy and safety of transcranial direct current stimulation in major neurodevelopmental disorders: ADHD, autism, and dyslexia. Brain Behav. 2022;12:e2724. https://doi.org/10.1002/brb3.2724.
67. Liu H, Wu M, Huang H, et al. Comparative efficacy of non-invasive brain stimulation on cognition function in patients with mild cognitive impairment: a systematic review and network meta-analysis. Ageing Res Rev. 2024;101:102508. https://doi.org/10.1016/j.arr.2024.102508.
68. Rossi S, Antal A, Bestmann S, et al. Safety and recommendations for TMS use in healthy subjects and patient populations, with updates on training, ethical and regulatory issues: expert guidelines. Clin Neurophysiol. 2021;132:269–306. https://doi.org/10.1016/j.clinph.2020.10.003.
69. Kang J, Lee H, Yu S, et al. Effects and safety of transcranial direct current stimulation on multiple health outcomes: an umbrella review of randomized clinical trials. Mol Psychiatry. 2024;29:3789. https://doi.org/10.1038/s41380-024-02624-3.

Autism Spectrum Disorder

9

Shi-Bin Wang, Linyan Fu, Jinming Liu, Min Dong,
and Fan Yang

Abstract

Autism spectrum disorder (ASD) is a neurodevelopmental disorder characterized by social communication deficits, repetitive behaviors, and restricted interests (RRBI). Currently, there is no single medication that alleviates the core symptoms of ASD, but noninvasive brain stimulation alternatives have been explored and developed. The evidence to date supports the efficacy of noninvasive neuromodulation techniques, including transcranial magnetic stimulation (TMS), theta burst stimulation (TBS), magnetic seizure therapy (MST), and transcranial direct current stimulation (tDCS), which will be discussed in this chapter.

Keywords

rTMS · dTMS · TBS · MST · tDCS

S.-B. Wang · L. Fu · M. Dong · F. Yang (✉)
Guangdong Mental Health Center, Guangdong Provincial People's Hospital
(Guangdong Academy of Medical Sciences), Southern Medical University,
Guangzhou, Guangdong Province, China
e-mail: yangfan@gdph.org.cn

J. Liu
Department of Maternal and Child Health, School of Public Health,
Sun Yat-sen University, Guangzhou, Guangdong Province, China

W. Zheng, Y. Ning (eds.), *TMS and tDCS for Psychiatric Disorders*,
https://doi.org/10.1007/978-981-96-8504-2_9

Abbreviations

ABA	applied behavior analysis
ABC	autism behavior checklist
ADHD	attention-deficit hyperactivity disorder
ALT	autism-like traits
APOE	apolipoprotein E
AQ	autism-spectrum quotient
ASD	autism spectrum disorder
ATEC	autism treatment evaluation checklist
BDNF	brain-derived neurotrophic factor
CARS	childhood autism rating scale
cTBS	the continuous theta-burst stimulation
DAN	the dorsal attention network
DLPFC	dorsolateral prefrontal cortex
DMN	default mode network
ECT	electroconvulsive therapy
FXS	fragile X syndrome
HD-tDCS	high-definition tDCS
HF	high frequency
iTBS	intermittent TBS
LF	low-frequency
LTP	long-term potentiation
M1	motor cortex
MEPs	motor-evoked potentials
MST	magnetic seizure therapy
mPFC	medial prefrontal cortex
NIBS	noninvasive brain stimulation
OCD	obsessive-compulsive disorder
pSTS	posterior superior temporal sulcus
RCTs	randomized controlled trials
RpSTS	the right posterior superior temporal sulcus
RRBI	repetitive and restricted behaviors and interests
rTMS	repetitive transcranial magnetic stimulation
rTPJ	the right temporoparietal junction
SRS-2	social responsiveness scale-2
SSRIs	selective serotonin reuptake inhibitors
sTMS	synchronized TMS
SMN	sensorimotor network
TBS	theta burst stimulation
TDC	typically developing children
tDCS	transcranial direct current stimulation

TEACCH	the treatment and education of autistic and communication-related handicapped children
tES	transcranial electrical stimulation
TMS	transcranial magnetic stimulation
VLPFC	the ventrolateral prefrontal cortex
VmPFC	the ventromedial prefrontal cortex
WHO	The World Health Organization (WHO)

9.1 Introduction

Autism spectrum disorder (ASD) is characterized by core impairments in social communication, repetitive and restricted behaviors and interests (RRBI). It is a neurodevelopmental disorder that emerges in early childhood, although the symptoms may not be immediately apparent.

Despite the influence of various factors, including the investigation period, research program, and tools, the prevalence of ASD has generally increased globally over the past 20–30 years. The World Health Organization (WHO) has suggested that the average prevalence of ASD is 1%. This disorder is widely considered a chronic disease with a poor long-term prognosis. Early diagnosis and intervention are crucial for improving the prognosis.

Pharmacotherapies are frequently utilized in individuals with ASD, yet these medications do not target the core deficits of the condition. The majority of evidence supporting the treatment of RRBI or comorbid disorders points to risperidone, aripiprazole, methylphenidate, and selective serotonin reuptake inhibitors (SSRIs). The adverse effects of medications used in ASD treatment should be systematically monitored. Education and behavioral intervention training are the most prevalent rehabilitation strategies in ASD treatment. Applied behavior analysis (ABA) and the treatment and education of autistic and communication-related handicapped children (TEACCH) are the most recognized and evidence-based therapies for ASD. It is crucial to bolster family support and empower parents to foster their child's development within the family unit. This can encompass techniques to encourage communication, manage behavior, and promote social interaction. With comprehensive guidance, parents can collaborate with teachers and doctors to implement these treatments.

Noninvasive brain stimulation (NIBS) techniques, such as transcranial magnetic stimulation (TMS) and transcranial direct current stimulation (tDCS), have recently been used to investigate the pathophysiology of various neuropsychiatric disorders, including ASD. These techniques are also being explored for the treatment of core and associated symptoms of ASD. This chapter will describe the effectiveness and safety of these NIBS treatments for ASD.

9.2 rTMS

9.2.1 Unilateral rTMS

9.2.1.1 Efficacy

Repetitive transcranial magnetic stimulation (rTMS) is a noninvasive brain stimulation technique that employs rapidly changing extracranial magnetic fields to induce local intracranial currents, which are sufficiently strong to depolarize a small group of neurons and thereby manipulate the electrophysiological activity of the cortex [1, 2]. Since 2008, high-frequency rTMS has been approved by the FDA for treating patients with treatment-resistant depression, as its efficacy and effectiveness have been demonstrated. Consequently, rTMS has been widely utilized as a therapeutic intervention for various psychiatric and neurological disorders [3–6].

Moreover, in recent years, an increasing number of studies have suggested that rTMS could be beneficial for treating the core symptoms of ASD [3, 7–9]. Studies have shown that the application of high-frequency rTMS to the left parietal cortex can improve social and language deficits in children with low-functioning ASD by enhancing remote brain connectivity and reorganization in these individuals [9]. Studies have also found that low-frequency rTMS targeting the dorsolateral prefrontal cortex (DLPFC) can improve repetitive and stereotyped behavior in individuals with autism [8]. Additionally, a study employing rTMS on the left dorsolateral prefrontal cortex to treat comorbid depression in autistic patients revealed that depressive symptoms ameliorated with treatment, and family members' evaluations of autism symptoms also showed improvement [3]. Although the optimal parameters for rTMS therapy have yet to be established, there remains significant potential for unilateral rTMS in treating the core symptoms of ASD.

9.2.1.2 Safety

In addition to efficacy, numerous clinical trials and meta-analyses have evaluated the safety of rTMS in both clinical practice and research, indicating that the incidence of adverse events is low across various populations and devices [1, 10, 11]. The International Federation of Clinical Neurophysiology has released safety guidelines. However, in comparison to the literature on adults, there is a relative scarcity of safety data for rTMS in the pediatric population. Available data suggest that children and adolescents have a safety profile similar to that of adults, with adverse event rates ranging from 3.4% to 10.11%. The incidence of adverse events varies depending on the patient population, the form of TMS, and the number of treatments. Consistent with the adult literature, the most commonly reported side effects in the pediatric literature are transient headache and neck pain [10]. However, due to the frequent use of psychotropic medications in individuals with ASD, and the increased prevalence of epilepsy within this population, individuals with ASD may have additional concerns regarding the relative risk of seizures [1]. This suggests that TMS should be administered with caution, with continuous monitoring for potential adverse effects and further investigation into the safety of TMS.

9.2.1.3 Treatment Regimen

Regarding research on unilateral rTMS therapy for ASD, most studies are categorized into two major stimulation frequency camps. The first camp aims to improve ASD symptoms by applying high-frequency rTMS to the temporoparietal cortex or the dorsolateral prefrontal cortex. Panerai et al. applied high-frequency (8 Hz) rTMS to the left anterior motor cortex of children with intellectual disability and ASD. The authors reported a significant improvement in eye-hand coordination, which was further enhanced when combined with behavioral eye-hand integration training [12]. A study employing 10 Hz high-frequency rTMS on the left dorsolateral prefrontal cortex demonstrated that both depressive symptoms and core symptoms of ASD improved with treatment [3].

The second camp aims to alleviate ASD symptoms using low-frequency rTMS. The earliest research on rTMS for ASD involved applying low-frequency stimulation to the dorsolateral prefrontal cortex (either on the left side or bilaterally in sequence) to decrease excitability in individuals with ASD. Following the protocol, behavioral assessments conducted after 18 sessions of transcranial magnetic stimulation indicated lower scores for irritability, hyperactivity, stereotypic behavior, and compulsions. Additionally, autonomic measures revealed a significant increase in cardiac septal variability, suggesting that 18 sessions of low-frequency rTMS for ASD enhanced cardiac vagal control and diminished sympathetic arousal [1, 13]. However, these experimental designs are all open-label (with a waiting list control group), so the results may be confounded by placebo effects or adaptation to the environment and protocol.

9.2.1.4 Clinical Recommendations

The application of unilateral rTMS, whether low-frequency or high-frequency, to stimulate the temporal parietal cortex or dorsolateral prefrontal cortex, may be effective. However, due to the relative scarcity of safety data for rTMS in pediatric populations, transcranial magnetic stimulation should be carefully considered in clinical practice and closely monitored for potential adverse reactions.

9.2.1.5 Bilateral rTMS

Efficacy

Regarding bilateral repetitive transcranial magnetic stimulation (rTMS), previous reports primarily involve direct stimulation of the bilateral dorsolateral prefrontal cortex, as well as separate stimulation of the left and right sides, followed by simultaneous stimulation of the bilateral prefrontal cortex. Wang et al. examined the impact of 12 sessions of rTMS on the autonomic activity of children with ASD in the dorsolateral prefrontal cortex (DLPFC). They observed a significant increase in the variability index of the patients' cardiac intervals, suggesting enhanced vagal nerve control and reduced sympathetic nerve excitability. Behavioral assessments indicated that the children's scores for irritability, hyperactivity, stereotyped behavior, and compulsive behavior decreased [14, 15]. Studies on rTMS have applied left-sided, right-sided, and bilateral stimulation sequentially. These studies have found that rTMS therapy targeting the prefrontal cortex can enhance brain function and social skills in individuals with autism [16–18]. At the same time, studies have found that this approach can improve executive function in patients with ASD, including enhancements in behavioral performance for visual attention tasks and reductions in omission error rates [19]. However, Manuel F et al. found no significant difference in the impact of the rMTS group and the sham stimulation group on executive function [20]. Therefore, the effectiveness of rTMS in treating executive function deficiencies associated with ASD in the DLPFC warrants further investigation.

Safety

The safety of bilateral rTMS treatment for ASD is comparable to that of unilateral rTMS stimulation. Existing studies have demonstrated that it has comparable and favorable safety profiles to those observed in other psychiatric populations. The most frequently reported side effects are transient headaches and neck pain [10], and the most severe side effect is an increased risk of epileptic seizures [1]. The incidence of adverse events varies depending on the patient population, the form of transcranial magnetic stimulation, and the frequency of treatment; therefore, close monitoring of possible adverse reactions is necessary.

Treatment Regimen

Studies on bilateral rTMS treatment for ASD can also be categorized into two camps based on the frequency of stimulation: those using low frequency and those using high frequency. The first camp advocates for low-frequency rTMS stimulation, which is also the frequency reported in most studies. These studies stimulated the dorsolateral prefrontal cortex using 0.5-1 Hz low-frequency rTMS, either simultaneously or alternately on the left and right sides, to explore changes in brain function and improvements in social and executive abilities [15, 16, 18, 19]. Another study utilized the second largest camp, employing high-frequency rTMS stimulation, to investigate its effects on executive function in individuals with ASD. Although no significant difference was found between the effects of active and sham stimulation rTMS on executive function, the study confirmed that repeated high-frequency rTMS targeting the DLPFC in young and middle-aged ASD patients has comparable safety profiles to those observed in other psychiatric populations [10].

Clinical Recommendations

The application of bilateral rTMS, whether low frequency or high frequency, to stimulate the DLPFC may be effective. The most noticeable effect, however, was the decrease in OCD-like (obsessive-compulsive disorder, OCD) symptoms at the end of the treatment, as measured by the Y-BOCS scale.

9.2.1.6 Accelerated rTMS

Accelerated TMS is an emerging treatment approach within the TMS field that can achieve similar, or even better, efficacy levels by reducing treatment time and enhancing response rates. Nonetheless, research into accelerated TMS remains in its infancy. While it may hold the promise of reducing treatment duration and quickly alleviating depressive symptoms in patients with depression, there are currently no established treatment protocols, and no studies have been conducted on its application in ASD. Consequently, further research is essential to elucidate its potential in treating ASD.

9.2.2 Deep TMS

9.2.2.1 Efficacy

Deep TMS employs H-Coils, which are flexible and come in various configurations to stimulate different brain regions associated with various neuropsychiatric disorders [21].

Only a few studies have reported on the effects of deep TMS in adults with ASD. The findings of these studies suggest that deep TMS applied to the mPFC may be beneficial for high-functioning ASD patients experiencing OCD or anxiety symptoms. A case report indicated that two weeks of daily excitatory stimulation (5 Hz) applied to the mPFC enhanced social functioning in a 20-year-old woman with high-functioning ASD [22]. A later research by the same study group discovered that stimulating the mPFC enhanced social relations and decreased socially related anxiety [23]. Another case report suggests that deep TMS may have modulated networks associated with mentalizing abilities and self-referential processes in two young patients diagnosed with high-functioning autism [24].

9.2.2.2 Safety

9.2.2.3 Treatment Regimen

The mPFC serves as a critical node within functional networks that are essential for mentalizing abilities, which are vital for social interactions and have been found to be impaired in adults with ASD.

One study examined the changes in emotional and cognitive processing following daily sessions of 5 Hz deep TMS H7 Coil over 5 weeks in two high-functioning ASD patients, and both patients exhibited improvement in various cognitive functions, with a global improvement of 20% for P1 and 30% for P2 in the neuropsychological battery [24]. Another study indicates that 28 adults with high-functioning ASD or Asperger's disorder completed two weeks of weekday treatment sessions (5 Hz, 10-second trains, 20-second inter-train intervals, and 1500 pulses per session). Participants in the deep TMS group, compared to those in the sham group, demonstrated a significant reduction in social-related symptoms posttreatment and at follow-up. Furthermore, a decrease in self-oriented (socially related) anxiety during challenging and emotional social situations was observed only in the deep TMS group from baseline to follow-up [23].

9.2.2.4 Clinical Recommendations

Deep TMS is increasingly being utilized in the treatment of both core and associated symptoms of ASD. Drawing from previous reports, we have noted alterations in the emotional, cognitive, and social domains among adult ASD patients, along with a significant reduction in symptoms akin to OCD. However, clinical recommendations remain unclear. Therefore, further research is necessary to determine if these positive outcomes can be sustained in the long term, and there is a pressing need for larger-scale, randomized controlled trials.

9.2.3 Priming TMS

Priming TMS is an innovative approach in rTMS, which involves preconditioning low-frequency (LF) stimulation trains with subthreshold stimulation at a high frequency (HF) [25], and there is a lack of research on the field of ASD.

9.2.4 Synchronized TMS

Synchronized TMS (sTMS) represents an innovative approach to noninvasive brain stimulation. Utilizing a trio of rotating neodymium magnets, sTMS can deliver extremely low-energy, sinusoidal magnetic fields that synchronize with an individual's intrinsic alpha frequency (IAF). However, scholarly investigations into its use for ASD are lacking.

9.2.5 TBS

9.2.5.1 iTBS

Efficacy

Theta burst stimulation (TBS) is a modified form of rTMS. Intermittent TBS (iTBS), which delivers bursts of TBS pulses for 2 seconds every 10 seconds, can enhance cortical excitability, yet it features a shorter stimulation duration (3 min) and a lower total number of TMS pulses and intensity [26]. Recent studies have investigated the efficacy of iTBS in ASD and support its effects on certain domains of ASD-associated deficits. Ni and colleagues applied iTBS over the bilateral posterior superior temporal sulcus (pSTS) or DLPFC to explore its impact in adults with ASD. The study demonstrated that a single session of active iTBS to the DLPFC led to immediate improvements on executive functioning tasks, while there were no significant differences in neuropsychological function following a single pSTS session [27]. Ni and colleagues also conducted a clinical trial on a group of intellectually capable children with autism and found that 8-week iTBS over the bilateral pSTS improved social relating and reduced repetitive behaviors in children with ASD. Higher IQ, better baseline social cognitive performance, and less attention-deficit hyperactivity disorder (ADHD) severity predicted a better response to pSTS stimulation for social relating [28]. Another study investigated the effects of iTBS on the pSTS in intellectually capable adults with ASD, revealing significant immediate impacts on parent-rated autistic symptoms among the participants. The outcomes of multisession iTBS on cognitive flexibility and clinical symptoms were influenced by baseline social-communicative impairment, baseline cognitive performance, concurrent psychotropic medication use, and baseline autistic symptoms [29]. Individuals with autism-like traits (ALT) are part of a subclinical group that shares social deficits with those diagnosed with ASD. Research has indicated that iTBS of the right pSTS can enhance the emotion perception abilities of ALT individuals by modulating the associated neural networks [30].

A case-control study investigated the use of iTBS to measure motor cortex (M1) plasticity in youth with ASD compared with typically developing children (TDC). It was found that both groups experienced a mean increase in excitability after 30 min of iTBS; however, the time course of excitability changes varied. Post-hoc testing revealed a significant decrease in amplitude in the ASD group at 20 min post-iTBS compared to the TDC, after correcting for multiple comparisons [31]. Abujadi and colleagues reported that iTBS in the right DLPFC could improve target executive function deficits and restricted, repetitive behaviors in male children and adolescents with ASD [32]. Oberman and coworkers observed an increased duration of response to the TBS paradigm among high-functioning adults with ASD, which may be associated with hyperplasticity [33].

Safety

The safety of iTBS in individuals with ASD has been investigated in previous studies. Meta-analysis has shown that iTBS is feasible and tolerable in capable youth and young adults with ASD [34]. A case-control study supported the finding that iTBS was well tolerated by all youths with ASD aged 11–18 years [31]. Self-reports indicated that iTBS was well tolerated. However, transient discomfort during stimulations due to muscle twitches around the eyes was reported by five participants in a study [30]. No other serious adverse effects, such as headaches or seizures, were reported in previous studies [32].

Treatment Regimen

The iTBS protocol includes 190 seconds of stimulation in a 2-second on and 8-second off pattern. Most studies have explored the effects of iTBS on the neuroplasticity of the pSTS, DLPFC, and M1, as well as the improvement in clinical symptoms. The stimulation camps include unilateral and bilateral stimulation. The study by Ni and colleagues shows promise for improvement in social symptoms and autistic symptoms through targeted iTBS pulses to the bilateral pSTS [28, 29]. Stimulation of the right pSTS could enhance emotion perception in individuals with ASD by significantly decreasing resting-state functional connectivity between the rpSTS and the left cerebellum [30]. Stimulation of the right DLPFC may enhance cognitive function and reduce repetitive behaviors in male children and adolescents with ASD [32]. These results suggest that further controlled interventional studies of iTBS for ASD are warranted.

Clinical Recommendations

Considering the improvement effect of iTBS on neuropsychological function, social symptoms, and autistic symptoms in ASD, iTBS could be a vital therapeutic strategy for the treatment of ASD. It is worth noting that individual factors modulating the response to iTBS should be explicitly considered. The level of clinical recommendation for iTBS was unclear.

9.2.5.2 Accelerated iTBS

Previous research has demonstrated that accelerated intermittent TBS has been shown to be an effective antidepressant treatment. The conclusions indicate that this treatment has demonstrated safety and excellent tolerability. The results encourage further research into the use of accelerated iTBS for the treatment of ASD patients with comorbid depression.

9.2.5.3 Continuation TBS

Efficacy

The continuous theta-burst stimulation (cTBS) protocol consists of 40 seconds of continuous stimulation. Various studies have examined the effectiveness of cTBS in individuals with ASD. Adults with ASD have shown a greater and more enduring inhibitory response to cTBS, which is likely attributed to abnormalities in cortical plasticity mechanisms associated with ASD [35]. A study investigated the mechanisms of plasticity and metaplasticity in humans with ASD and fragile X syndrome (FXS) following cTBS over the primary motor cortex (M1). It found that after a 40-second cTBS train, individuals with ASD exhibited a significantly longer duration of suppression in motor-evoked potentials (MEPs) amplitude compared to healthy controls, whereas individuals with FXS demonstrated a significantly shorter duration [36]. Jannati et al. found that motor-evoked potential (MEP) at T15 post-cTBS is a significant biomarker for adults with ASD, and its utility is modulated by brain-derived neurotrophic factor (BDNF) and apolipoprotein E (APOE) polymorphisms [35].

However, Yeh et al. did not find a significant impact of cTBS over the left DLPFC on white matter macrostructure and microstructure, as well as on connections, in children and emerging adults with autism [37]. Furthermore, an 8-week randomized, double-blind, sham-controlled trial was conducted to investigate the efficacy of inhibitory cTBS over the left DLPFC in individuals with ASD. The findings do not support the notion that cTBS over the left DLPFC is more effective than sham stimulation for children, adolescents, and adults with ASD [38]. An additional study, where TBS was applied to children with ASD, demonstrated an increase in the duration of response across childhood and revealed a subgroup of children who showed paradoxical facilitation to the typically suppressive cTBS paradigm. The authors suggested that their findings may reflect abnormalities in GABAergic inhibitory control in those individuals who showed paradoxical facilitation [39].

Safety

The safety of cTBS has been evaluated in a previous study, which indicated that all participants tolerated cTBS procedures without complications or unexpected side effects [35].

Treatment Regimen

Previous studies have explored the impact of cTBS over the DLPFC on core symptoms in ASD and found no significant effects [37]. In another study, the variations in symptoms and executive function were explored using cTBS over the left DLPFC in individuals with ASD. The active group did not demonstrate superiority over the Sham group in any clinical or neuropsychological metrics at Week 8 or Week 12 [38]. Oberman et al. explored M1 plasticity and metaplasticity using cTBS in individuals with ASD and found that these individuals exhibited a significantly longer duration of suppression in motor-evoked potentials (MEPs) amplitude [36].

Clinical Recommendations

Previous studies have found that cTBS has greater and more enduring effects on motor cortical excitability in children with ASD, with aberrant responses becoming more pronounced as they age [40, 41]. It is suggested that cTBS induces inhibitory neuromodulation at the stimulated site. However, some findings indicate that earlier positive open-label trial results may be attributable to generalized placebo effects [38]. The clinical recommendation level for iTBS was unclear.

9.2.5.4 Bilateral TBS

Efficacy

Studies have found that iTBS applied to the bilateral DLPFC and bilateral posterior superior temporal sulcus (pSTS) is associated with improvements in autistic symptoms, executive functioning, social interaction, and reductions in repetitive behaviors among individuals with ASD [27–29].

Safety

Previous studies have demonstrated that a course of at least 1 month of regular and repeated bilateral iTBS is feasible and acceptable for intellectually capable children and adolescents with ASD [28].

Treatment Regimen

Studies have investigated the therapeutic effect of iTBS on the bilateral pSTS using various designs, such as single-session [27] and multisession [29]. Multisession iTBS over the bilateral pSTS could improve parent-rated autistic symptoms in ASD, while there were no significant differences in neuropsychological function following a single pSTS session.

Table 9.1 TMS in ASD

TMS	Clinical recommendations				Levels + References
	Target	Frequency	Length (min)[a]	Duration (sessions)	
Unilateral rTMS	unclear	unclear	unclear	unclear	unclear
Bilateral rTMS	unclear	unclear	unclear	unclear	unclear
Accelerated rTMS	unclear	unclear	unclear	unclear	unclear
Deep TMS	unclear	unclear	unclear	unclear	unclear
Priming TMS	unclear	unclear	unclear	unclear	unclear
Synchronized TMS	unclear	unclear	unclear	unclear	unclear
iTBS	unclear	unclear	unclear	unclear	unclear
Accelerated iTBS	unclear	unclear	unclear	unclear	unclear
cTBS	unclear	unclear	unclear	unclear	unclear
Bilateral TBS	unclear	unclear	unclear	unclear	unclear
MST	unclear	unclear	unclear	unclear	unclear

Clinical Recommendations

It has been demonstrated that iTBS over the pSTS is practicable and safe in individuals with ASD, and it may favorably increase patient compliance and accessibility [28]. The clinical recommendation level for iTBS was unclear.

9.2.6 MST

Magnetic seizure therapy (MST) employs HF magnetic stimulation to induce generalized seizures [42] and alleviates adverse neurocognitive effects by specifically targeting localized brain regions and inducing seizures in superficial cortical areas [43]. MST results in shorter recovery times and lower levels of cognitive impairments in individuals with major depressive disorder compared to ECT [44]. Consequently, MST has been proposed as a potential alternative to ECT [45]. Emerging evidence supports the clinical efficacy of MST in treating depression [46], bipolar depression [44], schizophrenia [42], and OCD [47], but its effectiveness for ASD remains unclear (Table 9.1).

9.3 tDCS

Transcranial direct current stimulation (tDCS) is a type of transcranial electrical stimulation (tES) that operates on the principle of stimulating the cerebral cortex using a weak, constant electrical current. Unlike transcranial magnetic stimulation (TMS), the current in tDCS is not strong enough to induce action potentials; instead, it modulates the likelihood of neuronal firing in the affected area [48]. The basic structure of tDCS comprises a stimulator, electrodes (an anode and a cathode), cables, conductive gel, and software.

The current intensity remains constant during brief ramping up and down phases, typically ranging from 0.5 to 2 mA, with a treatment session lasting between 5 and 30 min [48]. Research into tDCS intervention for autism has been steadily growing in recent years, with the majority using conventional tDCS to assess its efficacy, while fewer studies have explored HD-tDCS (High-definition tDCS).

9.3.1 Conventional tDCS

9.3.1.1 Efficacy

Most studies have reported that conventional tDCS is effective in alleviating symptoms in autism. In terms of its mechanism, conventional tDCS may modify local and global brain network dynamics through processes such as the regulation of neural synapses [48, 49]. Thus, targeting various sites can result in different effects. The most frequently targeted area is the left DLPFC. This region is closely associated with executive function, and certain neuropsychological studies have found that stimulating this area can enhance attention, working memory, and cognitive flexibility [50–52]. Results from randomized controlled trials have shown that stimulating this area effectively reduces scores on symptom assessment tools (CARS, ATEC, SRS-2, etc.) for individuals with autism [53–56] and even demonstrates a cumulative efficacy over time [57]. The stimulation of the right temporoparietal junction (rTPJ) may enhance social functioning and empathy in individuals with autism [58]. Additionally, the activation of the ventromedial prefrontal cortex (vmPFC) has been shown to improve emotion recognition [59], and the cerebellum has been associated with increased brain complexity [60].

9.3.1.2 Safety

Conventional tDCS has not typically exhibited adverse effects in individuals with autism in most studies. During the treatment period, the most commonly reported effect was itchiness [56, 61]. Posttreatment, the predominantly reported effects were mood changes, particularly irritability [57]. Mood changes (42.9%), irritability (35.7%), tingling (28.6%), and itching (28.6%) were the most commonly observed effects in a recent review [62]. Clinically, a review has indicated that children may exhibit greater scalp sensitivity, which could be mitigated by employing a lower concentration of NaCl solution (15–140 mmol/L) to minimize the likelihood of side effects [63].

9.3.1.3 Treatment Regimen

There are two types of stimulation: anodal stimulation, which facilitates activity in targeted brain regions, and cathodal stimulation, which suppresses it. As previously mentioned, stimulation targets for autism typically include the DLPFC, rTPJ, vmPFC, and cerebellum. Among various stimulation strategies, the most common plan for autism involves placing the anodal electrode over the left DLPFC and the cathodal electrode on the right shoulder. The stimulation current is usually kept at a stable level between 1 and 2 mA, with 30-second ramp-up and ramp-down phases.

However, it is recommended that the dose for children and adolescents be lower, ranging from 0.5 to 1 mA. The stimulation time typically lasts between 10 and 40 min, often set at 20 min, and the stimulation frequency is between 5 and 20 sessions per day, which has been shown to yield the best therapeutic outcomes.

9.3.1.4 Clinical Recommendations

A large number of randomized controlled trials and systematic reviews have confirmed the efficacy of stimulation targeting the left DLPFC, but other stimulation targets, such as the rTPJ, are still in the stage of case study or pilot study. Additionally, it is recommended that randomized controlled trials (RCTs) involving tDCS in children and adolescents systematically evaluate side effects, as they may be more sensitive and potentially unblind the treatment more easily.

9.3.2 HD-tDCS

9.3.2.1 Efficacy

Compared to conventional tDCS, HD-tDCS devices enable more specific stimulation of the targeted cortical region. However, research on their efficacy for autism is limited. Only two studies have reported that stimulation of the right temporoparietal junction (rTPJ) improved autism-spectrum quotient (AQ) scores and emotional face processing in individuals with high autistic traits [64, 65], and a pilot study found significant effects when stimulating the ventrolateral prefrontal cortex (VLPFC) [66].

9.3.2.2 Safety

Only a few recent studies have reported short-term adverse effects, the most common being itching and scalp pain.

9.3.2.3 Treatment Regimen

The most common target for stimulation is the rTPJ, with the stimulation current and duration typically set at 2 mA and 20 min, respectively. The stimulation frequency is usually once per day, and the number of sessions typically ranges from 5 to 20.

9.3.2.4 Clinical Recommendations

RCTs that investigate the treatment efficacy of HD-tDCS for autism are scarce, with most studies concentrating on individuals with high autistic traits. The clinical recommendation level for tDCS was unclear (Table 9.2).

Table 9.2 tDCS in ASD

	Clinical recommendations					
tDCS	Anode position	Cathode position	Intensity (mA)	Length (min)	Duration (sessions)	Levels +References
Conventional tDCS	unclear	unclear	unclear	unclear	unclear	unclear
HD-tDCS	unclear	unclear	unclear	unclear	unclear	unclear

9.4 TMS Vs. tDCS

9.4.1 Efficacy

9.4.1.1 Cognitive Effects

TMS may enhance cognitive function by affecting the activity or connectivity of neural networks. For instance, it can stimulate the release of dopamine and boost the functional neuroanatomical connections between the DLPFC and subcortical structures, such as the anterior cingulate cortex and the striatum [67]. Additionally, TMS has been shown to improve neural efficiency in cognitive processing speed and cortical plasticity, potentially accelerating cognitive processes [67].

According to a systematic review and meta-analysis, stimulation of the DLPFC through tDCS can yield partial cognitive benefits in various disorders, including depression and schizophrenia. These benefits manifest as improvements in attention/vigilance, working memory, executive function, processing speed, verbal fluency, verbal learning, and social cognition [68]. Another study indicated that tDCS alone did not exhibit superior cognitive enhancement effects compared to the sham stimulation group. Nonetheless, this suggests that tDCS has the potential to improve cognition, but it requires more focused targeting of specific areas [67].

The mechanism of tDCS for cognitive facilitation is similar to that of TMS, affecting the target regions and the connected neural networks. For instance, it has been reported that the application of tDCS to the DLPFC can increase the release of striatal dopamine.

A meta-analysis revealed that both TMS (effect size [ES] = 0.17, $p = 0.015$) and tDCS (ES = 0.17, $p = 0.021$) exhibited small yet significant effects on working memory. Additionally, tDCS outperformed sham stimulation in attention/vigilance (ES = 0.20, $p = 0.020$) [67]. Another study indicated that both tDCS and TMS can yield cognitive benefits, and the combination of these two interventions appears to create a synergistic effect [68]. When tDCS is utilized as a prestimulation followed by repetitive transcranial magnetic stimulation (rTMS), it can induce more enduring changes in cortical excitability and plasticity.

Thus, both TMS and tDCS can enhance cognitive function in patients with ASD to some degree, particularly in areas such as working memory and attention/vigilance. tDCS appears to have a more pronounced effect on improving attention/vigilance, whereas TMS might enhance cognitive function by affecting the activity and connectivity of neural networks. The concurrent application of these two techniques could potentially yield a synergistic effect, but further research is required to ascertain their respective benefits and the most effective ways to apply them in treating ASD.

9.4.1.2 Emotional and Behavioral Effects

Studies have demonstrated that TMS can influence emotional recognition in patients with ASD. For instance, research has found that the accuracy of emotional recognition in ASD patients who received TMS treatment was improved. Specifically, when TMS was applied to the right posterior superior temporal sulcus (rpSTS), it enhanced the emotional perception of ASD patients by regulating the relevant neural networks [69]. TMS has also been proposed as a treatment for the depressive aspects of ASD and is considered a well-tolerated treatment modality. In a study, 25 sessions of rTMS were administered to 10 individuals with a moderate burden of depressive and autistic symptoms. The results indicated an improvement in the depressive symptoms of ASD and suggested a potential impact on the behavioral aspects of the disorder [70].

tDCS has been utilized to enhance the behavioral and functional outcomes of individuals with ASD. In a particular study, 26 children diagnosed with ASD underwent tDCS treatment. The evaluation was conducted using the autism behavior checklist (ABC) scores, and the results indicated significant improvements in the ABC scores of the experimental group following tDCS treatment [69]. tDCS contributes to emotional regulation and behavioral aspects in patients with ASD by modulating activity in the DLPFC. Another study revealed that the severity of autistic symptoms, theory of mind, emotional regulation strategies, and emotion-behavior functions were enhanced in ASD patients who received tDCS treatment [71].

Both TMS and tDCS have demonstrated potential for improving the emotional and behavioral symptoms of patients with ASD. TMS, in particular, has shown efficacy in enhancing emotional recognition and alleviating depressive symptoms, whereas tDCS has demonstrated benefits in improving behavioral and functional outcomes [72].

9.4.1.3 Biological Effects

TMS has the capability to influence the neuroplasticity of patients with ASD by altering synaptic strength, akin to the long-term potentiation (LTP) effect [73]. TMS can modulate cortical excitability. For instance, studies conducted by Oberman et al. revealed that following iTBS, the corticospinal excitability in the ASD group was enhanced more persistently. Conversely, after cTBS, the inhibition was more persistent, indicating that individuals with ASD may exhibit enhanced plasticity [1]. Research by Sokhadze et al. showed that after TMS treatment, the prefrontal P3a amplitude in ASD patients decreased, and repetitive behaviors were reduced [74].

tDCS significantly influences the regulation of brain activity and cortical functional connectivity in children with ASD. Studies have indicated that tDCS, when compared to sham stimulation, can decrease resting-state functional connectivity [74]. Furthermore, tDCS has the potential to enhance functional connectivity within the Default Mode Network (DMN), which is vital for processing self-related and other-related information, retrieving episodic and autobiographical memories, and understanding the mental states of others. It may also improve the responsiveness of individuals with ASD to sensory inputs, thereby enhancing their multisensory integration and social abilities. Additionally, tDCS can increase the functional connectivity of key regions within the Dorsal Attention Network (DAN), potentially improving attention control and top-down goal-directed processing in individuals with ASD [74].

Both TMS and tDCS can influence the neuroplasticity and cortical excitability of patients with ASD to some extent, but their mechanisms of action and stimulation targets differ. TMS affects neural activity through magnetic fields, whereas tDCS influences neural excitability via weak direct current. Notably, tDCS exhibits effects in regulating the functional connectivity of the DMN, Sensorimotor Network (SMN), and DAN, which are associated with the core symptoms of ASD. Consequently, both TMS and tDCS have the potential to improve the biological effects in patients with ASD to some degree. However, their distinct mechanisms of action and stimulation targets result in varying effects on different biological aspects of ASD. Further research is required to ascertain the optimal stimulation parameters and methods, as well as their comparative advantages in the treatment of ASD.

9.4.2 Safety

The side effects of TMS typically include headaches, scalp discomfort, and, in rare cases, seizures. Research has indicated that TMS appears to be safe when used at low frequencies in individuals with ASD and can enhance various patient-oriented outcomes [75].

tDCS is relatively safe and well-tolerated. The side effects are typically mild, such as a tingling sensation, itching, or a slight headache. However, it is important to note that tDCS is largely unregulated and lacks certification. Home-use devices operated by nonmedical professionals or services provided by nonprofessional institutions cannot ensure the safety controls found in a clinical setting. In 19 tDCS studies involving over 500 participants, the participants received 1 to 10 tDCS treatments, each lasting 5 to 30 min at an intensity of 1 to 2 mA. Thirteen of these studies reported no adverse events (either severe or non-severe) [76]. Another review indicated that tDCS is a simple, safe, convenient, cost-effective, and well-tolerated method that has been used for cognitive modification and the alleviation of cognitive symptoms in healthy individuals for over 20 years [73].

Both TMS and tDCS are generally well-tolerated by patients with ASD. When comparing the safety profiles of TMS and tDCS, the side effects of tDCS are typically less severe than those of TMS. Nevertheless, the use of tDCS devices outside of clinical settings, such as at home, carries potential risks due to the lack of widespread regulation and certification for these devices. In contrast, TMS is subject to stricter safety protocols when administered in a clinical environment, whereas tDCS may not have equivalent safeguards when used in domestic or nonprofessional contexts. Therefore, it is essential that both treatments be conducted under the supervision of qualified medical personnel to guarantee patient safety.

9.5 Conclusion

The mechanisms of autism spectrum disorder are believed to be associated with neurodevelopment. Consequently, NIBS techniques, such as repetitive rTMS and tDCS, have been utilized as alternative or adjunctive treatments targeting these brain network mechanisms of ASD. However, it is important to note that NIBS techniques are still used off-label for ASD.

rTMS and tDCS, as noninvasive brain stimulation techniques for the treatment of ASD, have shown potential in treating the disorder. Currently, due to limited evidence, the widespread off-label use of these techniques in the ASD population is not yet endorsed. Future research should focus on conducting larger randomized controlled trials, incorporating neuroimaging to develop biomarkers for predicting treatment responses, and refining treatment parameters. Additionally, the safety, efficacy, and optimal application methods of these techniques in the treatment of ASD warrant further exploration.

Acknowledgments We thank Professor Yu Jin from Sun Yat-sen University and Professor Hongbo He from the Guangdong Mental Health Center, Guangdong Provincial People's Hospital (Guangdong Academy of Medical Sciences), and Southern Medical University for their support in drafting the manuscript.

Disclosure/Conflicts of Interest The authors declare no conflicts of interest in conducting this study or preparing the manuscript.

Financial Support This study was funded by the National Center for Mental Health of China which initiated the "Project on integrated development system of community prevention and rehabilitation of mental disorders and home care", the National Natural Science Foundation of China (30,600,207 and 81,171,293), the China Disabled Persons' Federation Special Project Foundation (2022CDPFHS-13), the Natural Science Foundation Program of Guangdong (2023A1515012442 and 2020A1515010942), the Science and Technology Planning Project of Guangdong (2011B031800020), the Medical Scientific Research Foundation of Guangdong (A20242171), and the Science and Technology Program of Guangzhou (2024A04J4576). The funders played no role in the study design, data collection and analysis, the decision to publish, or the preparation of the manuscript.

References

1. Oberman LM, et al. Transcranial magnetic stimulation in autism spectrum disorder: challenges, promise, and roadmap for future research. Autism Res. 2016;9(2):184–203.
2. Wagner T, Valero-Cabre A, Pascual-Leone A. Noninvasive human brain stimulation. Annu Rev Biomed Eng. 2007;9:527–65.
3. Gwynette MF, et al. Treatment of adults with autism and major depressive disorder using transcranial magnetic stimulation: an open label pilot study. Autism Res. 2020;13(3):346–51.
4. Levkovitz Y, et al. Efficacy and safety of deep transcranial magnetic stimulation for major depression: a prospective multicenter randomized controlled trial. World Psychiatry: Off J World Psychiatric Assoc (WPA). 2015;14(1):64–73.
5. Gogulski J, et al. Personalized repetitive transcranial magnetic stimulation for depression. Biol Psychiatry Cogn Neurosci Neuroimag. 2023;8(4):351–60.
6. Lefaucheur J-P, et al. Evidence-based guidelines on the therapeutic use of repetitive transcranial magnetic stimulation (rTMS): an update (2014-2018). Clin Neurophysiol. 2020;131(2):474–528.
7. Sokhadze E, et al. Low-frequency repetitive transcranial magnetic stimulation (rTMS) affects event-related potential measures of novelty processing in autism. Appl Psychophysiol Biofeedback. 2010;35(2):147–61.

8. Sokhadze EM, et al. Effects of low frequency repetitive transcranial magnetic stimulation (rTMS) on gamma frequency oscillations and event-related potentials during processing of illusory figures in autism. J Autism Dev Disord. 2009;39(4):619–34.
9. Yang Y, et al. Repetitive transcranial magnetic stimulation modulates long-range functional connectivity in autism spectrum disorder. J Psychiatr Res. 2023;160:187–94.
10. Oberman LM, Francis SM, Lisanby SH. The use of noninvasive brain stimulation techniques in autism spectrum disorder. Autism Res. 2024;17(1):17–26.
11. Smith JR, et al. Treatment response of transcranial magnetic stimulation in intellectually capable youth and young adults with autism Spectrum disorder: a systematic review and meta-analysis. Neuropsychol Rev. 2023;33(4):834–55.
12. Panerai S, et al. Effects of repetitive transcranial magnetic stimulation in performing eye-hand integration tasks: four preliminary studies with children showing low-functioning autism. Autism. 2014;18(6):638–50.
13. Casanova MF, et al. Effects of weekly low-frequency rTMS on autonomic measures in children with autism spectrum disorder. Front Hum Neurosci. 2014;8:851.
14. Croarkin PE, Wall CA, Lee J. Applications of transcranial magnetic stimulation (TMS) in child and adolescent psychiatry. Int Rev Psychiatry. 2011;23(5):445–53.
15. Wang Y, et al. Heart rate variability and skin conductance during repetitive TMS course in children with autism. Appl Psychophysiol Biofeedback. 2016;41(1):47–60.
16. Kang J, et al. Effects of 1Hz repetitive transcranial magnetic stimulation on autism with intellectual disability: a pilot study. Comput Biol Med. 2022;141:105167.
17. Kang J, et al. Impact of repetitive transcranial magnetic stimulation on the directed connectivity of autism EEG signals: a pilot study. Med Biol Eng Comput. 2022;60(12):3655–64.
18. Kang JN, et al. Effects of repetitive transcranial magnetic stimulation on children with low-function autism. CNS Neurosci Ther. 2019;25(11):1254–61.
19. Sokhadze EM, et al. Prefrontal neuromodulation using rTMS improves error monitoring and correction function in autism. Appl Psychophysiol Biofeedback. 2012;37(2):91–102.
20. Casanova MF, et al. Autism spectrum disorders: linking neuropathological findings to treatment with transcranial magnetic stimulation. Acta Paediatr. 2015;104(4):346–55.
21. Harmelech T, et al. Deep TMS H7 coil: features, applications & future. Expert Rev Med Devices. 2021;18(12):1133–44.
22. Enticott PG, et al. Deep repetitive transcranial magnetic stimulation associated with improved social functioning in a young woman with an autism Spectrum disorder. J ECT. 2011;27(1):41–3.
23. Enticott PG, et al. A double-blind, randomized trial of deep repetitive transcranial magnetic stimulation (rTMS) for autism Spectrum disorder. Brain Stimul. 2014;7(2):206–11.
24. Avirame K, et al. Enhanced cognition and emotional recognition, and reduced obsessive compulsive symptoms in two adults with high-functioning autism as a result of deep transcranial magnetic stimulation (dTMS): a case report. Neurocase. 2017;23(3–4):187–92.
25. Arıkan MK, İlhan R, Esmeray T, Laçin Çetin H, Aytar EK, Aktas H, et al. Deep transcranial magnetic stimulation effects on the electrophysiological arameters in obsessive-compulsive disorder. Clin EEG Neurosci. 2022;53(6):484–90.
26. Huang YZ, Edwards MJ, Rounis E, Bhatia KP, Rothwell JC. Theta burst stimulation of the human motor cortex. Neuron. 2005;45:201–6.
27. Ni HC, Hung J, Wu CT, Wu YY, Chang CJ, Chen RS, Huang YZ. The impact of single session intermittent theta-burst stimulation over the dorsolateral prefrontal cortex and posterior superior temporal sulcus on adults with autism Spectrum disorder. Front Neurosci. 2017;11:255.
28. Ni HC, Chen YL, Chao YP, Wu CT, Wu YY, Liang SH, Chin WC, Chou TL, Gau SS, Huang YZ, Lin HY. Intermittent theta burst stimulation over the posterior superior temporal sulcus for children with autism spectrum disorder: a 4-week randomized blinded controlled trial followed by another 4-week open-label intervention. Autism Int J Res Prac. 2021;25:1279–94.
29. Ni HC, Lin HY, Chen YL, Hung J, Wu CT, Wu YY, Liang HY, Chen RS, Gau SS, Huang YZ. 5-day multi-session intermittent theta burst stimulation over bilateral posterior superior temporal sulci in adults with autism-a pilot study. Biom J. 2022;45:696–707.

30. Liu P, Xiao G, He K, Zhang L, Wu X, Li D, Zhu C, Tian Y, Hu P, Qiu B, Ji GJ, Wang K. Increased accuracy of emotion recognition in individuals with autism-like traits after five days of magnetic stimulations. Neural Plast. 2020;2020:9857987.
31. Pedapati EV, Gilbert DL, Erickson CA, Horn PS, Shaffer RC, Wink LK, Laue CS, Wu SW. Abnormal cortical plasticity in youth with autism Spectrum disorder: a transcranial magnetic stimulation case-control pilot study. J Child Adolesc Psychopharmacol. 2016;26:625–31.
32. Abujadi C, Croarkin PE, Bellini BB, Brentani H, Marcolin MA. Intermittent theta-burst transcranial magnetic stimulation for autism spectrum disorder: an open-label pilot study. Revista Brasileira de psiquiatria, vol. 40. Sao Paulo, Brazil; 1999. p. 309–11.
33. Oberman L, Eldaief M, Fecteau S, Ifert-Miller F, Tormos JM, Pascual-Leone A. Abnormal modulation of corticospinal excitability in adults with Asperger's syndrome. Eur J Neurosci. 2012;36:2782–8.
34. Smith JR, DiSalvo M, Green A, Ceranoglu TA, Anteraper SA, Croarkin P, Joshi G. Treatment response of transcranial magnetic stimulation in intellectually capable youth and young adults with autism Spectrum disorder: a systematic review and meta-analysis. Neuropsychol Rev. 2023;33:834–55.
35. Jannati A, Ryan MA, Block G, Kayarian FB, Oberman LM, Rotenberg A, Pascual-Leone A. Modulation of motor cortical excitability by continuous theta-burst stimulation in adults with autism spectrum disorder. Clin Neurophysiol. 2021;132:1647–62.
36. Oberman LM, Ifert-Miller F, Najib U, Bashir S, Heydrich JG, Picker J, Rotenberg A, Pascual-Leone A. Abnormal mechanisms of plasticity and Metaplasticity in autism Spectrum disorders and fragile X syndrome. J Child Adolesc Psychopharmacol. 2016;26:617–24.
37. Yeh CH, Lin PC, Tseng RY, Chao YP, Wu CT, Chou TL, Chen RS, Gau SS, Ni HC, Lin HY. Lack of effects of eight-week left dorsolateral prefrontal theta burst stimulation on white matter macro/microstructure and connection in autism. Brain Imaging Behav. 2024;18:794–807.
38. Ni HC, Chen YL, Chao YP, Wu CT, Chen RS, Chou TL, Gau SS, Lin HY. A lack of efficacy of continuous theta burst stimulation over the left dorsolateral prefrontal cortex in autism: a double blind randomized sham-controlled trial. Autism Res: Off J Int Society Autism Res. 2023;16:1247–62.
39. Oberman LM, Pascual-Leone A. Hyperplasticity in autism Spectrum disorder confers protection from Alzheimer's disease. Med Hypotheses. 2014;83:337–42.
40. Oberman LM, Pascual-Leone A, Rotenberg A. Modulation of corticospinal excitability by transcranial magnetic stimulation in children and adolescents with autism spectrum disorder. Front Hum Neurosci. 2014;8:627.
41. Jannati A, Block G, Ryan MA, Kaye HL, Kayarian FB, Bashir S, Oberman LM, Pascual-Leone A, Rotenberg A. Continuous theta-burst stimulation in children with high-functioning autism Spectrum disorder and typically developing children. Front Integr Neurosci. 2020;14:13.
42. Zhang X-Y, Chen H-D, Liang W-N, Yang X-H, Cai D-B, Huang X, et al. Adjunctive magnetic seizure therapy for schizophrenia: a systematic review. Front Psych. 2021;12:813590.
43. Deng Z-D, Lisanby SH, Peterchev AV. Electric field strength and focality in electroconvulsive therapy and magnetic seizure therapy: a finite element simulation study. J Neural Eng. 2011;8(1):016007.
44. Tang VM, Blumberger DM, Throop A, McClintock SM, Voineskos D, Downar J, et al. Continuation magnetic seizure therapy for treatment-resistant ynipolar or bipolar depression. J Clin Psychiatry. 2021a;82(6):20m13677.
45. Chen M, Yang X, Liu C, Li J, Wang X, Yang C, et al. Comparative efficacy and cognitive function of magnetic seizure therapy vs. electroconvulsive therapy for major depressive disorder: a systematic review and meta-analysis. Transl Psychiatry. 2021;11(1):437.
46. Daskalakis ZJ, Dimitrova J, McClintock SM, Sun Y, Voineskos D, Rajji TK, et al. Magnetic seizure therapy (MST) for major depressive disorder. Neuropsychopharmacology. 2020;45(2):276–82.
47. Tang VM, Blumberger DM, Weissman CR, Dimitrova J, Throop A, McClintock SM, et al. A pilot study of magnetic seizure therapy for treatment-resistant obsessive-compulsive disorder. Depress Anxiety. 2021b;38(2):161–71.

48. Oberman LM, Francis SM, Lisanby SH. The use of noninvasive brain stimulation techniques in autism spectrum disorder [J]. Autism Research: Off J Int Soc Autism Res. 2024;17(1):17–26.
49. Zhou T, Kang J, Li Z, et al. Transcranial direct current stimulation modulates brain functional connectivity in autism [J]. NeuroImage Clin. 2020;28:102500.
50. Rothärmel M, Moulier V, Vasse M, et al. A prospective open-label pilot study of transcranial direct current stimulation in high-functioning autistic patients with a Dysexecutive syndrome [J]. Neuropsychobiology. 2019;78(4):189–99.
51. Chen L, Du B, Li K, et al. The effect of tDCS on inhibitory control and its transfer effect on sustained attention in children with autism spectrum disorder: an fNIRS study [J]. Brain Stimul. 2024;17(3):594–606.
52. Ratsapbhayakul T, Keeratitanont K, Chonprai C, et al. Anodal transcranial direct-current stimulation and non-verbal intelligence in autism spectrum disorder: a randomized controlled trial [J]. Dev Med Child Neurol. 2024;66(9):1244–54.
53. Amatachaya A, Auvichayapat N, Patjanasoontorn N, et al. Effect of anodal transcranial direct current stimulation on autism: a randomized double-blind crossover trial [J]. Behav Neurol. 2014;2014:173073.
54. Amatachaya A, Jensen MP, Patjanasoontorn N, et al. The short-term effects of transcranial direct current stimulation on electroencephalography in children with autism: a randomized crossover controlled trial [J]. Behav Neurol. 2015;2015:928631.
55. Hadoush H, Nazzal M, Almasri NA, et al. Therapeutic effects of bilateral anodal transcranial direct current stimulation on prefrontal and motor cortical areas in children with autism Spectrum disorders: a pilot study [J]. Autism Res: Off J Int Soc Autism Res. 2020;13(5):828–36.
56. Han YMY, Chan MMY, Shea CKS, et al. Neurophysiological and behavioral effects of multisession prefrontal tDCS and concurrent cognitive remediation training in patients with autism spectrum disorder (ASD): a double-blind, randomized controlled fNIRS study [J]. Brain Stimul. 2022;15(2):414–25.
57. Auvichayapat P, Intayot K, Udomchat C, et al. Long-term effects of transcranial direct current stimulation in the treatment of autism spectrum disorder: a randomized controlled trial [J]. Dev Med Child Neurol. 2023;65(6):811–20.
58. Esse Wilson J, Quinn DK, Wilson JK, et al. Transcranial direct current stimulation to the right Temporoparietal junction for social functioning in autism Spectrum disorder: a case report [J]. J ECT. 2018;34(1):e10–e3.
59. Salehinejad MA, Paknia N, Hosseinpour AH, et al. Contribution of the right temporoparietal junction and ventromedial prefrontal cortex to theory of mind in autism: a randomized, sham-controlled tDCS study [J]. Autism Res: Off J Int Soc Autism Res. 2021;14(8):1572–84.
60. Hadoush H, Hadoush A. Modulation of resting-state brain complexity after bilateral cerebellar anodal transcranial direct current stimulation in children with autism Spectrum disorders: a randomized controlled trial study [J]. Cerebellum (London, England). 2023;22(6):1109–17.
61. Chan MMY, Choi CXT, Tsoi TCW, et al. Effects of multisession cathodal transcranial direct current stimulation with cognitive training on sociocognitive functioning and brain dynamics in autism: a double-blind, sham-controlled, randomized EEG study [J]. Brain Stimul. 2023;16(6):1604–16.
62. Lee JC, Kenney-Jung DL, Blacker CJ, et al. Transcranial direct current stimulation in child and adolescent psychiatric disorders [J]. Child Adolesc Psychiatr Clin N Am. 2019;28(1):61–78.
63. Osório AAC, Brunoni AR. Transcranial direct current stimulation in children with autism spectrum disorder: a systematic scoping review [J]. Dev Med Child Neurol. 2019;61(3):298–304.
64. Donaldson PH, Kirkovski M, Rinehart NJ, et al. Autism-relevant traits interact with temporoparietal junction stimulation effects on social cognition: a high-definition transcranial direct current stimulation and electroencephalography study [J]. Eur J Neurosci. 2018;47(6):669–81.
65. Qiao Y, Hu Q, Xuan R, et al. High-definition transcranial direct current stimulation facilitates emotional face processing in individuals with high autistic traits: a sham-controlled study [J]. Neurosci Lett. 2020;738:135396.
66. Parmar D, Enticott PG, Albein-Urios N. Anodal HD-tDCS for cognitive inflexibility in autism spectrum disorder: a pilot study [J]. Brain Stimul. 2021;14(5):1298–300.

67. Begemann MJ, et al. Efficacy of non-invasive brain stimulation on cognitive functioning in brain disorders: a meta-analysis. Psychol Med. 2020;50(15):2465–86.
68. Li X, et al. Cognitive enhancing effect of rTMS combined with tDCS in patients with major depressive disorder: a double-blind, randomized, sham-controlled study. BMC Med. 2024;22(1):253.
69. Massoni L. Transcranial magnetic stimulation (TMS) and transcranial direct current stimulation (TDCS) in autism Spectrum disorder. Med Discover. 2024;3(1):1107.
70. Hameed MQ, et al. Transcranial magnetic and direct current stimulation in children. Curr Neurol Neurosci Rep. 2017;17(2):11.
71. Luckhardt C, et al. Brain stimulation by tDCS as treatment option in autism Spectrum disorder-a systematic literature review. Prog Brain Res. 2021;264:233–57.
72. Qiu J, et al. Transcranial direct current stimulation (tDCS) over the left dorsal lateral prefrontal cortex in children with autism Spectrum disorder (ASD). Neural Plast. 2021;2021:6627507.
73. Wang CSM, et al. Comparing transcranial direct current stimulation (tDCS) with other non-invasive brain stimulation (NIBS) in the treatment of Alzheimer's disease: a literature review. J Med Biol Eng. 2023;43:362–75.
74. Kang J, et al. Effects of transcranial direct current stimulation on brain activity and cortical functional connectivity in children with autism spectrum disorders. Front Psych. 2024;15:1407267.
75. Casanova MF, et al. Transcranial magnetic stimulation in autism Spectrum disorders: neuropathological underpinnings and clinical correlations. Semin Pediatr Neurol. 2020;35:100832.
76. Bikson M, et al. Safety of transcranial direct current stimulation: evidence based update 2016. Brain Stimul. 2016;9(5):641–61.

Anxiety Disorder

10

Wuyou Bao, Shiqi Yuan, Peiying Li, and Bin Zhang

Abstract

Anxiety disorders rank as the ninth leading cause of health-related disability and are characterized by high prevalence, chronicity, and comorbidity (Vos T, Abajobir AA, Abate KH, The Lancet 390:1211–59, 2017). According to the Diagnostic and Statistical Manual of Mental Disorders, Fifth Edition (DSM-5), anxiety disorders encompass generalized anxiety disorder, social anxiety disorder, panic disorder, specific phobias, and agoraphobia(Edition F, Am Psychiatric Assoc 21:591–643, 2013) While some individuals benefit from pharmacotherapy or psychotherapy, (Cuijpers P, Cristea IA, Karyotaki E, et al., World Psychiatry 15(3):245–58, 2016; Slee A, Nazareth I, Bondaronek P, et al., Lancet 393(10173):768–77, 2019) many individuals still suffer from treatment resistance and poor disease prognosis. Over the past two decades, numerous clinical trials have been conducted to explore more effective treatment options for anxiety disorders. Among these, transcranial magnetic stimulation (TMS) and transcranial direct current stimulation (tDCS) have emerged as promising, safe, and noninvasive neuromodulation techniques. However, current studies lack consensus on the optimal treatment protocols for TMS and tDCS. Further research is essential to refine stimulation parameters and enhance the clinical efficacy of these approaches for anxiety disorders. This chapter provides an overview of TMS and tDCS as potential treatments for anxiety disorders.

W. Bao · P. Li
Tianjin Anding Hospital, Tianjin Medical University, Tianjin, China

S. Yuan
Brain Hospital of Guangzhou Medical University, Guangzhou, China

B. Zhang (✉)
Tianjin Anding Hospital, Tianjin Medical University, Tianjin, China

Mental Health Center of Tianjin University, Tianjin, China

W. Zheng, Y. Ning (eds.), *TMS and tDCS for Psychiatric Disorders*,
https://doi.org/10.1007/978-981-96-8504-2_10

Keywords

Anxiety Disorders · Efficacy · Generalized Anxiety Disorder (GAD) · Panic Disorder (PD) · Neuromodulation · Transcranial Magnetic Stimulation (tDCS) · Transcranial Direct Current Stimulation, Efficacy, and Safety.

Abbreviations

ACC	Anterior Cingulate Cortex
amPFC	Anterior Medial Prefrontal Cortex
BAI	Beck Anxiety Inventory
BD	Bipolar Disorder
CAS	Clinical Anxiety Scale
cTBS	Continuous Theta Burst Stimulation
deep TMS	Deep Transcranial Magnetic Stimulation
DLPFC	Dorsolateral Prefrontal Cortex
DSM-5	The Diagnostic and Statistical Manual of Mental Disorders, Fifth Edition
ECT	Electroconvulsive Therapy
EEG	Electroencephalogram
fMRI	Functional Magnetic Resonance Imaging
FPz	Frontal Pole Midline
GAD	Generalized Anxiety Disorder
GAD-7	Generalized Anxiety Disorder 7-item Scale
HAMA	Hamilton Anxiety Rating Scale
HD-tDCS	High-Definition Transcranial Direct Current Stimulation
HRSA	Hamilton Rating Scale for Anxiety
IAF	Individual Peak Alpha Frequency
iTBS	Intermittent Theta Burst Stimulation
MDD	Major Depressive Disorder
mPFC	Medial Prefrontal Cortex
MST	Magnetic Seizure Therapy
NIBS	Noninvasive Brain Stimulation
OCD	Obsessive-Compulsive Disorder
Oz	Occipital Midline
PD	Panic Disorder
PDSS	Panic Disorder Severity Scale
PFC	Prefrontal Cortex
priming TMS	Priming Transcranial Magnetic Stimulation
PTSD	Post-Traumatic Stress Disorder
rTMS	Repetitive Transcranial Magnetic Stimulation
SAD	Social Anxiety Disorder
SCARED	Screen for Child Anxiety-Related Emotional Disorders
SMA	Supplementary Motor Area

SSRIs	Selective Serotonin Reuptake Inhibitors
STAI	State-Trait Anxiety Inventory
synchronized TMS	Synchronized Transcranial Magnetic Stimulation
TBS	Theta Burst Stimulation
tDCS	Transcranial Direct Current Stimulation
TMS	Transcranial Magnetic Stimulation

10.1 Introduction

Anxiety disorders rank as the second most prevalent psychiatric condition after depression and are a leading cause of disability worldwide [1]. The Diagnostic and Statistical Manual of Mental Disorders, Fifth Edition (DSM-5) categorizes anxiety disorders as including generalized anxiety disorder (GAD), social anxiety disorder (SAD), panic disorder (PD), specific phobias, and agoraphobia [2]. These disorders are characterized by pervasive feelings of anxiety and fear, often accompanied by significant behavioral disturbances. Among the primary subtypes, lifetime prevalence rates are estimated at 13% for SAD, 6.2% for GAD, 5.2% for PD, and 2.6% for agoraphobia [5]. Notably, a meta-analysis has shown that the prevalence of anxiety increased substantially during the COVID-19 pandemic, with approximately 25% of adults reporting anxiety symptoms between the onset of the pandemic in 2019 and August 2020 [6]. Furthermore, anxiety disorders frequently co-occur with or exacerbate various medical conditions, including cardiovascular diseases, gastrointestinal disorders, pulmonary diseases, cancer, chronic pain, and migraines [7].

Current clinical management of anxiety disorders primarily relies on pharmacotherapy, psychotherapy, and physical interventions. Pharmacological treatments commonly include anxiolytics (e.g., benzodiazepines) and antidepressants (e.g., selective serotonin reuptake inhibitors, SSRIs), which, while effective in symptom management, are often associated with significant side effects. Despite their efficacy for many patients, a subset of individuals exhibit poor responses to these treatments, experience unfavorable prognoses, or discontinue therapy due to adverse effects. Given these limitations, recent research has increasingly focused on developing novel therapeutic approaches to enhance treatment outcomes.

Over the past two decades, numerous clinical trials have explored innovative treatment strategies to improve both safety and efficacy in managing anxiety disorders. Among these, transcranial magnetic stimulation (TMS) and transcranial direct current stimulation (tDCS) have emerged as promising noninvasive neuromodulation techniques. These modalities are well-tolerated and capable of directly modulating neural activity within brain regions implicated in emotional and cognitive regulation, such as the dorsolateral prefrontal cortex (DLPFC), insula, and amygdala.

Accumulating evidence supports the potential efficacy of TMS and tDCS in alleviating anxiety symptoms. For instance, high-frequency TMS targeting the left DLPFC or low-frequency TMS targeting the right DLPFC has been shown to significantly reduce symptoms in patients with GAD and SAD. Similarly, tDCS has demonstrated anxiolytic effects when the cathodal electrode is positioned over the right DLPFC. However, findings across studies remain inconsistent, particularly concerning treatment efficacy and optimal stimulation protocols.

The clinical application of TMS and tDCS for anxiety disorders faces several challenges. Variability in stimulation parameters—such as frequency, intensity, treatment duration, and target locations—contributes to heterogeneity in treatment outcomes. Additionally, individual patient characteristics, including disease duration, symptom severity, and comorbidities, require further investigation to optimize treatment efficacy. Larger, well-designed randomized controlled trials are essential to refine these techniques, establish optimal stimulation parameters and targets, and assess their long-term efficacy and safety.

Overall, TMS and tDCS represent significant advancements in the treatment of anxiety disorders, offering promising alternatives to traditional pharmacological and psychotherapeutic interventions. As these techniques continue to evolve and as more high-quality studies are conducted, noninvasive neuromodulation may become an effective adjunct to conventional treatments, expanding the therapeutic landscape for anxiety disorders. This chapter will provide a comprehensive overview of the application of TMS and tDCS in anxiety disorder treatment.

10.2 TMS

TMS is a noninvasive neuromodulation technique that regulates neural activity in specific brain regions through electromagnetic induction. As a safe and effective alternative or adjunctive treatment, TMS has emerged as a key research focus in the treatment of anxiety disorders, demonstrating substantial therapeutic potential. The mechanism of action involves high-frequency stimulation to enhance neuronal excitability or low-frequency stimulation to inhibit hyperactive brain regions, thereby modulating neural networks involved in emotion and cognition. Depending on the stimulation pattern, TMS can be classified into traditional repetitive transcranial magnetic stimulation (rTMS), theta burst stimulation (TBS), priming transcranial magnetic stimulation (priming TMS), synchronized transcranial magnetic stimulation (synchronized TMS), and deep transcranial magnetic stimulation (deep TMS). While numerous studies have validated the efficacy of TMS in treating anxiety disorders, further research is required to standardize treatment parameters and optimize target regions.

10.2.1 rTMS

10.2.1.1 Unilateral rTMS

Unilateral rTMS is a noninvasive brain stimulation technique that modulates neural activity within specific regions of one hemisphere. It is characterized by high target specificity and adjustable stimulation parameters, making it a promising therapeutic approach for anxiety disorders.

Efficacy

Bystritsky and colleagues were the first to investigate the application of rTMS in the treatment of GAD. Their study included 10 patients who underwent low-frequency (1 Hz) stimulation over the right DLPFC for three weeks, totaling six sessions. Prior to treatment, functional magnetic resonance imaging (fMRI) was used to identify the most hyperactive regions in the frontal cortex, with the peak voxel in the right frontal lobe serving as the individualized rTMS stimulation target. Symptom severity was assessed using the Hamilton Anxiety Rating Scale (HAMA). The study reported a significant reduction in HAMA scores from baseline to posttreatment, with 60% (six patients) achieving significant clinical improvement and remission within three weeks. An additional two patients experienced a reduction in HAMA scores exceeding 50% but did not meet remission criteria [8].

Diefenbach et al. conducted a randomized, double-blind, sham-controlled trial involving 25 patients diagnosed with GAD. Participants were randomly assigned to either an active or sham rTMS group. The active group received low-frequency (1 Hz) stimulation targeting the right DLPFC for 15 min per session over a total of 30 sessions. While both the active and sham groups exhibited significant improvements in Hamilton Anxiety Rating Scale (HARS) scores following treatment, only the active group demonstrated sustained symptom reduction during follow-up. Moreover, the active stimulation group displayed higher response and remission rates compared to the sham group, with significant group-by-time interactions observed for anxiety, worry, and depressive symptoms [9].

In a separate randomized, double-blind, sham-controlled clinical trial, Dilkov et al. evaluated the efficacy of high-frequency (20 Hz) rTMS targeting the right DLPFC over a six-week treatment period in patients with GAD. After 25 sessions, the active rTMS group exhibited clinically significant reductions in HARS scores compared to the sham group. Additionally, during the 2-week and 4-week follow-up periods, the active group maintained or slightly improved its symptom reduction, suggesting sustained treatment effects [10].

Mantovani et al. conducted a randomized, double-blind trial examining the effects of rTMS in patients with PD and comorbid major depressive disorder. Twenty-five participants were randomly assigned to receive either active or sham 1 Hz rTMS targeting the right DLPFC. The active rTMS group exhibited significantly greater improvements in panic symptoms compared to the sham group, though no significant differences were observed in depressive symptoms. At week 4, the response rate for PD was 50% in the active group compared to 8% in the sham group. By week eight, the response rate had increased to 67% for PD and 50% for depressive symptoms. Repeated measures analysis of variance confirmed significant improvements in both PD and depression, with these effects maintained at a six-month follow-up [11].

A separate randomized, double-blind, sham-controlled study assessed the efficacy of rTMS in 15 patients with SSRI-refractory PD. Participants were assigned to either an active treatment group, receiving 1 Hz rTMS over the right DLPFC, or a sham stimulation group, which continued SSRI treatment alongside sham stimulation. During treatment, the sham group demonstrated statistically significant reductions in HAMA and Beck Anxiety Inventory (BAI) scores, whereas no comparable changes were observed in the rTMS group. However, at both the 2-week and 4-week follow-ups, no significant differences in overall symptom scores were detected between the two groups. Notably, both groups exhibited significant improvements in Panic Disorder Severity Scale (PDSS) scores by the end of treatment, but the differences in average PDSS scores between the two groups were not statistically significant during follow-up [12].

Overall, unilateral rTMS represents a promising therapeutic option for patients with anxiety disorders. However, further research is needed to establish optimal stimulation parameters, including frequency, intensity, treatment duration, and target regions, to enhance clinical efficacy and treatment outcomes.

Safety

The safety of unilateral rTMS in patients with anxiety disorders has been extensively studied alongside its efficacy. Overall, unilateral rTMS is considered a safe neuromodulation technique for this population. Most reported side effects are mild and transient, including headaches, dizziness, and localized scalp discomfort. However, Dilkov et al. documented a case in which a patient with GAD experienced a generalized tonic-clonic seizure during high-frequency (20 Hz) unilateral rTMS targeting the right DLPFC [10]. Although the patient fully recovered and completed the study, this incident underscores the need for vigilant monitoring of potential adverse effects associated with rTMS and highlights the importance of further research to ensure its safety profile.

Treatment Regimen

The most commonly employed rTMS treatment regimen for anxiety disorders involves high-frequency stimulation of the right DLPFC. Diefenbach et al. demonstrated significant symptom improvement in patients with GAD using a 1 Hz rTMS protocol [9], with subsequent studies replicating these findings [8, 13]. Conversely, high-frequency (20 Hz) unilateral rTMS targeting the right DLPFC has also been shown to significantly alleviate anxiety symptoms, with the active rTMS group outperforming the sham group. These conflicting findings regarding the efficacy of different unilateral rTMS protocols highlight the need for further investigation to establish optimal treatment parameters.

Clinical Recommendations

Although unilateral rTMS is not currently included in clinical treatment guidelines for anxiety disorders, emerging evidence suggests it may offer therapeutic benefits (Table 10.1). Under specific conditions, unilateral rTMS may be considered an experimental treatment for anxiety disorders. However, its efficacy and safety require further validation through rigorous clinical research. Before clinical implementation, the potential benefits and risks should be carefully weighed, and unilateral rTMS should not yet be regarded as a standard treatment approach.

Table 10.1 Optimization strategies of TMS in anxiety disorder

TMS	Clinical recommendations				Levels + references
	Target	Frequency	Length (min)[a]	Duration (sessions)	
Unilateral rTMS	Unclear	Unclear	Unclear	Unclear	Unclear
Bilateral rTMS	Unclear	Unclear	Unclear	Unclear	Unclear
Accelerated rTMS	Unclear	Unclear	Unclear	Unclear	Unclear
Deep TMS	Unclear	Unclear	Unclear	Unclear	Unclear
Priming TMS	Unclear	Unclear	Unclear	Unclear	Unclear
Synchronized TMS	Unclear	Unclear	Unclear	Unclear	Unclear
iTBS	Unclear	Unclear	Unclear	Unclear	Unclear
Accelerated iTBS	Unclear	Unclear	Unclear	Unclear	Unclear
cTBS	Unclear	Unclear	Unclear	Unclear	Unclear
Bilateral TBS	Unclear	Unclear	Unclear	Unclear	Unclear
MST	Unclear	Unclear	Unclear	Unclear	Unclear

Abbreviations: *cTBS* continuous theta burst stimulation, *iTBS* intermittent theta burst stimulation, *MST* magnetic seizure therapy, *rTMS* repetitive transcranial magnetic stimulation, *TBS* theta burst stimulation, *TMS* transcranial magnetic stimulation

10.2.1.2 Bilateral rTMS

Bilateral rTMS typically involves the application of high-frequency stimulation (commonly >5 Hz) to one hemisphere, such as the left DLPFC, to enhance neural activity, while low-frequency stimulation (commonly 1 Hz) is administered to the contralateral hemisphere, such as the right DLPFC, to suppress excessive neural activity. Alternatively, bilateral stimulation may involve simultaneous application to both the left and right DLPFC.

Efficacy

Lu et al. conducted a study involving 35 patients with GAD recruited from an outpatient clinic. Participants received low-frequency rTMS using a 70 mm "figure-eight" coil, targeting the right DLPFC followed by the left DLPFC, five times per week over a 2-week period. Following treatment, HARS scores showed a significant reduction from baseline, demonstrating a clinically meaningful alleviation of anxiety symptoms [13].

White et al. investigated the effects of rTMS in 13 adult outpatients diagnosed with comorbid major depressive disorder (MDD) and GAD. The treatment protocol consisted of low-frequency (1 Hz) stimulation to the right DLPFC, followed by high-frequency (10 Hz) stimulation to the left DLPFC, delivered over 24 to 36 sessions across 5 to 6 weeks. By the end of the treatment, 11 of the 13 patients (84.6%) experienced significant clinical improvement, with their GAD-7 scores decreasing to below 5 [14].

Clarke et al. examined the effects of unilateral and bilateral rTMS in patients with depression, both with and without comorbid anxiety. A total of 103 patients received sequential bilateral rTMS, while 149 underwent unilateral rTMS. The bilateral rTMS protocol consisted of 15 min of intermittent 10 Hz stimulation to the left DLPFC, followed by 15 min of continuous 1 Hz stimulation to the right DLPFC. In contrast, the unilateral rTMS protocol involved 15 min of 1 Hz stimulation applied to the right DLPFC. Both groups demonstrated significant posttreatment reductions in anxiety scores, but there was no statistically significant difference between them [15].

Chen et al. analyzed treatment data from 697 patients with MDD across five clinical sites. Among them, 179 received high-frequency rTMS to the left DLPFC, 218 underwent low-frequency rTMS to the right DLPFC, and 300 received sequential bilateral rTMS. All three protocols effectively alleviated anxiety symptoms, demonstrating comparable efficacy across treatments. Notably, patients with both high and low baseline anxiety scores exhibited significant symptom reductions [16].

Croarkin et al. retrospectively analyzed the outcomes of bilateral rTMS treatment in 59 adolescents (aged 12–17) diagnosed with MDD based on DSM-5 criteria. The bilateral rTMS protocol consisted of 1 Hz stimulation to the right DLPFC, followed by 20 Hz stimulation to the left DLPFC, administered over 36 sessions. The analysis revealed significant improvements in both depressive and anxiety symptoms, with 32 out of 57 patients (56.1%) meeting response criteria and 29 out of 57 patients (50.9%) achieving remission [17].

While bilateral rTMS has been explored as a potential treatment for anxiety disorders, the current body of evidence is too limited to conclusively establish its efficacy. Further research should employ rigorous methodologies is necessary to generate more generalizable findings.

Safety

The most frequently reported adverse effects of bilateral rTMS include headaches, mild scalp tenderness, and dizziness. Importantly, there are no documented reports in the literature of severe adverse events such as epileptic seizures, suicide attempts, completed suicides, or other serious complications. Future research should focus on systematically monitoring potential adverse effects and further evaluating the safety profile of bilateral rTMS to ensure its clinical viability.

Treatment Regimen

Bilateral rTMS protocols for anxiety disorders typically involve pairing low-frequency stimulation of the right DLPFC with high-frequency stimulation of the left DLPFC, or alternatively, applying low-frequency stimulation to both DLPFCs simultaneously to achieve therapeutic benefits. However, research in this area remains limited, and further large-scale studies are needed to identify the most effective treatment parameters.

Clinical Recommendations

Bilateral rTMS is not currently included in clinical treatment guidelines for anxiety disorders; however, preliminary evidence suggests it may have therapeutic potential (Table 10.1). Under specific conditions, bilateral rTMS may be considered an experimental treatment for anxiety disorders. Nevertheless, its efficacy and safety require further validation through well-designed clinical trials. Before clinical implementation, its potential benefits and risks should be carefully evaluated, and it should not yet be regarded as a standard treatment approach.

10.2.1.3 Accelerated rTMS

Accelerated rTMS refers to the administration of two or more rTMS sessions within a single day.

Efficacy

Senda et al. conducted a multiphase study to evaluate treatment outcomes in 48 patients diagnosed with depression, anxiety disorders, and/or PTSD. The first phase involved virtual reality therapy, followed by 25 sessions of accelerated rTMS targeting the left DLPFC in phase 2A, and an additional 25 sessions targeting the medial prefrontal cortex (mPFC) in phase 2B. Participants with GAD demonstrated significant reductions in anxiety symptoms following accelerated rTMS compared to baseline [18].

Holtzheimer III et al. investigated the effects of accelerated rTMS in 14 patients with MDD. The protocol involved 10 Hz rTMS targeting the left DLPFC over two consecutive days, with a total of 15 sessions (5 sessions on the first day and 10 on the second). Compared to baseline, HAMA scores showed significant reductions on days 3, 21, and 42 following treatment [19].

McGirr et al. conducted a study in which 27 patients with MDD underwent 10 Hz high-frequency rTMS applied to the left DLPFC. rTMS sessions were administered twice daily over 2 weeks. Following this accelerated treatment regimen, patients showed significant improvements in both depression and anxiety symptom scores [20].

While accelerated rTMS has been widely explored for the treatment of depressive disorders, research on its efficacy in anxiety disorders remains limited. However, preliminary findings are promising. Further investigation, particularly through sham-controlled trials, is necessary to refine accelerated rTMS protocols and establish its effectiveness in treating anxiety disorders.

Safety

The primary adverse effects associated with accelerated rTMS include headaches, scalp discomfort, fatigue, and emotional fluctuations. Caulfield et al. conducted a systematic review of 50 studies involving 741 participants who collectively underwent 15,693 rTMS sessions. The review reported no occurrences of epileptic seizures and found that the incidence of side effects was comparable to that of traditional once-daily rTMS protocols. These findings suggest that administering multiple rTMS sessions per day does not lead to cumulative adverse effects or an increased incidence of side effects [21]. In a separate study by Holtzheimer III et al., one participant discontinued treatment due to heightened suicidal ideation [19]. However, no other studies have reported severe adverse reactions, including suicide.

Treatment Regimen

Accelerated rTMS has demonstrated significant improvements in anxiety symptoms through stimulation of the left DLPFC. Additionally, some studies suggest that targeting the anterior medial prefrontal cortex (amPFC) may also yield therapeutic benefits. However, the optimal treatment protocol for anxiety disorders using accelerated rTMS has yet to be determined and warrants further research.

Clinical Recommendations

Accelerated rTMS is not currently included in clinical treatment guidelines for anxiety disorders. Nevertheless, emerging evidence suggests it may hold potential for alleviating anxiety symptoms (Table 10.1). Under specific conditions, accelerated rTMS may be considered an experimental treatment option for anxiety disorders; however, its efficacy and safety require further validation through well-designed clinical trials. Before clinical implementation, its potential benefits and risks should be carefully evaluated, and it should not yet be regarded as a standard treatment approach.

10.2.2 Deep TMS

Deep TMS employs specially designed H-coils capable of generating a more powerful magnetic field. Unlike conventional TMS, which primarily stimulates superficial brain regions at a depth of approximately 2–3 centimeters, deep TMS can penetrate to depths of 6–8 centimeters while maintaining a safe stimulation intensity. This allows for effective modulation of neurons within deeper brain structures. Furthermore, deep TMS has the unique capability to simultaneously target multiple subcortical and cortical regions, broadening its therapeutic potential.

10.2.2.1 Efficacy

In a study by Moraga-Amaro et al., 55 patients with GAD and 13 healthy individuals experiencing occupational stress were recruited. Using H4 and H2 coils, sequential stimulation was applied to the bilateral insula and the bilateral prefrontal cortex (PFC), with insula stimulation preceding PFC stimulation. High-frequency stimulation was administered to both regions 10 times per day, with each session lasting 30 min. The results showed significant reductions in anxiety, as evidenced by lower scores on the HARS, the State-Trait Anxiety Inventory (STAI), and the Clinical Anxiety Scale (CAS) when compared to baseline measurements [22].

Hovav et al. conducted a retrospective analysis of data from 10 patients, aged 20 to 74. The treatment protocol involved high-frequency (18 Hz) stimulation of the left PFC using an H1 coil, followed by low-frequency (1 Hz) stimulation of the right PFC. Following 20 sessions across 4 weeks, patients showed significant improvements in both depression and anxiety scores [23].

In another case study, Maiquez et al. treated a 10-year-old boy with autism, severe anxiety, and OCD who had shown limited response to psychotherapy and medication. Due to the prominence of anxiety symptoms, low-frequency (1 Hz) stimulation was initially applied to the right PFC using an H7 coil. Once obsessive symptoms subsided, stimulation was redirected to the mPFC and anterior cingulate cortex (ACC). After 20 sessions of low-frequency stimulation targeting the PFC, the patient's Screen for Child Anxiety Related Emotional Disorders (SCARED) scale scores decreased from 52 to 28, with the improvement sustained for two months [24].

Although research into the use of deep TMS for anxiety disorders is still in its early stages, preliminary results suggest that it holds significant therapeutic potential. However, due to the small sample sizes in current studies, more large-scale trials are needed to draw robust conclusions.

10.2.2.2 Safety

To date, no reports have documented of epileptic seizures, suicide attempts, completed suicides, or other severe adverse reactions associated with deep TMS treatment. However, continuous monitoring for potential adverse effects is necessary to ensure patient safety. Further research is required to validate the long-term safety profile of deep TMS.

10.2.2.3 Treatment Regimen

There is currently no consensus regarding the optimal treatment protocols for deep TMS. For instance, Moraga-Amaro et al. employed H4 and H2 coils for stimulating the bilateral insula and bilateral PFC, respectively [22]. In contrast, Maiquez et al. used H1 coils for stimulation of the left and right prefrontal areas [24]. The most effective protocol for deep TMS in treating anxiety disorders remains unclear, highlighting the need for further research to refine and standardize treatment strategies.

10.2.2.4 Clinical Recommendations

While deep TMS has not yet been included in clinical treatment guidelines for anxiety disorders, certain studies suggest it may be beneficial for alleviating anxiety symptoms (Table 10.1). Under specific conditions, deep TMS can be considered an experimental treatment option for anxiety disorders. However, its efficacy and safety need further validation through additional research. Prior to application, the potential benefits and risks should be carefully evaluated, and it should not be regarded as a standard treatment approach.

10.2.3 Priming TMS

Priming TMS is an emerging therapeutic technique within the field of TMS. This approach involves preconditioning a low-frequency stimulation sequence with high-frequency subthreshold stimulation [25]. Research suggests that this method significantly enhances the neural response to subsequent low-frequency stimulation, potentially improving therapeutic efficacy [26].

10.2.3.1 Efficacy

While definitive studies on the use of priming TMS for anxiety disorders remain limited, preliminary evidence suggests its potential efficacy in reducing anxiety symptoms, particularly in patients with OCD.

Vidya et al. conducted a prospective, hospital-based, double-blind, randomized, sham-controlled trial involving 30 patients with OCD. Participants were randomly assigned to either a priming TMS group or a non-priming TMS group. The priming TMS protocol consisted of 10 min of 6 Hz stimulation targeting the supplementary motor area (SMA), followed by 20 min of 1 Hz stimulation. Both groups underwent a total of 10 treatment sessions. Results showed that patients in the priming TMS group experienced significantly greater reductions in HARS scores compared to those in the non-priming TMS group, highlighting the potential therapeutic benefits of this approach [27].

10.2.3.2 Safety

To date, no severe adverse effects—such as seizures, suicide attempts, completed suicides, or other major complications—have been reported in association with priming TMS. However, ongoing vigilance and further studies are essential to comprehensively assess its long-term safety profile.

10.2.3.3 Treatment Regimen

The optimal protocol for priming TMS in the treatment of anxiety disorders has yet to be established. Further research is needed to develop evidence-based guidelines and standardized treatment protocols for its clinical application.

10.2.3.4 Clinical Recommendations

Priming TMS is not currently included in clinical treatment guidelines for anxiety disorders. However, preliminary findings suggest that it may offer therapeutic benefits in alleviating anxiety symptoms (Table 10.1). Under specific conditions, priming TMS may be considered an experimental treatment option for anxiety disorders. Nevertheless, further research is required to validate its efficacy and safety. Before clinical application, a careful assessment of its potential benefits and risks is necessary, and it should not yet be regarded as a standard treatment approach.

10.2.4 Synchronized TMS

Synchronized TMS is an emerging therapeutic modality. This technique utilizes rotating spherical rare-earth (neodymium) magnets positioned along the midline sagittal plane of the scalp, delivering stimulation that is synchronized with the individual's alpha brainwave frequency.

10.2.4.1 Efficacy

Synchronized TMS has shown potential as a treatment for anxiety disorders, although further research is necessary to confirm its efficacy.

Philip et al. conducted a multicenter study evaluating the effectiveness of synchronized TMS in patients with MDD. The findings indicated that individuals with more severe depression and heightened anxiety symptoms responded more favorably to synchronized TMS, suggesting its promise as a therapeutic option for anxiety disorders [28].

10.2.4.2 Safety

Reported adverse events associated with synchronized TMS include tinnitus, visual disturbances, difficulty with defecation, nonspecific gastrointestinal discomfort, back pain, pain in the hands, arms, shoulders, legs, or feet, dizziness, headache, paresthesia, insomnia, and fatigue. However, the overall safety profile of synchronized TMS remains incompletely understood, necessitating further investigation and careful monitoring in future studies.

10.2.4.3 Treatment Regimen

Synchronized TMS typically begins with determining the individual's peak alpha frequency (IAF) using a single-channel electroencephalogram (EEG) recording. Electrodes are placed at the Frontal Pole midline (FPz) and Occipital midline (Oz) positions, following the international 10–20 system. The synchronized TMS device consists of three sagittally aligned cylindrical neodymium magnets, magnetized along their diameters, positioned between the participant's forehead and the top of the head in close proximity to the scalp.

During treatment, participants remain in a supine position with their eyes closed while staying awake throughout the procedure. However, the optimal treatment protocol for synchronized TMS in managing anxiety disorders remains undefined, necessitating further research to establish evidence-based guidelines.

10.2.4.4 Clinical Recommendations

Synchronized TMS is not currently included in clinical treatment guidelines for anxiety disorders. However, preliminary studies suggest it may have potential in alleviating anxiety symptoms (Table 10.1). Under specific conditions, synchronized TMS may be considered an experimental treatment option for anxiety disorders. Nevertheless, its efficacy and safety require further validation through additional research. Before clinical implementation, the potential benefits and risks should be carefully assessed, and it should not yet be regarded as a standard treatment approach.

10.2.5 TBS

TBS is an advanced rTMS technique that significantly reduces stimulation duration compared to traditional rTMS. TBS exerts a powerful and lasting effect on synaptic plasticity and comprises two distinct protocols: intermittent TBS (iTBS) and continuous TBS (cTBS). While iTBS enhances cortical excitability, cTBS suppresses it, highlighting their complementary roles in modulating cortical activity.

10.2.5.1 iTBS

iTBS is a rapid and efficient stimulation protocol designed to enhance neuronal excitability and plasticity by delivering bursts of intermittent stimulation to targeted neural circuits.

Efficacy

Although iTBS has been widely studied in the context of depression, research on its efficacy for anxiety disorders remains limited, necessitating further investigation.

In a study by Zhang et al., 37 patients experiencing symptoms of both anxiety and depression underwent 10 sessions of iTBS targeting the left DLPFC. The treatment resulted in significant reductions in both anxiety and depressive symptoms [29].

Similarly, Deppermann et al. conducted a randomized study involving 44 patients who received either active or sham iTBS over the left DLPFC. While both groups, in combination with psychoeducation, exhibited significant reductions in scores on the HAMA, the Panic and Agoraphobia Scale, and the Cardiac Anxiety Questionnaire compared to baseline, no statistically significant differences were found between the active and sham iTBS groups. These findings highlight the need for further research to clarify the specific benefits of active iTBS [30].

A retrospective analysis by Lee et al. examined data from 42 patients with MDD who underwent iTBS at a tertiary medical center. The study revealed a significant reduction in GAD-7 scores among patients with anxious depression following at least 20 sessions of iTBS [31].

In another study, Zhang et al. randomly assigned 90 patients to receive either twice-daily iTBS, once-daily iTBS, or sham iTBS. Improvements in anxiety and other clinical symptoms were observed in both the once-daily and twice-daily iTBS groups. However, response and remission rates remained relatively low, with no significant differences observed between the treatment groups at any time point. Similarly, no significant differences were found between the active iTBS and sham iTBS groups, underscoring the need for further investigation into the clinical efficacy of iTBS [32].

Safety

The primary side effects of iTBS include headaches, dizziness, neck and shoulder pain, nausea, paresthesia, and facial twitching. Additionally, a case report has documented the induction of seizures in a child with schizophrenia following iTBS treatment [33]. While a separate case series has highlighted the potential for treatment-induced manic symptoms [34]. These findings underscore the importance of continued monitoring to fully assess and ensure the safety of iTBS.

Treatment Regimen

Currently, the left DLPFC is the most commonly targeted region for iTBS, with typical stimulation frequencies of 20 Hz or 50 Hz and treatment durations ranging from 10 to 30 min. However, the optimal protocol for using iTBS in the treatment of anxiety disorders remains undetermined. Further research is needed to refine therapeutic strategies and establish evidence-based guidelines.

Clinical Recommendations

iTBS is not yet included in clinical treatment guidelines for anxiety disorders. However, preliminary studies suggest it may have potential in alleviating anxiety symptoms (Table 10.1). Under specific conditions, iTBS may be considered an experimental treatment option for anxiety disorders. Nevertheless, its efficacy and safety require further validation through additional research. Before clinical application, a careful assessment of its potential benefits and risks is necessary, and it should not yet be regarded as a standard treatment approach.

10.2.5.2 Accelerated iTBS

Accelerated iTBS is a modified approach to TBS designed to achieve therapeutic effects more rapidly. Whereas conventional iTBS is typically administered once daily over several weeks, accelerated iTBS delivers multiple treatment sessions within a condensed timeframe, often completing the full course in just a few days.

Efficacy

Accelerated iTBS has been extensively studied for the treatment of depressive disorders; however, its application in anxiety disorders remains limited, necessitating further research to establish its efficacy in managing anxiety symptoms.

Tang et al. conducted a retrospective analysis of patients previously treated for affective disorders, including 27 individuals with depression or comorbid GAD who received either accelerated iTBS or cTBS. Both treatment modalities led to significant improvements in depressive and anxiety symptoms following intervention [35].

Goodman et al. performed an open-label clinical trial assessing the effects of extended accelerated iTBS treatment in patients with moderate to severe depression who required electroconvulsive therapy (ECT). The study comprised three phases: acute treatment, tapering, and relapse prevention. Across all phases, patients exhibited significant reductions in GAD-7 scores compared to baseline [36].

Similarly, Quinn et al. conducted a non-blind, single-arm prospective cohort study involving 25 elderly patients with depression who underwent 45 sessions of fMRI-guided accelerated iTBS. Significant reductions in GAD-7 scores were observed both mid-treatment and posttreatment compared to baseline [37].

In contrast, Cantù et al. evaluated six patients with MDD or bipolar disorder (BD) who underwent accelerated iTBS targeting the left DLPFC. The protocol involved three sessions per day over 6 days. Unlike previous studies, their results indicated a progressive increase in HAMA scores during and after treatment, with no symptom relief. This highlights variability in treatment outcomes and underscores the need for further investigation [38].

Safety

Adverse effects associated with accelerated iTBS are generally mild and include fatigue, headaches, muscle twitching, discomfort, sleep disturbances, and dizziness. However, Goodman et al. reported two cases of suicidal ideation or attempts during treatment [36]. To date, no severe adverse events such as epileptic seizures, completed suicides, or other major complications have been reported. Nevertheless, continuous monitoring and further research are necessary to thoroughly assess the safety profile of accelerated iTBS.

Treatment Regimen

Although accelerated iTBS shows promise in treating anxiety disorders, optimal treatment parameters remain undefined. Common stimulation protocols involve delivering a triplet-pulse burst at 30 Hz or 50 Hz, repeated at a frequency of 5 Hz. Each stimulation burst lasts 2 s, followed by an 8-s interval, with sessions typically consisting of either 600 or 1800 pulses per session [36, 38]. Despite these standardized parameters, further research is required to refine treatment protocols and determine the most effective stimulation settings for anxiety disorders.

Clinical Recommendations

Accelerated iTBS has not yet been incorporated into clinical treatment guidelines for anxiety disorders. However, preliminary studies suggest it may hold potential for alleviating anxiety symptoms (Table 10.1). Under specific conditions, accelerated iTBS can be considered an experimental treatment option for anxiety disorders; however, its efficacy and safety require further validation through additional research. Before clinical application, its potential benefits and risks should be carefully weighed, and it should not be regarded as a standard treatment approach.

10.2.5.3 Continuation TBS

cTBS is an advanced neuromodulation technique that modulates neuronal activity through rapid sequences of electromagnetic pulses. It has been explored as a potential intervention for various psychiatric disorders, including anxiety disorders.

Efficacy

Although research on cTBS for anxiety disorders remains limited, existing evidence suggests its potential therapeutic benefits. Further studies, including placebo-controlled trials, are needed to enhance our understanding of its efficacy and underlying mechanisms.

Li et al. conducted a randomized study involving 120 patients with GAD, who were assigned to receive cTBS targeting the right DLPFC, 1 Hz low-frequency rTMS, or sham cTBS. Each cTBS session consisted of 600 pulses delivered over 40 s, whereas iTBS sessions involved 1200 pulses over 20 min. After 20 sessions, both the cTBS and 1 Hz rTMS groups demonstrated significantly lower HAMA scores compared to the sham group. Notably, the cTBS group exhibited a greater reduction in anxiety symptoms than the 1 Hz rTMS group, both immediately posttreatment and at the 1-month follow-up. Additionally, the cTBS group demonstrated a significantly higher remission rate [39].

In a study by José et al., 27 GAD patients underwent 35 sessions of cTBS targeting the right DLPFC. Following treatment, anxiety symptoms, as measured by the Beck Anxiety Inventory (BAI), decreased by an average of 67.86%, with mean scores dropping from 26.74 to 8.59 [40].

Similarly, Littman et al. recruited nine patients with severe depression or bipolar depression from a private TMS clinic to receive off-label cTBS targeting the right DLPFC. Following treatment, anxiety symptoms in eight out of nine patients decreased by more than 50% on the HAMA scale [41].

Additionally, multiple studies on MDD, BD, and OCD have reported that cTBS can effectively alleviate anxiety symptoms, further supporting its potential as a therapeutic intervention [42–49].

Safety

The adverse effects of cTBS are generally mild and transient. Common side effects include fatigue, transient headaches, dizziness, facial muscle twitching, mild discomfort, somnolence, sleep disturbances, and scalp itching. Notably, no cases of epileptic seizures, suicide attempts, completed suicides, or other severe adverse reactions have been reported in association with cTBS. However, continuous monitoring of potential adverse effects remains essential to further evaluate its safety profile.

Treatment Regimen

The right DLPFC is the most commonly targeted brain region for cTBS in anxiety disorders. Standard stimulation parameters involve delivering three pulses at 50 Hz, repeated every 200 milliseconds (equivalent to a frequency of 5 Hz). Each session typically consists of 600 pulses administered over 40 s, with a 15-min interval between consecutive sets of 600 pulses. Further research is needed to determine the optimal cTBS protocol for anxiety disorders and refine its clinical application.

Clinical Recommendations

Currently, cTBS has not been incorporated into clinical treatment guidelines for anxiety disorders. However, preliminary evidence suggests it may hold promise in alleviating anxiety symptoms (Table 10.1). Under specific conditions, cTBS can be considered an experimental treatment option for anxiety disorders. However, its efficacy and safety require further validation through additional research. Before clinical implementation, the potential benefits and risks should be carefully weighed, and it should not yet be regarded as a standard treatment approach.

10.2.5.4 Bilateral TBS

Bilateral TBS is an advanced neuromodulation approach designed to target both hemispheres of the brain, either simultaneously or sequentially. This technique aims to enhance therapeutic outcomes by leveraging the functional connectivity between hemispheric regions implicated in mood and anxiety regulation.

Efficacy

The efficacy of bilateral TBS in treating anxiety disorders has yielded mixed results, with some studies demonstrating significant reductions in anxiety symptoms, while others report no clear advantage over sham or alternative treatments. Further research is required to establish consistent conclusions regarding its therapeutic potential.

Stubbeman et al. conducted a study involving 65 patients diagnosed with MDD and comorbid anxiety. Using navigation-guided 20 Hz TBS to stimulate the bilateral DLPFC, the study observed a significant reduction in BAI scores, which decreased by 62%—from 29.4 to 11.2—following treatment, suggesting a substantial alleviation of anxiety symptoms [50].

Chen et al. performed a three-arm, single-blind, randomized controlled multicenter trial to evaluate the effects of bilateral TBS on anxiety. Participants were randomly assigned to one of three groups: a control group receiving 10 Hz rTMS applied to the left DLPFC, a low-intensity bilateral group receiving cTBS and iTBS applied to the right and left DLPFC, respectively, and a high-intensity bilateral group using the same stimulation protocol but at higher intensities. All groups underwent 20 treatment sessions. While significant reductions in BAI scores were observed across all groups from baseline to weeks 4 and 8, no significant differences in anxiety reduction were found between the three treatment modalities [51].

Tavares et al. conducted a 6-week, single-center, double-blind, randomized, parallel-group, sham-controlled trial to assess the efficacy of bilateral TBS in anxiety treatment. The protocol involved cTBS applied to the right DLPFC, followed by iTBS applied to the left DLPFC. However, no significant differences in HAMA scores were observed between the active and sham groups, further highlighting the need for additional investigation [52].

Given these mixed findings, the potential of bilateral TBS in anxiety treatment remains uncertain. While some studies report substantial symptom improvement, others do not demonstrate superiority over control conditions. Further research, including larger, well-controlled trials, is essential to refine treatment protocols and determine the clinical utility of bilateral TBS for anxiety disorders.

Safety

Bilateral TBS is generally well-tolerated, with commonly reported side effects including headaches, fatigue, muscle twitching, insomnia, nausea, and variations in systolic blood pressure. Importantly, no severe adverse events—such as seizures, suicide attempts, completed suicides, or other major complications—have been reported. However, continued monitoring and further studies are necessary to fully characterize the safety profile of bilateral TBS and identify any potential long-term risks.

Treatment Regimen

The most commonly used bilateral TBS protocol involves applying iTBS to the left DLPFC, followed by cTBS to the right DLPFC. However, the optimal stimulation parameters for treating anxiety disorders remain undefined. Further research is needed to refine and optimize bilateral TBS treatment strategies, including variations in stimulation intensity, frequency, and session duration, to maximize therapeutic efficacy.

Clinical Recommendations

Bilateral TBS has not yet been incorporated into clinical treatment guidelines for anxiety disorders. Nevertheless, preliminary studies suggest it may have potential as an experimental intervention for alleviating anxiety symptoms (Table 10.1). Given the current lack of standardized protocols and mixed efficacy results, bilateral TBS should not be considered a standard treatment approach. Instead, its application should be carefully evaluated on a case-by-case basis, with potential benefits and risks weighed before implementation. Further rigorous research is required to establish its role in clinical practice and determine whether it can be integrated into evidence-based treatment frameworks for anxiety disorders.

10.2.6 Magnetic Seizure Therapy (MST)

MST is an emerging neuromodulation technology that employs short-duration, high-frequency magnetic pulses to noninvasively penetrate the scalp and skull. By inducing controlled seizures through high-intensity magnetic fields, MST aims to achieve therapeutic effects. As a novel intervention, most MST research to date has focused on depressive disorders, with no dedicated studies explicitly examining its efficacy for anxiety disorders.

However, MST may hold promise for highly treatment-resistant cases of anxiety disorders due to its capacity to generate focal seizures that can induce therapeutic change, as observed in depressive disorders [53]. Notably, MST has demonstrated the ability to alleviate anxiety symptoms in patients with depression. For instance, Wang et al. reported that accelerated MST significantly improved both depressive and anxiety symptoms in depressed patients [54].

Despite these findings, MST has not yet been incorporated into clinical treatment guidelines for anxiety disorders. While preliminary evidence suggests its potential to mitigate anxiety symptoms (Table 10.1), further research is necessary to determine its efficacy and establish its role as a viable treatment option for anxiety disorders.

10.3 tDCS

tDCS is a noninvasive neuromodulation technique that involves the application of weak direct currents (typically 1–2 mA, with a maximum limit of 4 mA) through electrodes placed on the scalp to modulate cortical excitability. tDCS primarily consists of two types: conventional tDCS and high-definition tDCS (HD-tDCS).

10.3.1 Conventional tDCS

Conventional tDCS administers a continuous, low-intensity direct current to specific brain regions, altering neural activity and modulating cortical excitability.

10.3.1.1 Efficacy

Several studies have investigated the efficacy of tDCS in treating anxiety disorders, yielding mixed findings.

Shiozawa et al. examined the effects of tDCS in a 58-year-old female patient with GAD. The cathode was placed over the right DLPFC to reduce neuronal excitability, while the anode was positioned over the contralateral deltoid muscle to minimize cortical excitability enhancement. The patient underwent 15 consecutive daily sessions (excluding weekends), with a current intensity of 2.0 mA applied for 30 min per session. The intervention resulted in a significant reduction in anxiety symptoms [55].

Lin et al. conducted a randomized controlled trial with 20 GAD patients, who were randomly assigned to either an active stimulation or sham stimulation group. In the active stimulation group, the cathode was placed over the right DLPFC and the anode over the left mastoid region, delivering 2.0 mA of direct current for 20 min per day over 10 days. The sham group received only brief stimulation at the beginning and end of each session (10 s). Patients in the active stimulation group exhibited significant improvements in HAMA scores at 2, 4, and 8 weeks ($p < 0.05$), whereas the sham group showed no significant changes throughout the study [56].

Contrasting results were reported by Lima et al., who conducted a double-blind, randomized, sham-controlled trial with 30 GAD patients. Participants in the active stimulation group received 2.0 mA of current for 20 min per day over five sessions, with the anode placed over the left DLPFC and the cathode over the right frontal cortex. The sham group underwent the same setup, but the current was turned off after 30 s. No significant improvements in anxiety, mood, stress, affect, or depressive symptoms were observed in the active stimulation group, although there was a notable reduction in physical stress symptoms [57].

Movahed et al. built upon previous findings by Shiozawa et al. and Bystritsky et al., conducting a randomized, single-blind, medication- and sham-controlled trial. Eighteen GAD patients were randomly assigned to receive either tDCS, medication, or sham stimulation. The tDCS protocol involved placing the cathode over the right frontal cortex and the anode over the contralateral deltoid muscle, with 2.0 mA of current administered for 20 min per session across 10 sessions. While both the tDCS and medication groups showed reductions in anxiety symptoms, these improvements were not statistically significant compared to the sham group [58].

Heeren et al. demonstrated that in individuals with SAD, anodal tDCS targeting the left DLPFC effectively attenuated attentional bias toward threats in female patients, thereby reducing their propensity to misinterpret neutral external cues as threatening [59].

Shiozawa et al. also explored the use of tDCS in a 44-year-old female patient with PD. The cathode was positioned over the right DLPFC, and the anode over the contralateral deltoid muscle, delivering 2.0 mA of current for 30 min across 10 consecutive days. This intervention significantly improved the patient's anxiety symptoms [60].

However, conflicting results were reported by Aksu et al., who conducted a triple-blind, randomized, sham-controlled trial involving 30 PD patients. The active stimulation group received tDCS with the anode placed over the left DLPFC (F3) and the cathode over the right DLPFC (F4) for 20 min per session over 10 consecutive days. While both the active and sham stimulation groups showed improvements in anxiety scores, no significant differences were observed between them, and the active stimulation group did not demonstrate superior therapeutic effects [61].

Overall, while tDCS has demonstrated potential benefits in clinical trials and case reports for anxiety disorders, the small sample sizes and heterogeneous findings underscore the need for larger, more rigorous studies to establish consistent and reliable conclusions.

10.3.1.2 Safety

Conventional tDCS is generally regarded as a safe neuromodulation technique. Data from human trials involving over 33,200 treatment sessions and more than 1000 participants receiving repeated treatments have reported no serious adverse effects or irreversible damage when adhering to standard tDCS protocols (≤40 min per session, ≤4 mA current intensity, ≤7.2 coulombs total charge) [62]. The most commonly reported side effects are mild and transient, including tingling or itching at the stimulation site, headaches, drowsiness, localized skin redness, and temporary changes in attention and mood.

A rare but documented side effect is a mild superficial electrolytic burn, with one case reporting a 1 × 1 cm light red mark on the forehead, which fully resolved within a week without complications in subsequent sessions. Short-term studies have also indicated that tDCS is safe and tolerable in children and adolescents with psychotic or neurodevelopmental disorders. However, given the potential impact on the developing brain, further systematic research is necessary to ensure its safety and efficacy in younger populations [62].

10.3.1.3 Treatment Regimen

The standard tDCS protocol for anxiety disorders typically employs a direct current intensity of 2.0 mA, delivered through anodal and cathodal electrodes. The cathode is generally positioned over the target brain region, commonly the right or left DLPFC, which plays a critical role in emotion regulation and cognitive control. The anode is often placed in extracranial or intracranial locations such as the contralateral deltoid, mastoid, or shoulder.

Each tDCS session typically lasts between 20 and 30 min, with repeated sessions administered over several days or weeks, depending on the study protocol. However, the optimal stimulation parameters—including electrode placement, current intensity, and session duration—remain uncertain for the effective treatment of anxiety

Table 10.2 Optimization strategies of tDCS in anxiety disorder

tDCS	Clinical Recommendations					Levels +References
	Anode position	Cathode position	Intensity (mA)	Length (min)	Duration (sessions)	
Conventional tDCS	Unclear	Unclear	Unclear	Unclear	Unclear	Unclear
HD-tDCS	Unclear	Unclear	Unclear	Unclear	Unclear	Unclear

Abbreviations: *HD-tDCS* high-definition transcranial direct current stimulation, *tDCS* transcranial direct current stimulation

disorders. Further rigorous research is necessary to establish standardized, evidence-based clinical guidelines.

10.3.1.4 Clinical Recommendations

At present, conventional tDCS is not included in clinical treatment guidelines for anxiety disorders. Nevertheless, some studies suggest that it may offer potential benefits for alleviating anxiety symptoms (Table 10.2). Under specific conditions, tDCS may be considered an experimental treatment option for anxiety disorders; however, its efficacy and safety require further validation through additional research. Before clinical application, the potential benefits and risks should be carefully evaluated, and tDCS should not be regarded as a first-line treatment approach.

10.3.2 HD-tDCS

HD-tDCS is considered a safer and more tolerable alternative to conventional tDCS. This is primarily due to its use of smaller, specialized electrode arrays, which enhance spatial resolution and precision. By enabling more targeted cortical stimulation, HD-tDCS minimizes undesired effects on surrounding brain regions, thereby improving the specificity of neural modulation.

10.3.2.1 Efficacy

Current research on HD-tDCS has primarily focused on conditions such as depression and PTSD, with limited investigation into its efficacy for anxiety disorders. As a result, definitive conclusions regarding its effectiveness in treating anxiety disorders remain inconclusive. However, emerging evidence suggests that HD-tDCS may have potential in alleviating anxiety symptoms, particularly in patients with depression.

Zhang et al. conducted a randomized, double-blind, active-controlled exploratory study involving 60 patients diagnosed with depression and anxiety. Participants were randomly assigned to either an active or sham stimulation group. In the active stimulation group, the anode was placed over the left DLPFC at the F3 electrode site, while four cathodes were positioned at the left Frontal-Central (FC1), left Anterior Frontal (AF3), left Inferior Frontal (F7), and left Frontal-Central (FC5) regions. A direct current of 2.0 mA was administered for 20 min per session. The sham stimulation group followed the same electrode placement and parameter settings, but the current was applied only for 30 s at the beginning and end of each session. Both groups underwent daily stimulation sessions, five times per week, for a total of 10 sessions. Posttreatment analysis revealed a significant reduction in anxiety symptoms, with BAI scores significantly lower in the active group compared to the sham group. These findings suggest that HD-tDCS may be a promising intervention for reducing anxiety symptoms; however, further research is needed to establish its efficacy as a targeted treatment for anxiety disorders [63].

10.3.2.2 Safety

Studies conducted on elderly individuals, healthy adults, and children have demonstrated that HD-tDCS is well-tolerated and has a favorable safety profile [64–66]. The most commonly reported side effects include mild and transient sensations such as tingling, itching, skin redness, a burning sensation, tinnitus, and headaches at the site of stimulation. Importantly, no severe adverse events have been reported in existing studies. However, ongoing research and systematic monitoring are necessary to further assess the long-term safety of HD-tDCS, particularly in clinical populations.

10.3.2.3 Clinical Recommendations

HD-tDCS has not yet been incorporated into clinical treatment guidelines for anxiety disorders. Nevertheless, preliminary research suggests that it may hold potential for alleviating anxiety symptoms (Table 10.2). Under specific conditions, HD-tDCS may be considered an experimental treatment option for anxiety disorders; however, its clinical efficacy and safety require further validation through well-designed, large-scale studies. Before clinical implementation, the potential benefits and risks should be carefully evaluated, and HD-tDCS should not be regarded as a standard treatment approach at this stage.

10.4 TMS versus tDCS

10.4.1 Efficacy

TMS and tDCS are prominent noninvasive brain stimulation (NIBS) techniques increasingly investigated for their therapeutic potential in anxiety disorders. Despite sharing a common noninvasive nature, these modalities differ significantly in their underlying mechanisms of action. TMS operates by generating brief electromagnetic pulses that induce changes in neuronal activity, with varying frequencies (e.g., low and high) exerting differential effects—low-frequency TMS typically decreases cortical excitability, while high-frequency TMS enhances it. In contrast, tDCS modulates cortical excitability via the application of a continuous direct current. Anodal tDCS generally increases excitability, whereas cathodal tDCS produces inhibitory effects [67].

The efficacy of TMS and tDCS in treating anxiety disorders appears to be disorder-specific, with mixed results reported across various studies. For instance, investigations involving patients with PD suggest that both active iTBS and tDCS do not outperform sham stimulation, implying a potential placebo effect or limited efficacy in this cohort. Conversely, patients with GAD appear to benefit more substantially from TMS interventions. Studies have demonstrated that both low-frequency and high-frequency TMS can significantly improve GAD symptoms, suggesting a disorder-specific efficacy profile. However, findings related to tDCS in GAD remain inconsistent; while some studies report significant symptom improvement, others fail to demonstrate efficacy beyond placebo. These discrepancies highlight the need for further research to elucidate the optimal parameters and mechanisms of tDCS for anxiety treatment.

Overall, TMS appears to have garnered more consistent support in clinical efficacy, particularly for GAD, while the therapeutic value of tDCS remains uncertain and warrants additional investigation.

10.4.2 Safety

Both TMS and tDCS are broadly recognized as safe and well-tolerated NIBS techniques, with reported adverse effects being generally mild and transient. In the case of TMS, common side effects include mild headaches and localized scalp discomfort, with no severe adverse reactions reported in the literature. Furthermore, TMS has demonstrated a favorable safety profile in pediatric and adolescent populations, extending its potential applicability [68, 69]. tDCS similarly boasts a high safety profile, with common adverse effects limited to minor itching, tingling, headaches, localized burning sensations, and discomfort. Compared to TMS, tDCS side effects are typically milder and of shorter duration [70]. Despite these positive findings, the safety of both techniques across diverse populations, particularly the elderly and individuals with comorbidities, remains an area requiring further scrutiny. Long-term studies are essential to comprehensively assess the safety of these modalities in broader patient demographics.

10.5 Conclusion

Although the efficacy of TMS and tDCS in treating anxiety disorders remains subject to ongoing debate, a substantial body of evidence suggests their potential therapeutic utility, particularly in patients with GAD. TMS has demonstrated a stronger empirical foundation for efficacy, while the effects of tDCS remain inconsistent and require further validation through rigorous study designs.

These two technologies offer distinct advantages, including noninvasiveness, a favorable safety profile, and minimal side effects. However, their clinical efficacy is influenced by various factors such as the specific type of anxiety disorder (e.g., GAD, PD), illness duration, and concurrent treatments (e.g., pharmacotherapy or psychotherapy).

Optimization of therapeutic outcomes will likely require fine-tuning of various parameters. For TMS, critical factors include stimulation frequency (high versus low), intensity, stimulation site (e.g., right or left dorsolateral prefrontal cortex), and duration of treatment. For tDCS, key variables include the selection of anodal or cathodal stimulation, electrode placement, and current intensity. Tailoring these parameters to individual patients' needs could significantly enhance clinical efficacy.

Future research should prioritize the identification of factors influencing individual treatment responses, such as neurobiological characteristics (e.g., cortical excitability and structural abnormalities), symptom profiles (e.g., presence of comorbid depressive symptoms), and genetic predispositions. Through the integration of personalized medicine approaches and large-scale research data, more precise and efficient treatment strategies can be developed.

From a clinical research perspective, there is a pressing need for large-scale, high-quality randomized controlled trials to establish the long-term efficacy and safety of these technologies and to clarify their applicability across diverse patient populations. Additionally, investigating the synergistic effects of combining NIBS with other interventions, such as pharmacotherapy and cognitive-behavioral therapy, remains an important area of exploration.

The continued advancement of neuroimaging techniques, such as fMRI and EEG, will provide valuable insights into the neural networks and functional changes associated with TMS and tDCS. Such technologies can offer a more reliable biological foundation for refining treatment protocols and enhancing therapeutic efficacy.

In conclusion, while current research presents several limitations, the future of TMS and tDCS in treating anxiety disorders appears promising. Through the development of more precise treatment designs and a deeper understanding of individualized response mechanisms, these technologies have the potential to provide safer, more effective, and personalized treatment options for patients with anxiety disorders.

Acknowledgments None.

Disclosure/Conflicts of Interest The authors declare no conflicts of interest in conducting this study or preparing the manuscript.

Financial Support None.

References

1. Vos T, Abajobir AA, Abate KH, et al. Global, regional, and national incidence, prevalence, and years lived with disability for 328 diseases and injuries for 195 countries, 1990–2016: a systematic analysis for the global burden of disease study 2016 [J]. Lancet. 2017;390(10100):1211–59.
2. Edition F. Diagnostic and statistical manual of mental disorders [J]. Am Psychiatric Assoc. 2013;21(21):591–643.
3. Cuijpers P, Cristea IA, Karyotaki E, et al. How effective are cognitive behavior therapies for major depression and anxiety disorders? A meta-analytic update of the evidence [J]. World Psychiatry. 2016;15(3):245–58.

4. Slee A, Nazareth I, Bondaronek P, et al. Pharmacological treatments for generalised anxiety disorder: a systematic review and network meta-analysis [J]. Lancet. 2019;393(10173):768–77.
5. Kessler RC, Petukhova M, Sampson NA, et al. Twelve-month and lifetime prevalence and lifetime morbid risk of anxiety and mood disorders in the United States [J]. Int J Methods Psychiatr Res. 2012;21(3):169–84.
6. Santabárbara J, Lasheras I, Lipnicki DM, et al. Prevalence of anxiety in the COVID-19 pandemic: an updated meta-analysis of community-based studies [J]. Prog Neuro-Psychopharmacol Biol Psychiatry. 2021;109:110207.
7. Kariuki-Nyuthe C, Stein DJ. Anxiety and related disorders and physical illness [J]. In: Sartorius N, RIG H, Maj M, editors. Comorbidity of mental and physical disorders. Berlin: Karger; 2015. p. 81–7.
8. Bystritsky A, Kaplan JT, Feusner JD, et al. A preliminary study of fMRI-guided rTMS in the treatment of generalized anxiety disorder [J]. J Clin Psychiatry. 2008;69(7):1092–8.
9. Diefenbach GJ, Bragdon LB, Zertuche L, et al. Repetitive transcranial magnetic stimulation for generalised anxiety disorder: a pilot randomised, double-blind, sham-controlled trial [J]. Br J Psychiatry. 2016;209(3):222–8.
10. Dilkov D, Hawken ER, Kaludiev E, et al. Repetitive transcranial magnetic stimulation of the right dorsal lateral prefrontal cortex in the treatment of generalized anxiety disorder: a randomized, double-blind sham controlled clinical trial [J]. Prog Neuro-Psychopharmacol Biol Psychiatry. 2017;78:61–5.
11. Mantovani A, Aly M, Dagan Y, et al. Randomized sham controlled trial of repetitive transcranial magnetic stimulation to the dorsolateral prefrontal cortex for the treatment of panic disorder with comorbid major depression [J]. J Affect Disord. 2013;144(1–2):153–9.
12. Prasko J, Zalesky R, Bares M, et al. The effect of repetitive transcranial magnetic stimulation (rTMS) add on serotonin reuptake inhibitors in patients with panic disorder: a randomized, double blind sham controlled study [J]. Neuroendocrinol Lett. 2007;28(1):33–8.
13. Lu R, Zhang C, Liu Y, et al. The effect of bilateral low-frequency rTMS over dorsolateral prefrontal cortex on serum brain-derived neurotropic factor and serotonin in patients with generalized anxiety disorder [J]. Neurosci Lett. 2018;684:67–71.
14. White D, Tavakoli S. Repetitive transcranial magnetic stimulation for treatment of major depressive disorder with comorbid generalized anxiety disorder [J]. Ann Clin Psychiatry. 2015;27(3):192–6.
15. Clarke E, Clarke P, Gill S, et al. Efficacy of repetitive transcranial magnetic stimulation in the treatment of depression with comorbid anxiety disorders [J]. J Affect Disord. 2019;252:435–9.
16. Chen L, Hudaib AR, Hoy KE, et al. Is rTMS effective for anxiety symptoms in major depressive disorder? An efficacy analysis comparing left-sided high-frequency, right-sided low-frequency, and sequential bilateral rTMS protocols [J]. Depress Anxiety. 2019;36(8):723–31.
17. Croarkin PE, Zuckerman S, Middleton VJ, et al. Clinical outcomes in adolescents undergoing sequential bilateral 1 Hz/20 Hz transcranial magnetic stimulation for treatment resistant depression [J]. Brain Stimul. 2024;17(2):431–3.
18. Senda M, Johnson K, Taylor I, et al. Sequenced pilot trial using virtual reality followed by two different accelerated TMS protocols for treating dysphoria [J]. Brain Stimulation: Basic Translat Clinical Res Neuromodulation. 2023;16(4):7.
19. Holtzheimer PE II, Mcdonald WM, Mufti M, et al. Accelerated repetitive transcranial magnetic stimulation for treatment-resistant depression [J]. Depress Anxiety. 2010;27(10):960–3.
20. Mcgirr A, Van Den Eynde F, Tovar-Perdomo S, et al. Effectiveness and acceptability of accelerated repetitive transcranial magnetic stimulation (rTMS) for treatment-resistant major depressive disorder: an open label trial [J]. J Affect Disord. 2015;173:216–20.
21. Caulfield K. A systematic review of the safety profile and key variables in accelerated rTMS [J]. Brain Stimulation: Basic Translat Clinical Res Neuromodulation. 2021;14(6):1718–9.
22. Moraga-Amaro R, Muñoz P, Villalobos T, et al. Real-world data of non-invasive stimulation of the human insula-prefrontal cortices using deep TMS to treat anxiety for occupational stress and generalized anxiety disorder [J]. Psychiatry Res. 2023;320:115036.

23. Hovav S, Kinback K. Deep TMS for comorbid major depressive disorder and anxiety–a brief report of patients in a real-world practice [J]. Brain Stimulation: Basic Translat Clinical Res Neuromodulation. 2014;7(5):e20.
24. Maiquez A, Moreno O, Jarvis A, et al. Deep-TMS in a 10-year-old with autism anxiety and OCD [J]. Brain Stimulation: Basic Translat Clinical Res Neuromodulation. 2021;14(5):1404.
25. Iyer MB, Schleper N, Wassermann EM. Priming stimulation enhances the depressant effect of low-frequency repetitive transcranial magnetic stimulation [J]. J Neurosci. 2003;23(34):10867–72.
26. Abraham WC, Bear MF. Metaplasticity: the plasticity of synaptic plasticity [J]. Trends Neurosci. 1996;19(4):126–30.
27. Vidya KL, Rao PG, Goyal N. Adjuvant priming repetitive transcranial magnetic stimulation for treatment-resistant obsessive-compulsive disorder: in search of a new paradigm! [J]. J ECT. 2022;38(1):e1–8.
28. Philip NS, Leuchter AF, Cook IA, et al. Predictors of response to synchronized transcranial magnetic stimulation for major depressive disorder [J]. Depress Anxiety. 2019;36(3):278–85.
29. Zhang Y, Li L, Bian Y, et al. Theta-burst stimulation of TMS treatment for anxiety and depression: a FNIRS study [J]. J Affect Disord. 2023;325:713–20.
30. Deppermann S, Vennewald N, Diemer J, et al. Does rTMS alter neurocognitive functioning in patients with panic disorder/agoraphobia? An fNIRS-based investigation of prefrontal activation during a cognitive task and its modulation via sham-controlled rTMS [J]. Biomed Res Int. 2014;2014:542526.
31. Lee NA, Kung S, Hu Y, et al. Depression and anxiety outcomes of adults undergoing intermittent theta burst stimulation (iTBS) treatment for major depressive disorder: a naturalistic study [J]. J Mood Anxiety Disorders. 2023;3:100028.
32. Zhang M, Li W, Ye Y, et al. Efficacy and safety of intermittent theta burst stimulation on adolescents and young adults with major depressive disorder: a randomized, double blinded, controlled trial [J]. J Affect Disord. 2024;350:214–21.
33. Purushotham A, Sinha VK, Goyal N, et al. Intermittent theta burst stimulation induced seizure in a child with schizophrenia: a case report [J]. Brain Stimul. 2018;11(6):1415–6.
34. Shere SS, Baliga SP, Mehta UM, et al. Treatment emergent affective switch in depression: a possible role of theta burst stimulation [J]. Brain Stimul. 2020;13(2):492–3.
35. Tang SJ, Holle J, Mor S, et al. Improvements in sleep quality in patients with major depressive and generalized anxiety disorders treated with individualized, parcel-guided transcranial magnetic stimulation [J]. Brain Behavior. 2024;14(10):e70088.
36. Goodman MS, Trevizol AP, Konstantinou GN, et al. Extended course accelerated intermittent theta burst stimulation as a substitute for depressed patients needing electroconvulsive therapy [J]. Neuropsychopharmacology. 2024.
37. Quinn DK, Upston J, Jones TR, et al. Electric field distribution predicts efficacy of accelerated intermittent theta burst stimulation for late-life depression [J]. Front Psych. 2023;14:1215093.
38. Cantù F, Schiena G, Sciortino D, et al. Use of 30-Hz accelerated iTBS in drug-resistant unipolar and bipolar depression in a public healthcare setting: a case series [J]. Front Psych. 2021;12:798847.
39. Li X, Zhang C, Tan J, et al. Clinical effects of continuous theta burst stimulation for generalized anxiety disorder and a mechanism involving α oscillations: a randomized controlled trial [J]. J Psychiatry Neurosci. 2022;47(2):E123–e33.
40. José C, Lua R, Iván C, et al. cTBS for generalized anxiety disorder in pharmaco-naïve patients [J]. Brain Stimulation: Basic Translat Clinical Res Neuromodulation. 2022;15(5):1325–6.
41. Littman L, Datto SM. The use of continuous theta burst stimulation for the treatment of anxiety: a case series [J]. Brain Stimulation: Basic Translat Clinical Res Neuromodulation. 2018;11(6):e13.
42. Zhao H, Jiang C, Zhao M, et al. Comparisons of accelerated continuous and intermittent theta burst stimulation for treatment-resistant depression and suicidal ideation [J]. Biol Psychiatry. 2024;96(1):26–33.

43. Mallik G, Mishra P, Garg S, et al. Safety and efficacy of continuous theta burst "intensive" stimulation in acute-phase bipolar depression: a pilot, exploratory study [J]. J ECT. 2023;39(1):28–33.
44. Guo Q, Wang K, Han H, et al. Continuous theta burst stimulation over the bilateral supplementary motor area in obsessive-compulsive disorder treatment: a clinical randomized single-blind sham-controlled trial [J]. Eur Psychiatry. 2022;65(1):e64.
45. Dutta P, Dhyani M, Garg S, et al. Efficacy of intensive orbitofrontal continuous theta burst stimulation (iOFcTBS) in obsessive compulsive disorder: a randomized placebo controlled study [J]. Psychiatry Res. 2021;298:113784.
46. Li P, Cheng J, Fan Q, et al. The functional connectivity predictor of therapeutic effect of continuous theta burst stimulation on obsessive-compulsive disorder: a preliminary study [J]. J Affect Disord. 2022;311:231–8.
47. Stubbeman WF, Converse J, Gencosmanoglu M, et al. Neuronavigated 20hz theta burst TMS for OCD comorbid with depression and anxiety sequentially targeting RDLPFC, LDLPFC, and ROFC [J]. Brain Stimulation: Basic Translat Clinical Res Neuromodulation. 2023;16(4):10.
48. Liu W, Shao H, Liao J, et al. Continuous theta-burst stimulation over the right orbitofrontal cortex in treatment-resistant obsessive-compulsive disorder treatment: a randomized sham-controlled trial [J]. Int J Gen Med. 2021;14:3109–18.
49. Mukherjee A, Kumre PK, Goyal N, et al. Adjunctive neuronavigated accelerated continuous theta-burst stimulation in obsessive-compulsive disorder: a randomized sham-controlled study [J]. CNS Spectr. 2023;28(4):495–504.
50. Stubbeman WF, Zarrabi B, Bastea S, et al. Bilateral neuronavigated 20Hz theta burst TMS for treatment refractory depression: an open label study [J]. Brain Stimul. 2018;11(4):953–5.
51. Chen L, Thomas EHX, Kaewpijit P, et al. Accelerated theta burst stimulation for the treatment of depression: a randomised controlled trial [J]. Brain Stimul. 2021;14(5):1095–105.
52. Tavares DF, Suen P, Rodrigues Dos Santos CG, et al. Treatment of mixed depression with theta-burst stimulation (TBS): results from a double-blind, randomized, sham-controlled clinical trial [J]. Neuropsychopharmacology. 2021;46(13):2257–65.
53. Kar SK, Sarkar S. Neuro-stimulation techniques for the Management of Anxiety Disorders: an update [J]. Clin Psychopharmacol Neurosci. 2016;14(4):330–7.
54. Wang J, Vila-Rodriguez F, Ge R, et al. Accelerated magnetic seizure therapy (aMST) for treatment of major depressive disorder: a pilot study [J]. J Affect Disord. 2020;264:215–20.
55. Shiozawa P, Leiva AP, Castro CD, et al. Transcranial direct current stimulation for generalized anxiety disorder: a case study [J]. Biol Psychiatry. 2014;75(11):e17–8.
56. Lin Y, Zhang C, Wang Y. A randomized controlled study of transcranial direct current stimulation in treatment of generalized anxiety disorder [J]. Brain Stimulation: Basic Translat Clinical Res Neuromodulation. 2019;12(2):403.
57. De Lima AL, Braga FMA, Da Costa RMM, et al. Transcranial direct current stimulation for the treatment of generalized anxiety disorder: a randomized clinical trial [J]. J Affect Disord. 2019;259:31–7.
58. Movahed FS, Goradel JA, Pouresmali A, et al. Effectiveness of transcranial direct current stimulation on worry, anxiety, and depression in generalized anxiety disorder: a randomized, single-blind pharmacotherapy and sham-controlled clinical trial [J]. Iranian J Psychiatry Behavioral Sci. 2018;12(2)
59. Heeren A, Billieux J, Philippot P, et al. Impact of transcranial direct current stimulation on attentional bias for threat: a proof-of-concept study among individuals with social anxiety disorder [J]. Soc Cogn Affect Neurosci. 2017;12(2):251–60.
60. Shiozawa P, Da Silva ME, Cordeiro Q. Transcranial direct current stimulation for treating depression in a patient with right hemispheric dominance: a case study [J]. J ECT. 2015;31(3):201–2.
61. Aksu S, Soyata AZ, Mursalova Z, et al. Transcranial direct current stimulation does not improve clinical and neurophysiological outcomes in panic disorder: a randomized sham-controlled trial [J]. Psychiatry Clin Neurosci. 2022;76(8):384–92.
62. Bikson M, Grossman P, Thomas C, et al. Safety of transcranial direct current stimulation: evidence based update 2016 [J]. Brain Stimul. 2016;9(5):641–61.

63. Zhang DW, Li GH, Wang ZB, et al. Exploratory randomized controlled trial of high-definition transcranial direct current stimulation combined with antidepressants in the treatment of depression with anxious distress features [J]. Chinese J Psychiatry. 2022;55(3):196–203.
64. El Jamal C, Harrie A, Rahman-Filipiak A, et al. Tolerability and blinding of high-definition transcranial direct current stimulation among older adults at intensities of up to 4 mA per electrode [J]. Brain Stimul. 2023;16(5):1328–35.
65. Turski CA, Kessler-Jones A, Chow C, et al. Extended multiple-field high-definition transcranial direct current stimulation (HD-tDCS) is well tolerated and safe in healthy adults [J]. Restor Neurol Neurosci. 2017;35(6):631–42.
66. Zewdie E, Ciechanski P, Kuo HC, et al. Safety and tolerability of transcranial magnetic and direct current stimulation in children: prospective single center evidence from 3.5 million stimulations [J]. Brain Stimul. 2020;13(3):565–75.
67. Nitsche MA, Paulus W. Transcranial direct current stimulation--update 2011 [J]. Restor Neurol Neurosci. 2011;29(6):463–92.
68. Qiu H, Liang K, Lu L, et al. Efficacy and safety of repetitive transcranial magnetic stimulation in children and adolescents with depression: a systematic review and preliminary meta-analysis [J]. J Affect Disord. 2023;320:305–12.
69. Allen CH, Kluger BM, Buard I. Safety of transcranial magnetic stimulation in children: a systematic review of the literature [J]. Pediatr Neurol. 2017;68:3–17.
70. Moffa AH, Brunoni AR, Nikolin S, et al. Transcranial direct current stimulation in psychiatric disorders: a comprehensive review [J]. Psychiatr Clin North Am. 2018;41(3):447–63.

Post-traumatic Stress Disorder

11

Sha Sha, Sixiang Liang, Yannan Deng, Leyi Zhu, Xin Chen, and Qinge Zhang

Abstract

Post-traumatic stress disorder (PTSD) represents a debilitating and chronic mental health disorder that can manifest following exposure to traumatic events. Recently, novel noninvasive neurostimulation techniques have emerged as potential therapeutic strategies for PTSD. Over the past few decades, numerous studies have explored the therapeutic effects of transcranial magnetic stimulation (TMS), transcranial direct current stimulation (tDCS), and electroconvulsive therapy (ECT) in the context of PTSD. Despite some promising results, findings have been inconsistent, and a definitive consensus regarding their efficacy remains elusive. Additional research is essential to substantiate the effectiveness of these approaches and to determine optimal treatment parameters for TMS and tDCS in PTSD therapy. This chapter provides a comprehensive review of innovative noninvasive neurostimulation methods, including TMS, tDCS, and ECT, in the treatment of PTSD. Given the diverse nature of TMS modalities, the chapter offers an in-depth exploration of repetitive TMS (r-TMS), deep TMS (D-TMS), priming TMS, synchronized TMS, theta burst stimulation (TBS), and magnetic seizure therapy (MST).

Keywords

Post-traumatic stress disorder (PTSD) · Repetitive transcranial magnetic stimulation (rTMS) · Deep TMS (dTMS) · Theta burst stimulation (TBS) · Transcranial direct current stimulation (tDCS) · Electroconvulsive therapy (ECT) · Magnetic seizure therapy (MST)

S. Sha · S. Liang · Y. Deng · L. Zhu · X. Chen · Q. Zhang (✉)
Beijing Key Laboratory of Mental Disorders, National Clinical Research Center for Mental Disorders & National Center for Mental Disorders, Beijing Anding Hospital, Capital Medical University, Beijing, China

W. Zheng, Y. Ning (eds.), *TMS and tDCS for Psychiatric Disorders*, https://doi.org/10.1007/978-981-96-8504-2_11

Abbreviations

A-rTMS	Accelerated Repetitive Transcranial Magnetic Stimulation
ACC	Anterior Cingulate Cortex
AITBS	Accelerated Intermittent Theta-Burst Stimulation
B-rTMS	Bilateral Repetitive Transcranial Magnetic Stimulation
BL ECT	Bilateral Electroconvulsive Therapy
BT ECT	Bitemporal Electroconvulsive Therapy
CPT	Cognitive Processing Therapy
cTBS	Continuation Theta Burst Stimulation
D-TMS	Deep Transcranial Magnetic Stimulation
DLPFC	Dorsolateral Prefrontal Cortex
dmPFC	Dorsomedial Prefrontal Cortex
ECT	Electroconvulsive Therapy
EEG	Electroencephalography
EMDR	Eye Movement Desensitization and Reprocessing
GAD	Generalized Anxiety Disorder
HD-tDCS	High-Definition Transcranial Direct Current Stimulation
HF-rTMS	High-Frequency Repetitive Transcranial Magnetic Stimulation
iTBS	Intermittent Theta Burst Stimulation
LF-rTMS	Low-Frequency Repetitive Transcranial Magnetic Stimulation
LMICs	Low- and Middle-Income Countries
LTC	Lateral Temporal Cortex
MDD	Major Depressive Disorder
MECT	Modified Electroconvulsive Therapy
MECTA-SR1	Modified Electroconvulsive Therapy Apparatus
MEP	Motor-Evoked Potential
mPFC	Medial Prefrontal Cortex
MST	Magnetic Seizure Therapy
NIBS	Noninvasive Brain Stimulation
PE	Prolonged Exposure Therapy
PTSD	Post-Traumatic Stress Disorder
r-TMS	Repetitive Transcranial Magnetic Stimulation
RCT	Randomized Controlled Trial
RMT	Resting Motor Threshold
RUL ECT	Right Unilateral Electroconvulsive Therapy
TBS	Theta Burst Stimulation
tDCS	Transcranial Direct Current Stimulation
TMS	Transcranial Magnetic Stimulation
U-rTMS	Unilateral Repetitive Transcranial Magnetic Stimulation
vmPFC	Ventromedial Prefrontal Cortex
VR	Virtual Reality

11.1 Introduction

Post-traumatic stress disorder (PTSD) is a mental health condition triggered by exposure to a traumatic event, characterized by delayed onset and persistence of symptoms, with a global lifetime prevalence of 3.9% [1]. Like other mental health disorders, PTSD arises from the complex interplay of social, psychological, and biological factors. While anyone can develop PTSD following a potentially traumatic event, individuals with prior traumatic experiences are at heightened risk. Approximately 70% of people worldwide will encounter a potentially traumatic event in their lifetime, yet only a minority (5.6%) will develop PTSD [2]. Despite the availability of effective treatments, only one in four individuals with PTSD in low- and middle-income countries (LMICs) seek any form of treatment [3].

PTSD is characterized by three core symptom clusters: re-experiencing the trauma, avoidance and emotional numbing, and hyperarousal. These symptoms often lead to significant functional impairment, comorbidities, and a major threat to social well-being, stability, and economic development. Hence, enhancing research on PTSD and providing effective treatment and support are critical to mitigating its societal impact. Trauma-focused therapies, including Cognitive Processing Therapy (CPT), Prolonged Exposure Therapy (PE), and Eye Movement Desensitization and Reprocessing (EMDR), are supported by empirical evidence [4], although these interventions often yield suboptimal outcomes [5]. Despite the effectiveness of pharmacotherapy and psychotherapy, approximately 20–30% of patients remain symptomatic [6].

Emerging studies indicate that specific neuroregulatory techniques, such as transcranial magnetic stimulation (TMS), theta burst stimulation (TBS), magnetic seizure therapy (MST), and transcranial direct current stimulation (tDCS), may activate or inhibit targeted neural circuits to alleviate PTSD symptoms. However, some of these techniques fail to stimulate deep brain regions, and existing research remains insufficient. This chapter will provide an in-depth exploration of these neuroregulatory approaches.

11.2 TMS

11.2.1 rTMS

Repetitive Transcranial Magnetic Stimulation (rTMS) is an advanced, noninvasive neuromodulation technique that alters brain activity through the application of repeated magnetic pulses, generated by fluctuations in the coil's magnetic field. This modulation can influence both cortical and subcortical brain regions. rTMS is typically classified into high-frequency (HF-rTMS; 5–20 Hz) and low-frequency (LF-rTMS; ≤1 Hz) categories. It has been proposed that HF-rTMS enhances cortical excitability, whereas LF-rTMS dampens cortical activity [7, 8]. While limited research exists, both HF- and LF-rTMS have demonstrated therapeutic potential in PTSD treatment, though there is no consensus regarding the optimal frequency for therapeutic efficacy.

11.2.1.1 Unilateral rTMS

Efficacy

Unilateral Repetitive Transcranial Magnetic Stimulation (U-rTMS) has emerged as a promising treatment modality for PTSD. U-rTMS has been shown to significantly reduce overall PTSD severity and alleviate core symptoms such as intrusive memories, avoidance behaviors, and hyperarousal [9–11].

Research indicates that U-rTMS targeting specific brain regions yields differential therapeutic effects. For instance, U-rTMS applied to the right dorsolateral prefrontal cortex (DLPFC) has been found to significantly reduce PTSD-associated anxiety and depression [12–14], while left DLPFC stimulation has been associated with improvements in comorbid depression [15, 16].

U-rTMS has also demonstrated potential cognitive benefits. Both left- and right-sided U-rTMS have shown trends toward cognitive improvement, with right-sided U-rTMS particularly enhancing verbal fluency [17].

Regarding the durability of effects, studies suggest that U-rTMS can sustain its therapeutic benefits over the long term, with improvements remaining significant from 2 weeks to 3 months posttreatment [12, 18, 19].

Safety

U-rTMS is generally regarded as a low-risk, well-tolerated treatment, with most adverse effects being mild and transient. No major adverse events, such as seizures, have been reported in current studies. Common side effects include temporary headache, syncope, scalp discomfort, and transient changes in hearing [20]. Some studies have also noted additional adverse events, such as dizziness, neck pain, sleepiness, and brief mood disturbances [6].

Treatment Regimen

The optimal parameters for U-rTMS treatment in PTSD have garnered considerable attention, particularly regarding the targeted brain regions. The DLPFC has emerged as a central area of focus, given its pivotal role in emotion regulation [21]. Typically, either the right or left DLPFC is targeted to modulate activity in the ventromedial prefrontal cortex (vmPFC), a key region involved in emotional processing.

Significant improvements in PTSD core symptoms have been observed with U-rTMS targeting both the left and right DLPFC [13, 14, 22]. However, there remains ongoing debate about whether left or right DLPFC yields better treatment outcomes in patients with PTSD. Studies over the past decades have produced inconsistent results. For instance, one randomized controlled trial (RCT) suggested that right DLPFC-targeted U-rTMS produces a larger effect compared to left DLPFC stimulation, whereas another review found no significant difference between the two approaches [9, 17]. This highlights the need for further research with larger sample sizes to resolve the issue of the optimal brain target in PTSD treatment.

An additional area of clinical debate is the choice between high- and low-frequency U-rTMS. Both approaches have shown mixed results in terms of efficacy. Some studies [13, 14, 18, 22] have demonstrated the effectiveness of low-frequency U-rTMS, while others [12, 15] have failed to show significant benefits. Similarly, while some research [12, 23] supports the effectiveness of high-frequency U-rTMS, other studies [18] have found no distinction between high-frequency stimulation and sham treatments.

Several meta-analyses [9, 10] suggest that high-frequency TMS is more effective than low-frequency TMS in the treatment of PTSD, producing a significantly larger therapeutic effect. These meta-analyses predominantly focus on studies utilizing U-rTMS, though a limited number also include bilateral rTMS and deep TMS, which may slightly confound the conclusions. Therefore, while high-frequency U-rTMS may appear more efficacious, the question of whether high- or low-frequency U-rTMS is superior remains unresolved and warrants further investigation.

Treatment parameters, such as session duration and frequency, though less frequently discussed, are also important potential moderators of treatment efficacy. Typical rTMS sessions range from 20 to 45 min, with treatment frequencies varying from two to five sessions per week [10, 24].

A previous study indicated no significant differences in effect size based on the number of treatment sessions or the total number of TMS pulses delivered over the course of treatment. Moreover, high-dose TMS protocols did not show superior efficacy compared to mid-dose protocols, suggesting that improving the efficiency of TMS treatments could make them more accessible to a broader patient population [9]. This meta-analysis focused mainly on U-rTMS studies, with a few exceptions involving other TMS modalities. Consequently, the impact of treatment parameters such as session duration and frequency on the effectiveness of U-rTMS remains uncertain.

Clinical Recommendations

The right DLPFC, as a potentially promising brain target, warrants increased clinical attention and further investigation, particularly in light of its superior effectiveness demonstrated in an RCT study. Additionally, high-frequency stimulation has been associated with potentially larger treatment effects, with some reviews advocating for its use. However, these recommendations remain inadequately substantiated by sufficient empirical evidence (see Table 11.1).

Consensus on the optimal parameters remains elusive, primarily due to the limited clinical evidence and the mixed results reported in current studies. To clarify the ideal treatment parameters and establish evidence-based clinical recommendations, further investigations are essential.

11.2.1.2 Bilateral rTMS

Efficacy

In the limited literature, bilateral rTMS (B-rTMS) targeting the DLPFC has been demonstrated to effectively reduce overall PTSD severity significantly [23].

However, there remains considerable debate regarding whether bilateral DLPFC stimulation is superior to unilateral DLPFC stimulation [20]. While some studies suggest that bilateral stimulation may lead to a greater reduction in PTSD severity

Table 11.1 TMS in PTSD

TMS	Clinical recommendations				Levels + references
	Target	Frequency	Length (min)[a]	Duration (sessions)	
rTMS	Right DLPFC	High(10 Hz)	20	10	B-grade Recommendation[a] [25]
Deep TMS	Unclear	Unclear	Unclear	Unclear	NA
Priming TMS	Unclear	Unclear	Unclear	Unclear	NA
Synchronized TMS	Unclear	Unclear	Unclear	Unclear	NA
TBS[b]	Unclear	Unclear	Unclear	Unclear	NA
MST	Unclear	Unclear	Unclear	Unclear	NA

[a]The WFSBP Guidelines mentions rTMS as B-grade recommendation for PTSD treatment
[b]TBS includes iTBS, accelerated iTBS, continuation TBS, and bilateral TBS

compared to unilateral stimulation (on either the right or left DLPFC), other studies report comparable effects between bilateral and unilateral right DLPFC stimulation [23, 26].

Further research is needed to clarify the optimal targeting strategy for DLPFC stimulation in the treatment of PTSD.

Safety

The side effect profile associated with B-rTMS is generally low, with most adverse events being mild and transient [6]. No major adverse events have been reported in existing trials. A few studies found that only a small number of patients with PTSD experienced transient headaches and discomfort, either during the initial sessions of B-rTMS or following treatment completion [23, 27].

Treatment Regimen

One study exploring bilateral DLPFC stimulation (20 Hz, 100% MT, 1200/2400 pulses per session, 10 sessions in total) not only significantly reduced PTSD symptoms but also achieved faster initial effects compared to unilateral right DLPFC stimulation [23].

Although most therapeutic B-rTMS studies have focused on the DLPFC, another study targeting the dorsomedial prefrontal cortex (dmPFC) reported improvements in overall severity in patients with comorbid PTSD and eating disorders (10/20/50 Hz, 120% MT, 3000/600 pulses per session, 20–30 sessions in total) [27].

Clinical Recommendations

Based on the limited studies available, high-frequency stimulation (10–20 Hz) targeting the DLPFC can be recommended for B-rTMS in the treatment of PTSD, given its efficacy in alleviating overall PTSD severity, as well as associated depression and anxiety [23, 26].

Additionally, the dmPFC is emerging as a novel stimulation target, with preliminary evidence suggesting its potential efficacy [27]. However, further large-scale studies are needed to validate its effectiveness.

11.2.1.3 Accelerated rTMS

Efficacy

Accelerated rTMS (A-rTMS) remains an experimental intervention with limited clinical evidence to support its widespread use. Although its potential has been explored in patients with Major Depressive Disorder (MDD) and Generalized Anxiety Disorder (GAD) [28–31], its application in the treatment of PTSD remains insufficiently investigated.

Among the few existing studies, trends suggesting anxiolytic and antidepressant effects have been observed in individuals with GAD, MDD, and Postpartum Depression following A-rTMS treatment [28, 29, 32–34]. Preliminary evidence also indicates that A-rTMS may reduce the intensity of suicidal ideation in suicidal inpatients [35]. However, whether similar therapeutic outcomes can be achieved in patients with PTSD remains uncertain and warrants further investigation in future studies.

Safety

In the limited available literature, A-rTMS has been reported to be safe and well-tolerated, with minimal adverse effects, such as scalp discomfort [33]. Nevertheless, additional studies are required to assess its safety profile specifically within the context of PTSD treatment.

Treatment Regimen

Given the scarcity of research on A-rTMS for PTSD, optimal treatment parameters remain undefined.

Brain targets, particularly the right DLPFC, have shown promising results with conventional rTMS for PTSD [17], underscoring the need for further clinical investigations to determine whether comparable outcomes can be achieved using A-rTMS protocols.

Clinical Recommendations

Further research is essential to establish the efficacy and safety of A-rTMS in the treatment of PTSD before any clinical recommendations can be made.

11.2.2 Deep TMS

11.2.2.1 Efficacy

Deep TMS (D-TMS) is a neuromodulation and neurostimulation technique that operates on the principle of electromagnetic induction within the brain [36]. The intensity and density of the induced electromagnetic field are sufficient to depolarize neurons, with repetitive transcranial magnetic stimulation pulses modulating cortical excitability. Depending on the stimulation parameters, this modulation can either reduce or enhance neuronal activity, with effects that may persist beyond the duration of the stimulation sequence [37]. While standard TMS typically employs

No. 8 coils, D-TMS can utilize various coil types, including H-type coils [38], C-type core coils [39], and round crown coils [40]. Among these, only the H-coil has been tested for safety and efficacy and is the sole coil type used in clinical trials [41]. The H-coil's ability to reach depths of up to 6 cm enables it to modulate both cortical activity and deep neural circuits [36].

Clinically, the efficacy of D-TMS for mental disorders characterized by hyperactivity or heightened arousal, such as PTSD, remains inconsistent [42]. Two clinical trials have investigated the use of H1 [43] and H7 coils [44] for D-TMS treatment in patients with PTSD. The first study [43] demonstrated significant therapeutic effects when D-TMS was combined with exposure therapy. Specifically, the group receiving combined D-TMS and exposure therapy showed greater improvements in Clinician-Administered PTSD Scale (CAPS) scores compared to those in the exposure therapy combined with the sham operation group or the D-TMS-only treatment group. However, the limited number of studies on sham-controlled D-TMS in PTSD populations means that these results are not definitive. In contrast, the second trial [44] reached an opposing conclusion, where both active and pseudo-D-TMS groups were treated simultaneously with exposure therapy. Surprisingly, the sham group showed greater improvement. The reasons for this discrepancy remain unclear. The authors [42] suggest that the differences may arise from the distinct brain regions targeted by the coils in each study: the former focused on the medial prefrontal cortex (mPFC), while the latter targeted the bilateral mPFC and anterior cingulate cortex (ACC). The therapeutic efficacy of various coil types in treating PTSD warrants further investigation.

11.2.2.2 Safety

The common side effects associated with D-TMS in patients with PTSD primarily include discomfort at the application site [43] and mild headache [43, 45]. Some patients may withdraw from trials due to increased discomfort or anxiety [43, 45]. Headaches are typically self-limiting and rarely necessitate pain medication [43]. Seizures are an infrequent occurrence, with one case of transient tonic-clonic generalized seizures reported after the eighth treatment. These seizures were self-limiting and did not require special intervention [43]. Although rare, instances of suicidal ideation and attempts have been noted, though they cannot be directly attributed to D-TMS therapy itself [45].

11.2.2.3 Treatment Regimen

Studies [12, 17, 18] have shown that rTMS applied to the right DLPFC yields more substantial improvements in PTSD compared to stimulation of the left DLPFC. A prior clinical trial using f8 coils on the left frontal cortex indicated only modest improvements in core PTSD symptoms [15]. These findings suggest that hemisphere targeting should be an important consideration when customizing D-TMS protocols for patients with PTSD.

Previous studies [43] commonly employed a treatment regimen of three sessions per week over 4 weeks. The most recent trial [44], utilizing H7 coils, targeted the bilateral mPFC and ACC, with treatment parameters including 18 Hz frequency, 2-s pulse trains, 20-s intertrain intervals, 80 pulses per group, and a resting motor threshold of 100%. However, these treatment protocols were administered in conjunction with exposure therapy, and the efficacy of D-TMS alone remains an area requiring further investigation.

11.2.2.4 Clinical Recommendations

In the context of MDD, D-TMS was approved by the U.S. FDA in 2013 for MDD treatment [46]. Due to its unique coil structure, D-TMS can stimulate deeper brain regions and larger volumes than traditional rTMS. However, even in MDD clinical studies, response and remission rates with D-TMS compared to sham treatments have not been consistently significant [47]. Currently, clinical research on PTSD remains limited. Notably, a study by Hung et al. [48] found that only the combination of D-TMS and antidepressants demonstrated superior efficacy over placebo drugs, while D-TMS alone had no significant effect. This insight may guide future clinical research, highlighting the need for additional randomized controlled trials to confirm the efficacy of D-TMS in PTSD treatment (see Table 11.1).

11.2.3 Priming TMS

11.2.3.1 Efficacy

As a variant of low-frequency TMS, priming transcranial magnetic stimulation involves a specific regimen wherein a short-term high-intensity, high-frequency TMS ("prime stimulation") is applied before administering low-frequency right DLPFC rTMS [49]. Previous studies have demonstrated that the antidepressant effects of priming with 1 Hz right rTMS were superior to those of 1 Hz right rTMS alone [50]. A recent study [51] found that applying iTBS 10 min after priming with continuation TBS (cTBS) resulted in a greater motor-evoked potential (MEP) change compared to iTBS following sham treatment. Although differing from the concepts under consideration, this finding offers valuable insights for incorporating priming TMS in clinical research design. To date, no studies have investigated the application of priming TMS in patients with PTSD.

11.2.3.2 Safety

A recent network meta-analysis [30] found that priming TMS produced a stronger antidepressant response than cTBS, with transcranial stimulation associated with a lower dropout rate compared to sham stimulation. However, due to the lack of clinical research on priming TMS for PTSD, its potential adverse effects remain unexplored. The safety of this technique requires further validation, supported by larger clinical trials.

11.2.3.3 Treatment Regimen

The treatment approach employed in the current study serves as a significant reference point. Unlike other TMS therapies, D-TMS itself serves as a form of priming therapy, using high-frequency stimulation for priming followed by low-frequency right DLPFC rTMS [49]. In the TBS treatment group, conventional iTBS was administered following a 10-min priming stimulation [51].

11.2.3.4 Clinical Recommendations

To date, randomized controlled trials evaluating priming TMS remain limited [50, 52], signaling the need for further investigation. Moreover, research into its specific applications for PTSD treatment is even more scarce, representing a notable gap in current research (see Table 11.1).

11.2.4 Synchronized TMS

11.2.4.1 Efficacy

Synchronous transcranial magnetic stimulation (synchronized TMS) utilizes low-field sinusoidal stimulation through a rotating spherical neodymium magnet positioned along the scalp's midline, synchronized with the individual's alpha EEG frequency [30, 53]. By harnessing the brain's natural resonance at the individual alpha frequency, synchronized TMS is thought to achieve antidepressant effects using lower intensity energy than traditional high-frequency rTMS [54]. Furthermore, research by the same team indicated that synchronized TMS may offer greater benefits to patients with more severe baseline depression and anxiety compared to pseudo-synchronized TMS [55], although this conclusion has not been universally corroborated across studies [30]. Most current studies focus on depression, providing some reference and guidance for PTSD treatment. A prospective, sham-controlled, multi-site pilot study by Noah et al. [56] involving 23 veterans with PTSD and MDD under pharmacological treatment showed that all participants experienced significant reductions in PTSD and MDD symptoms. Notably, active stimulation significantly alleviated moderate to severe PTSD symptoms, including "threshold PTSD" symptom count, PCL score, and QIDS score.

11.2.4.2 Safety

Similar to priming TMS, the safety profile of synchronized TMS requires further investigation, particularly through randomized controlled trials and comparative studies with other TMS modalities. Previous studies suggest that synchronized TMS may cause side effects such as headache and nausea. However, relevant research remains limited [56].

11.2.4.3 Treatment Regimen

Currently, there is no authoritative recommendation for the use of synchronized TMS in PTSD treatment. However, the design of prior studies, such as Noah et al.'s [56], offers valuable insights. In their experiment, participants received either active or sham synchronized TMS for 4 weeks (five sessions per week), followed by an optional non-blind synchronized TMS for an additional 4 weeks.

11.2.4.4 Clinical Recommendations

The existing clinical evidence primarily involves populations with comorbid PTSD and MDD [56]. Given the high comorbidity of these conditions, this body of research offers significant implications for PTSD treatment. Further development in synchronized TMS research, particularly involving larger sample sizes and more specific patient populations, is urgently needed (see Table 11.1).

11.2.5 TBS

Theta burst stimulation (TBS) is a noninvasive brain stimulation technique employed to modulate neural circuits involved in psychiatric and neurological disorders, offering a novel approach to the treatment of PTSD [57]. TBS modalities primarily include intermittent theta-burst stimulation (iTBS), accelerated intermittent theta-burst stimulation (AITBS), and bilateral TBS.

iTBS, a relatively recent form of rTMS, is characterized by shorter treatment durations compared to traditional rTMS protocols. It involves the delivery of bursts consisting of three high-frequency (50 Hz) TMS pulses every 200 ms for a duration of 2 s, repeated every 10 s in an intermittent manner [58].

AITBS is an advanced variant of noninvasive brain stimulation (NIBS) utilizing transcranial magnetic stimulation (TMS) to target specific brain regions. This method delivers multiple TBS sessions per day, each comprising a series of brief magnetic pulses at a theta frequency (approximately 5 Hz) [59].

11.2.5.1 iTBS

Efficacy

iTBS has demonstrated efficacy in treating depression and shows potential for PTSD intervention. Several studies have reported significant improvements in PTSD symptoms, as well as in social, occupational, and depressive functioning, within 2 weeks, with continued progress observed at one-month follow-ups [60]. When compared to conventional rTMS, iTBS may offer a similarly effective yet more time-efficient treatment option [61]. However, research on iTBS in PTSD remains limited, and the robustness of these findings requires further investigation.

Safety

Safety assessments of iTBS in PTSD have been conducted in multiple clinical trials, with side effects consistent with those observed in rTMS studies, such as mild headaches and discomfort at the treatment site. Although the risk of seizures is low, it remains a consideration, necessitating vigilant monitoring for patient safety [60]. Overall, current data support iTBS as a safe and well-tolerated treatment modality for PTSD [61].

Treatment Regimen

The typical iTBS treatment regimen for PTSD involves delivering magnetic pulses to the right DLPFC. Protocols vary across studies, but a common approach includes daily sessions for 10–15 days, with each session lasting approximately 3–10 min [60, 61]. To achieve optimal therapeutic effects, a cumulative dose of iTBS over an extended period may be required.

The DLPFC has been identified as a key target for iTBS in PTSD therapy [60–62].

In studies investigating iTBS, the stimulation intensity is generally set at 80% of the resting motor threshold (RMT), with 50 Hz bursts delivered at a frequency of 5 Hz [58]. Each iTBS session typically involves the administration of 1800 pulses over 9.5 min [60].

Clinical Recommendations

Targeting the right DLPFC for iTBS treatment in PTSD is recommended, as this region has demonstrated the most consistent efficacy in alleviating PTSD symptoms. The stimulation intensity should be set at 80% of the RMT, with bursts of 50 Hz delivered at a frequency of 5 Hz. This protocol offers a balanced approach, optimizing both efficacy and safety. A minimum of 10–15 treatment sessions over several weeks is typically required to achieve significant symptom improvements, with extended treatment durations potentially necessary to sustain therapeutic benefits.

iTBS shows considerable promise as an effective and time-efficient treatment for PTSD. However, further investigation is essential to refine treatment parameters, assess long-term efficacy, and identify potential biomarkers predictive of treatment response.

11.2.5.2 Accelerated iTBS

Efficacy

Research focused on patients with PTSD suggests that AITBS may also be an effective treatment modality [60, 63]. Notably, case reports indicate that AITBS targeting the DLPFC appears to yield better outcomes for individuals with MDD comorbid with PTSD [59]. Nonetheless, research in this area remains scarce, and higher-level evidence is lacking.

Safety

AITBS is generally considered a safe and well-tolerated treatment. Commonly reported side effects include mild to moderate discomfort at the stimulation site, headaches, and transient increases in anxiety or depressive symptoms. These side effects are typically short-lived and manageable, with most patients tolerating the treatment without significant adverse events.

Notably, the safety of AITBS for PTSD has not been fully established across all patient populations. For example, patients with conditions such as epilepsy or metal implants in the head may face an increased risk of adverse events and should undergo careful evaluation prior to receiving AITBS treatment. Additionally, the long-term safety and efficacy of AITBS for PTSD remain undetermined and require further study.

Treatment Regimen

The AITBS treatment regimen for PTSD generally involves delivering brief bursts of high-frequency magnetic pulses to the right DLPFC or bilateral DLPFC, utilizing TMS to induce synaptic plasticity, which may contribute to symptom alleviation [59].

Treatment protocols vary according to symptom severity and individual response. However, a typical regimen consists of daily sessions over 10–15 days, each lasting 3–10 min. The stimulation follows a theta-burst pattern, with bursts of 50 Hz repeated at 5 Hz (with a 200 ms interval between bursts).

The intensity of the magnetic pulses is usually calibrated to the motor threshold, sufficient to induce a muscle twitch in the contralateral hand or forearm, ensuring adequate stimulation of neural activity in the targeted brain region.

Unlike conventional TMS, AITBS employs the theta-burst pattern, which is believed to induce synaptic plasticity more effectively, potentially enhancing its therapeutic efficacy for PTSD.

Clinical Recommendations

AITBS offers a promising treatment alternative for individuals with PTSD, with preliminary evidence indicating its potential efficacy in reducing symptoms and demonstrating good tolerability. However, further research is required to establish its safety and effectiveness in larger, more diverse patient populations and to explore the underlying neurobiological mechanisms responsible for its effects.

11.2.5.3 Bilateral TBS

Efficacy

Early studies have examined the efficacy of bilateral TBS in mitigating PTSD symptoms [60]. Bilateral sequential TBS has been shown to effectively alleviate post-traumatic stress and associated depressive symptoms [64]. Additionally, several studies have reported improvements in both depression and PTSD symptoms following TBS treatment, with bilateral stimulation potentially yielding more substantial responses compared to single-site stimulation [65]. These findings suggest that bilateral TBS could be a viable treatment strategy for individuals with comorbid PTSD and MDD.

Safety

The safety of bilateral TBS for PTSD has been evaluated in multiple studies, with preliminary evidence indicating that TBS is relatively safe, causing minimal side effects such as mild headaches and discomfort at the stimulation site [60, 64].

While these findings are encouraging, further research is necessary to fully assess the potential risks and long-term effects of TBS. Additionally, individual variations in tolerability, as well as contraindications such as the presence of metallic implants or certain medical conditions, should be carefully considered.

Treatment Regimen

The optimal treatment regimen for bilateral TBS in PTSD remains an area of active investigation. The bilateral sequential TBS protocol typically consists of 67 iTBS trains, delivered to each hemisphere at 8-s intervals. Treatment is administered 5 days per week for 2 weeks, totaling 10 sessions. The TBS protocol involves delivering 20 columns of stimulation at 5 Hz, each containing three 50 Hz pulses repeated at 5 Hz. Previous studies have reported a treatment schedule consisting of 5 consecutive treatment days followed by 2 days of rest, extending over a 3-week period [60, 64].

The choice of treatment regimen may vary depending on factors such as the severity of PTSD symptoms, comorbid conditions, and patient preferences. Thus, it is crucial for clinicians to tailor treatment plans to the individual, closely monitoring patient response and tolerability.

Clinical Recommendations

Bilateral theta-burst stimulation holds significant promise as a therapeutic option for individuals with PTSD, particularly those with comorbid MDD. Preliminary evidence supports its effectiveness in alleviating PTSD symptoms, with good tolerability and minimal side effects. However, additional research is needed to better understand the optimal treatment regimen, assess potential risks, and evaluate the long-term effects of this innovative treatment modality.

11.2.6 MST

Magnetic seizure therapy (MST) is an emerging noninvasive brain stimulation technique that utilizes high-intensity TMS to induce a generalized seizure under anesthesia. Derived from electroconvulsive therapy (ECT), MST differs in several significant ways. Unlike ECT, which targets deeper brain structures, MST focuses on more localized brain regions and induces seizures in the superficial cortical layers, potentially reducing the impact on deeper structures such as the hippocampus [66, 67]. Additionally, MST employs pulsed magnetic fields, while ECT uses electric currents. Although both therapies are effective in treating severe psychiatric disorders, MST is generally regarded as safer, with fewer cognitive and memory side effects, whereas ECT may cause transient cognitive impairments but remains highly effective in a broader range of conditions. Both treatments have shown efficacy in treating severe depressive disorder and other psychiatric conditions, with MST being a newer and less explored option compared to ECT. The choice between the two treatments should be informed by individual patient factors and preferences, in consultation with a qualified psychiatrist [20].

11.2.6.1 Efficacy

Although MST has demonstrated potential in treating a variety of psychiatric disorders, including MDD and schizophrenia, its effectiveness in treating PTSD has not been extensively researched. However, the underlying mechanisms by which MST induces neural plasticity and alters resting-state functional connectivity suggest that it may be a promising tool for modulating the fear extinction network, which plays a pivotal role in PTSD treatment [66]. Furthermore, emerging evidence supports the clinical efficacy of MST in MDD [68], bipolar depression [69], and schizophrenia [70], hinting at potential benefits for patients with PTSD. In studies examining MST for MDD, patients have reported substantial improvements in depressive symptoms and reductions in suicidal ideation, with fewer cognitive side effects compared to ECT [20]. These results imply that MST may provide a comparable therapeutic benefit for PTSD, particularly in cases where other treatments have been ineffective.

Notably, no studies have specifically investigated the efficacy of MST for PTSD. However, ongoing clinical trials and future research are essential to explore the potential application of MST in PTSD treatment further.

11.2.6.2 Safety

MST has demonstrated a promising safety and tolerability profile for treating various psychiatric conditions, including MDD and schizophrenia. However, its safety in the treatment of PTSD has not been extensively studied. MST is generally considered more targeted than ECT, which results in a reduced side effect profile. Common side effects of MST, such as headache, nausea, and transient memory loss, are typically mild and resolve quickly after treatment [70].

When comparing the cognitive side effects of MST and ECT, MST has been shown to cause less cognitive impairment [71]. This is particularly relevant for patients with PTSD, who may already be facing cognitive challenges due to the effects of their trauma. Despite its generally favorable side effect profile, MST carries potential risks, including the risk of seizure induction, as it involves anesthesia and the induction of generalized seizures. Therefore, careful patient evaluation and monitoring are crucial to ensure safety during MST.

11.2.6.3 Treatment Regimen

Since specific studies on MST for PTSD are lacking, the treatment regimen for this application has not been definitively established. However, based on existing studies of MST for other psychiatric conditions, some general guidelines can be inferred. MST typically targets the cortex, particularly the vertex or prefrontal cortex, depending on the desired therapeutic effect and protocol [70]. For PTSD, the specific stimulation targets may need to focus on the neural circuits and brain regions implicated in trauma-related responses, such as the prefrontal cortex and amygdala.

The intensity of MST stimulation is high enough to induce a seizure under anesthesia. While the exact intensity may vary depending on the patient's tolerance and the therapeutic goals, high-frequency MST (e.g., 100 Hz) has been found to be effective in alleviating symptoms of depression [68]. The optimal frequency for

PTSD treatment, however, remains unknown and will require further research to determine.

MST treatment typically involves a series of sessions over several weeks, with the number and duration of treatments adjusted based on patient response and symptom severity. In general, longer MST treatment courses are associated with higher remission rates in conditions like depression [66].

11.2.6.4 Clinical Recommendations

While MST has not yet been specifically studied for PTSD, its mechanisms of action and the promising results observed in other psychiatric conditions suggest that it could be a viable treatment option for PTSD. However, further research is needed to explore the safety, efficacy, and optimal treatment regimens for MST in PTSD, to determine its potential as a therapeutic tool for this disorder (See Table 11.1).

11.3 tDCS

11.3.1 Conventional tDCS

Transcranial direct current stimulation (tDCS) is a noninvasive neuromodulatory technique that has shown promise in modulating neuronal activity through the application of a low-intensity direct current delivered via saline-soaked electrodes. This method can influence cognitive processes such as learning and memory [72–75], making it a potential adjunct to treatments for various psychiatric disorders, including depression, schizophrenia, and PTSD [76, 77]. Conventional tDCS is receiving growing attention as an adjunctive treatment for PTSD, owing to its capacity to target specific brain regions associated with PTSD symptomatology [78].

11.3.1.1 Efficacy

In recent studies, conventional tDCS has demonstrated efficacy in reducing PTSD symptoms. Small-sample, randomized double-blind trials have shown that tDCS can alleviate hyperarousal, negative cognitive and emotional symptoms, as well as symptoms of depression and anxiety in individuals with PTSD [76]. Research has further highlighted that patients with PTSD often exhibit reduced volume and activation in regions such as the ventromedial prefrontal cortex (vmPFC) during fear extinction processes [79]. As a result, studies have suggested that tDCS, when applied during the consolidation phase of fear extinction, can enhance extinction memory and potentially improve the treatment of PTSD [80].

Advancements in technology have led to the development of novel approaches that combine tDCS with other therapeutic modalities, such as computerized cognitive training and virtual reality (VR). One study involved a computerized cognitive program administered over 5 weeks following tDCS treatment in a small group of patients with PTSD. The results indicated significant improvements not only in cognitive and emotional performance indicators but also in neurophysiological changes, as evidenced by electroencephalography (EEG) recordings taken before and after treatment [81].

In the context of PTSD treatment, combining tDCS with VR has yielded even more significant improvements. A double-blind randomized trial found that the combination of tDCS and VR resulted in a significant reduction in self-reported PTSD symptom severity. Similarly, a single-blind trial demonstrated that the combined treatment group showed a decrease in physiological arousal and clinically significant reductions in PTSD symptoms compared to the VR-only group with sham stimulation [82, 83].

In conclusion, tDCS has demonstrated promising efficacy in treating PTSD symptoms, along with associated emotional and cognitive sub-symptoms. However, given the heterogeneous nature of PTSD and the variability in patient responses, further research is necessary to optimize treatment parameters and determine long-term outcomes.

11.3.1.2 Safety

Conventional tDCS is widely regarded as a safe neuromodulatory intervention with minimal serious adverse effects. In a study involving 12 male veterans with war-zone-related PTSD, common side effects included mild headaches, skin irritation at electrode sites, and transient tingling sensations, typically manageable with over-the-counter analgesics. Although rare, issues such as itching, burning, and erythema have been reported in prior studies, yet larger-scale investigations in diverse populations have not corroborated these observations [82, 84].

Consequently, patients should be informed about potential side effects prior to treatment, with careful monitoring both during and following the intervention.

11.3.1.3 Treatment Regimen

The optimal protocol for conventional tDCS in the treatment of PTSD remains undetermined and warrants further investigation. However, several general parameters can be inferred from existing studies, including stimulation target, intensity, frequency, duration, and treatment session count.

Research in PTSD, encompassing both animal models and human studies, has revealed dysregulations in neural circuits, notably hyperactivity in the amygdala and dorsal anterior cingulate cortex—regions involved in fear processing—alongside diminished activity in the vmPFC, a key area responsible for inhibiting fear responses [85–87]. As such, the vmPFC is a primary target for tDCS in PTSD, particularly for augmenting extinction learning [80].

Another frequently targeted area is the bilateral DLPFC, with its therapeutic efficacy demonstrated in a double-blind, randomized controlled trial. In this study, 40 patients with PTSD were randomly assigned to receive either 10 sessions of active tDCS at 2 mA to the right (cathode) and left (anode) DLPFC or sham stimulation. The active treatment group showed significant improvements across a spectrum of PTSD symptoms as well as mood and cognitive sub-symptoms [76]. Additional research has indicated that left prefrontal anodal stimulation enhances working memory performance, with the effect being dependent on stimulation polarity and site specificity [88].

The standard tDCS protocol (<40 min, 4 mA, 7.2 Coulombs) has not been associated with serious adverse events or irreversible damage, even after over 33,200 sessions and 1000 repeated sessions across various subjects [89]. The majority of studies employ 2 mA as the stimulation intensity, with session durations typically ranging from 20 to 30 min [76, 83]. An effective treatment regimen typically involves administering five to ten sessions over the course of 1 to 2 weeks. The frequency of stimulation varies across studies, with some conducting sessions daily for 1–2 weeks, while others employ intermittent schedules; however, research has predominantly concentrated on continuous stimulation rather than intermittent protocols [60]. Evidence suggests that administering 6–10 sessions over 2–3 weeks yields relatively favorable outcomes [76, 83].

11.3.1.4 Clinical Recommendations

Based on existing evidence, typical clinical protocols for conventional tDCS in the treatment of PTSD involve delivering 2 mA tDCS to the DLPFC or vmPFC for 20–30 min per session. The number of sessions generally ranges from 10 to 20, administered over a period of 2–3 weeks. Clinical trials have indicated that targeting the DLPFC or vmPFC with 2 mA tDCS can effectively reduce PTSD symptoms, particularly those related to physiological arousal and re-experiencing [78, 82].

When combined with exposure therapy (such as VR exposure) or cognitive training, conventional tDCS has shown potential for enhancing treatment efficacy. The integration of tDCS with VR, specifically, has been noted as technically feasible and may evolve into a valuable clinical adjunct [81, 83].

11.3.2 HD-tDCS

High-Definition Transcranial Direct Current Stimulation (HD-tDCS) represents an advancement in noninvasive brain stimulation techniques, utilizing low-intensity electric currents to modulate neuronal excitability. Unlike traditional pad-based tDCS, which employs a single large anode and cathode, 4 × 1 ring HD-tDCS delivers more focal brain stimulation. This configuration surrounds a central electrode with four "ring" electrodes, ensuring that the modulation of cortical excitability is confined to the region enclosed by the ring electrodes. The peak electric field intensity is localized to the central electrode, optimizing the precision of stimulation and enhancing the current density within the target region, while minimizing effects on nontarget areas. As such, HD-tDCS offers a promising approach for more targeted and effective modulation of neuronal activity in the treatment of PTSD.

11.3.2.1 Efficacy

Research has explored the efficacy of HD-tDCS in PTSD, particularly for patients exhibiting symptoms such as re-experiencing. Due to its ability to precisely and deeply stimulate specific brain regions, HD-tDCS [84] may offer superior therapeutic effects compared to conventional tDCS. This is supported by previous studies demonstrating the effectiveness of HD-tDCS in treating other psychiatric conditions. For example, in aphasia treatment, HD-tDCS significantly improved naming accuracy and response time by precisely targeting higher cortical areas, outperforming traditional tDCS techniques [90].

However, patients with PTSD exhibiting specific symptom profiles may require longer or more intensive treatments, or more precise targeting, to achieve optimal therapeutic outcomes. Given the small sample sizes and issues with reproducibility in current studies, the findings related to HD-tDCS in PTSD have not yet reached the threshold for robust evidence-based recommendations. Nevertheless, the potential of HD-tDCS for treating PTSD, especially in patients with distinct symptomatology, remains an area of active interest. Future studies, with larger sample sizes and rigorous methodologies, are needed to further validate and refine its therapeutic applications.

11.3.2.2 Safety

HD-tDCS is widely recognized as a safe and well-tolerated intervention for PTSD, with the most commonly reported mild adverse effects being itching and tingling in studies involving patients with PTSD [84]. No significant or severe adverse events have been linked to HD-tDCS. However, studies involving other populations, such as the elderly, frequently report sensations of tingling and burning, with the majority of cases being mild, although 4% of individuals report "severe" sensations [91]. These observations may have relevance for PTSD populations as well.

11.3.2.3 Treatment Regimen

Drawing from the theoretical frameworks and neuroanatomical models of PTSD, it is postulated that targeting the LTC with HD-tDCS could lead to enhanced therapeutic outcomes. Given that the LTC and insula may engage in interactions with the fear circuit, creating a feedback loop that exacerbates the maladaptive symptoms of PTSD, and considering that the amygdala, hippocampus, and ventromedial prefrontal cortex constitute the "fear circuit" [92, 93], which is considered the physiological foundation of PTSD, focused and precise stimulation of the LTC may more effectively address this deeply located brain region, which is difficult to target with noninvasive techniques, thereby promoting symptom alleviation.

The literature indicates that administering HD-tDCS at 2 mA for 20 min across consecutive working days (Monday to Friday), totaling ten sessions, can yield positive effects in patients with PTSD [84]. These results suggest that further investigation into the optimization of these treatment parameters is warranted.

11.3.2.4 Clinical Recommendations

Currently, while limited research exists on HD-tDCS for PTSD, the treatment has demonstrated some efficacy. Additional clinical trials and replication studies are essential to validate and expand these findings and to identify the most effective treatment parameters (see Table 11.2).

Table 11.2 tDCS in PTSD

tDCS	Clinical recommendations					Levels + references
	Anode position	Cathode position	Intensity (mA)	Length (min)	Duration (sessions)	
Conventional tDCS	Unclear	Unclear	Unclear	Unclear	Unclear	NA
HD-tDCS	Unclear	Unclear	Unclear	Unclear	Unclear	NA

Abbreviations: *HD-tDCS* high-definition transcranial direct current stimulation

11.4 TMS vs. tDCS

11.4.1 Efficacy

A systematic review of 981 patients from 21 randomized controlled trials [94] compared the effectiveness of various physical therapies in alleviating PTSD symptoms and comorbid conditions. Among the treatments evaluated, dual tDCS was found to improve PTSD symptoms more effectively than HF-rTMS, which, in turn, was more beneficial than LF-rTMS.

TMS has been shown to improve core PTSD symptoms, including intrusive memories, avoidance behaviors, and hyperarousal [9–11]. The therapeutic effects of rTMS on brain regions are frequency-dependent: low-frequency rTMS at 1 Hz applied to the DLPFC reduces overactivity in the region, alleviating symptoms such as hypervigilance [8, 95], while HF-rTMS targeting the DLPFC enhances emotion regulation and mitigates excessive fear responses in patients with PTSD [96, 97]. Additionally, TMS has demonstrated potential to improve cognitive function, with both left-sided and right-sided rTMS associated with cognitive enhancements, particularly with right-sided rTMS, which significantly improves verbal fluency [17].

tDCS has been shown to significantly reduce PTSD symptoms, including hyperarousal, negative cognitive and mood changes, and anxiety [76]. Notably, dual tDCS has been linked to a reduction in anxiety symptoms in patients with PTSD during the follow-up phase, suggesting that the therapeutic effects on anxiety may persist long-term [98].

While the effects of noninvasive stimulation are generally less durable for persistent disorders [99], Xu Guobin's meta-analysis confirmed the long-term stability of symptom relief in patients with PTSD treated with rTMS [10]. Previous studies have indicated that the effects of follow-up treatment remain significant for periods ranging from 2 weeks to 3 months [12, 17, 18]. Conversely, Liu's study found that only the therapeutic effect of dual tDCS on anxiety symptoms remained significant during the follow-up period (from half a month to 6 months, with a mean duration of 2.2 months) [94]. These observations suggest cautious optimism regarding the long-term efficacy of neuromodulatory treatments for PTSD.

11.4.2 Safety

Both TMS and tDCS are considered low-risk interventions, with side effects primarily involving physical discomfort, with transient mood disturbances observed in TMS treatments [6]. No significant difference was observed in the incidence of side effects between the two groups [94].

11.5 ECT

11.5.1 Efficacy

To date, PTSD has not been considered an indication for ECT [100]. However, mounting evidence suggests that ECT may be beneficial for patients with refractory PTSD [101], particularly those who have failed at least four different antidepressants and 12 sessions of cognitive behavioral therapy. ECT has also shown promise in treating PTSD individuals with comorbid MDD [100, 102] and in providing maintenance therapy for individuals with chronic PTSD (symptoms persisting for more than 3 months) [101].

Current evidence indicates that ECT can alleviate core PTSD symptoms, including intrusion, avoidance, and hyperarousal. However, it remains controversial whether improvements in PTSD symptom severity are linked to concurrent improvements in depressive symptoms [100]. The comorbidity rate of PTSD and MDD is high, with nearly half of patients with PTSD meeting the diagnostic criteria for MDD [103], suggesting a significant overlap between depression and PTSD. As a result, distinguishing the specific effects of ECT on these two symptom clusters in clinical studies presents challenges [104]. A recent meta-analysis suggested that comorbid depression, given its high prevalence, may serve as an independent indication for ECT trials in patients with PTSD [105], though these observations require further validation. Notably, no significant differences in cardiovascular mortality or all-cause mortality have been observed between PTSD and MDD individuals treated with ECT and those without these comorbidities. However, ECT was independently associated with reduced cardiovascular and all-cause mortality, as well as a lower risk of suicide in patients with both depression and PTSD, showing more robust outcomes compared to antidepressant treatment alone [102].

A previous study found that the intention-to-treat response rate for ECT in patients with PTSD resistant to both pharmacological and psychotherapeutic interventions was 70%. Although the sample size was small, this suggests that ECT may be particularly suitable for PTSD complicated by psychotic features [101].

In conclusion, while the acute-phase efficacy of MECT in improving core PTSD symptoms appears promising, further supporting evidence is required to substantiate this claim. From a long-term clinical outcome perspective, ECT has been shown to independently reduce the risk of suicide, cardiovascular diseases, and all-cause mortality in PTSD individuals with comorbid MDD, with effects stronger than those achieved by antidepressant therapy alone.

11.5.2 Safety

The safety profile of ECT has been broadly validated in the general population, with most side effects primarily affecting cognition [106]. Anterograde amnesia is typically transient, while retrograde amnesia can be long-lasting or, in some cases, permanent. However, due to the limited number of clinical studies involving ECT for PTSD, specific safety data in this context remain sparse [107].

11.5.3 Treatment Regimen

There are relatively few clinical studies evaluating ECT in patients with PTSD. In one prospective study [101], pretreatment medications included pentobarbital (4–5 mg/kg) and succinylcholine (1 mg/kg), with ECT performed using the MECTA-SR1 device. The protocol involved an 800 mA current amplitude, 1.5 ms pulse width, and 50 Hz pulse frequency. Stimulation duration was adjusted to ensure full seizure induction, with EEG monitoring defined as lasting at least 25 s. The treatment regimen consisted of 6 sessions, with double temporal electrodes applied twice a week.

In a recent meta-analysis [105], four studies utilized right unilateral ECT (RUL ECT), while one study employed bilateral ECT (BL ECT). For patients who did not respond adequately to initial RUL ECT, BL or bitemporal ECT (BT ECT) was administered. The onset threshold for ECT varied from 100% to 600% over the course of 6–14 treatments.

11.5.4 Clinical Recommendations

While PTSD has not been officially included in the indications for MECT, a notable proportion of patients (approximately 10%) have received ECT for both major depression and PTSD. This indicates that a significant number of patients with PTSD are already being treated with ECT. Future studies should include larger sample sizes, randomized controlled trials, and more refined patient populations to further evaluate the effectiveness and safety of MECT in treating PTSD [108].

11.6 Conclusion

This chapter provides an overview of TMS, tDCS, ECT, and their contemporary variations in the management of PTSD.

Regarding rTMS, U-rTMS, particularly when high-frequency stimulation is applied to the right DLPFC, has shown promising results in alleviating core PTSD symptoms. B-rTMS, which targets both the right and left DLPFC with high-frequency stimulation, has also proven effective. However, the mixed outcomes reported across existing studies have led to uncertainty regarding the optimal parameters for these two protocols. In contrast, A-rTMS, a more recent treatment modality, offers the advantage of greater time efficiency compared to U-rTMS and B-rTMS. However, the notable lack of research on its application in patients with PTSD limits the ability to establish its efficacy and safety for this condition.

Among the novel TMS protocols, D-TMS, which stimulates deeper brain regions, shows significant potential, particularly when combined with exposure therapy for PTSD treatment. However, larger and more rigorously controlled studies are necessary to validate its efficacy and safety. The therapeutic benefits of priming TMS and synchronized TMS in PTSD treatment remain unclear due to limited available evidence. TBS, although still in the early stages of research, has exhibited potential for achieving comparable effectiveness while offering a more time-efficient treatment option compared to rTMS. Preliminary evidence also suggests that accelerated iTBS and bilateral TBS may reduce PTSD symptoms.

Furthermore, tDCS has shown promise in alleviating PTSD symptoms, enhancing fear-extinction learning, and improving cognitive functions. High-definition tDCS (HD-tDCS), which delivers more focused stimulation, is currently under investigation to gather further clinical evidence and establish its efficacy.

Extensive evidence supports ECT as a potentially effective treatment for several PTSD-related conditions, including refractory PTSD, PTSD with comorbid MDD, and maintenance therapy for chronic PTSD.

MST, derived from ECT, offers superior targeting precision, resulting in fewer cognitive side effects compared to traditional ECT. However, the efficacy of MST for PTSD remains unproven due to a lack of clinical evidence.

In terms of safety, both TMS and tDCS are generally regarded as low-risk treatments, characterized by mild and transient side effects. These modalities notably lack the cognitive side effects often associated with ECT. However, there is a potential risk of seizures, particularly in individuals with preexisting risk factors, warranting careful monitoring and individualized risk assessments.

Overall, current evidence highlights the promising potential of TMS and tDCS in PTSD treatment. However, due to the heterogeneity of PTSD and observed variations in treatment parameters and responses across trials, future studies should aim to validate the effectiveness of these techniques, clarify optimal treatment protocols, and investigate their long-term efficacy and safety. Larger, well-controlled studies are essential to confirm the preliminary findings and establish robust, long-term efficacy and safety profiles for neurostimulation techniques in PTSD treatment.

Given that PTSD is a highly heterogeneous condition that may co-occur with distinct psychiatric diagnoses, developing a universally effective treatment for this patient population remains challenging. Therefore, exploring the underlying neurobiological mechanisms of neurostimulation is essential for the development of more precise, targeted stimulation protocols that can address specific symptoms and facilitate individualized treatment approaches.

Acknowledgments We acknowledge the contributions of each author and the guidance from Dr. Wei Zheng.

Disclosure/Conflicts of Interest The authors declare no conflicts of interest in conducting this study or preparing the manuscript.

Financial Support This study was funded by the High-level Talents—the Second Batch of Scientific Backbone (KY480-11).

References

1. World Health Organization. Post-traumatic stress disorder; 2024. https://www.who.int/news-room/fact-sheets/detail/post-traumatic-stress-disorder
2. Kessler RC, Aguilar-Gaxiola S, Alonso J, Benjet C, Bromet EJ, Cardoso G, Degenhardt L, de Girolamo G, Dinolova RV, Ferry F, Florescu S, Gureje O, Haro JM, Huang Y, Karam EG, Kawakami N, Lee S, Lepine JP, Levinson D, et al. Trauma and PTSD in the WHO world mental health surveys. Eur J Psychotraumatol. 2017;8(sup5):1353383. https://doi.org/10.1080/20008198.2017.1353383.
3. Charlson F, van Ommeren M, Flaxman A, Cornett J, Whiteford H, Saxena S. New WHO prevalence estimates of mental disorders in conflict settings: a systematic review and meta-analysis. Lancet. 2019;394(10194):240–8. https://doi.org/10.1016/s0140-6736(19)30934-1.
4. Schrader C, Ross A. A review of PTSD and current treatment strategies. Mo Med. 2021;118(6):546–51.
5. Alexandra Kredlow M, Fenster RJ, Laurent ES, Ressler KJ, Phelps EA. Prefrontal cortex, amygdala, and threat processing: implications for PTSD [Review]. Neuropsychopharmacology. 2022;47(1):247–59. https://doi.org/10.1038/s41386-021-01155-7.
6. Gouveia FV, Davidson B, Meng Y, Gidyk DC, Rabin JS, Ng E, Abrahao A, Lipsman N, Giacobbe P, Hamani C. Treating post-traumatic stress disorder with neuromodulation therapies: transcranial magnetic stimulation, transcranial direct current stimulation, and deep brain stimulation. Neurotherapeutics. 2020;17(4):1747–56. https://doi.org/10.1007/s13311-020-00871-0.
7. Brunoni AR, Sampaio-Junior B, Moffa AH, Aparício LV, Gordon P, Klein I, Rios RM, Razza LB, Loo C, Padberg F, Valiengo L. Noninvasive brain stimulation in psychiatric disorders: a primer. Revista brasileira de psiquiatria (Sao Paulo, Brazil: 1999). 2019;41(1):70–81. https://doi.org/10.1590/1516-4446-2017-0018.
8. Lefaucheur J-P, Aleman A, Baeken C, Benninger DH, Brunelin J, Lazzaro VD, Filipović SR, Grefkes C, Hasan A, Hummel FC, Jääskeläinen SK, Langguth B, Leocani L, Londero A, Nardone R, Nguyen J-P, Nyffeler T, Oliveira-Maia AJ, Oliviero A, et al. Evidence-based guidelines on the therapeutic use of repetitive transcranial magnetic stimulation (rTMS): an update (2014–2018). Clin Neurophysiol. 2020;131(2):474–528. https://doi.org/10.1016/j.clinph.2019.11.002.
9. Harris A, Reece J. Transcranial magnetic stimulation as a treatment for posttraumatic stress disorder: a meta-analysis. J Affect Disord. 2021;289:55–65. https://doi.org/10.1016/j.jad.2021.04.003.
10. Xu G, Li G, Yang Q, Li C, Liu C. Explore the durability of repetitive transcranial magnetic stimulation in treating post-traumatic stress disorder: an updated systematic review and meta-analysis. Stress Health. 2024;40(1):e3292. https://doi.org/10.1002/smi.3292.
11. Yan T, Xie Q, Zheng Z, Zou K, Wang L. Different frequency repetitive transcranial magnetic stimulation (rTMS) for posttraumatic stress disorder (PTSD): A systematic review and meta-analysis. J Psychiatr Res. 2017;89:125–35. https://doi.org/10.1016/j.jpsychires.2017.02.021.
12. Cohen H, Kaplan Z, Kotler M, Kouperman I, Moisa R, Grisaru N. Repetitive transcranial magnetic stimulation of the right dorsolateral prefrontal cortex in posttraumatic stress disorder: a double-blind, placebo-controlled study. Am J Psychiatry. 2004;161(3):515–24. https://doi.org/10.1176/appi.ajp.161.3.515.

13. Kozel FA, Trees KV, Larson V, Phillips S, Hashimie J, Gadbois B, Johnson S, Gallinati J, Barrett B, Toyinbo P, Weisman M, Centorino M, Gibson C-A, Catalano G. One hertz versus ten hertz repetitive TMS treatment of PTSD: A randomized clinical trial. Psychiatry Res. 2019;273:153–62. https://doi.org/10.1016/j.psychres.2019.01.004.
14. Watts BV, Landon B, Groft A, Young-Xu Y. A sham controlled study of repetitive transcranial magnetic stimulation for posttraumatic stress disorder. Brain Stimul. 2012;5(1):38–43. https://doi.org/10.1016/j.brs.2011.02.002.
15. Rosenberg PB, Mehndiratta RB, Mehndiratta YP, Wamer A, Rosse RB, Balish M. Repetitive transcranial magnetic stimulation treatment of comorbid posttraumatic stress disorder and major depression. J Neuropsychiatry Clin Neurosci. 2002;14(3):270–6. https://doi.org/10.1176/jnp.14.3.270.
16. Philip NS, Ridout SJ, Albright SE, Sanchez G, Carpenter LL. 5-Hz transcranial magnetic stimulation for comorbid posttraumatic stress disorder and major depression. J Trauma Stress. 2016;29(1):93–6. https://doi.org/10.1002/jts.22065.
17. Boggio PS, Rocha M, Oliveira MO, Fecteau S, Cohen RB, Campanhã C, Ferreira-Santos E, Meleiro A, Corchs F, Zaghi S, Pascual-Leone A, Fregni F. Noninvasive brain stimulation with high-frequency and low-intensity repetitive transcranial magnetic stimulation treatment for posttraumatic stress disorder. J Clin Psychiatry. 2010;71(8):992–9. https://doi.org/10.4088/JCP.08m04638blu.
18. Leong K, Chan P, Ong L, Zwicker A, Willan S, Lam R, McGirr A. A randomized sham-controlled trial of 1-Hz and 10-Hz repetitive transcranial magnetic stimulation (rTMS) of the right dorsolateral prefrontal cortex in civilian post-traumatic stress disorder: un essai randomisé contrôlé simulé de stimulation magnétique transcrânienne repetitive (SMTr) de 1 Hz et 10 Hz du cortex préfrontal dorsolatéral droit dans le trouble de stress post-traumatique chez des civils. Can J Psychiatry. 2020;65(11):770–8. https://doi.org/10.1177/0706743720923064.
19. Thierrée S, Raulin-Briot M, Legrand M, Gouge AL, Vancappel A, Tudorache A-C, Brizard B, Clarys D, Caille A, El-Hage W. Combining trauma script exposure with rTMS to reduce symptoms of post-traumatic stress disorder: randomized controlled trial. Neuromodulation. 2022;25(4):549–57. https://doi.org/10.1111/ner.13505.
20. Becker CR, Milad MR. Contemporary approaches toward neuromodulation of fear extinction and its underlying neural circuits. Curr Top Behav Neurosci. 2023;64:353–87. https://doi.org/10.1007/7854_2023_442.
21. Wassermann EM, Lisanby SH. Therapeutic application of repetitive transcranial magnetic stimulation: a review. Clin Neurophysiol. 2001;112(8):1367–77. https://doi.org/10.1016/s1388-2457(01)00585-5.
22. Kozel FA, Motes MA, Didehbani N, DeLaRosa B, Bass C, Schraufnagel CD, Jones P, Morgan CR, Spence JS, Kraut MA, Hart J. Repetitive TMS to augment cognitive processing therapy in combat veterans of recent conflicts with PTSD: a randomized clinical trial. J Affect Disord. 2018;229:506–14. https://doi.org/10.1016/j.jad.2017.12.046.
23. Ahmadizadeh M-J, Rezaei M. Unilateral right and bilateral dorsolateral prefrontal cortex transcranial magnetic stimulation in treatment post-traumatic stress disorder: a randomized controlled study. Brain Res Bull. 2018;140:334–40. https://doi.org/10.1016/j.brainresbull.2018.06.001.
24. Koek RJ, Roach J, Athanasiou N, Wout-Frank M v t, Philip NS. Neuromodulatory treatments for post-traumatic stress disorder (PTSD). Prog Neuro-Psychopharmacol Biol Psychiatry. 2019;92:148–60. https://doi.org/10.1016/j.pnpbp.2019.01.004.
25. Bandelow B, Allgulander C, Baldwin DS, Costa DL d C, Denys D, Dilbaz N, Domschke K, Hollander E, Kasper S, Möller H-J, Eriksson E, Fineberg NA, Hättenschwiler J, Kaiya H, Karavaeva T, Katzman MA, Kim Y-K, Inoue T, Lim L, et al. World Federation of Societies of Biological Psychiatry (WFSBP) guidelines for treatment of anxiety, obsessive-compulsive and posttraumatic stress disorders – version 3. Part II: OCD and PTSD. World J Biol Psychiatry. 2023;24(2):118–34. https://doi.org/10.1080/15622975.2022.2086296.

26. Tseng P-T, Zeng B-Y, Wang H-Y, Zeng B-S, Liang C-S, Chen Y-CB, Stubbs B, Carvalho AF, Brunoni AR, Su K-P, Tu Y-K, Wu Y-C, Chen T-Y, Li D-J, Lin P-Y, Chen Y-W, Hsu C-W, Hung K-C, Shiue Y-L, Li C-T. Efficacy and acceptability of noninvasive brain stimulation for treating posttraumatic stress disorder symptoms: a network meta-analysis of randomized controlled trials. Acta Psychiatr Scand. 2024;150(1):5–21. https://doi.org/10.1111/acps.13688.
27. Woodside DB, Colton P, Lam E, Dunlop K, Rzeszutek J, Downar J. Dorsomedial prefrontal cortex repetitive transcranial magnetic stimulation treatment of posttraumatic stress disorder in eating disorders: an open-label case series. Int J Eat Disord. 2017;50(10):1231–4. https://doi.org/10.1002/eat.22764.
28. Croarkin PE, Dojnov A, Middleton VJ, Bowman J, Kriske J, Donachie N, Siddiqi SH, Downar J. Accelerated 1 Hz dorsomedial prefrontal transcranial magnetic stimulation for generalized anxiety disorder in adolescents and young adults: A case series. Brain Stimul. 2024;17(2):269–71. https://doi.org/10.1016/j.brs.2024.02.018.
29. Dilkov D, Hawken ER, Kaludiev E, Milev R. Repetitive transcranial magnetic stimulation of the right dorsal lateral prefrontal cortex in the treatment of generalized anxiety disorder: a randomized, double-blind sham controlled clinical trial. Prog Neuro-Psychopharmacol Biol Psychiatry. 2017;78:61–5. https://doi.org/10.1016/j.pnpbp.2017.05.018.
30. Julian M, Vijeinika V, Ben C, René H, Cynthia HYF, Allan HY. Comparative efficacy and acceptability of non-surgical brain stimulation for the acute treatment of major depressive episodes in adults: systematic review and network meta-analysis. BMJ. 2019;364:l1079. https://doi.org/10.1136/bmj.l1079.
31. Sonmez AI, Camsari DD, Nandakumar AL, Voort JLV, Kung S, Lewis CP, Croarkin PE. Accelerated TMS for depression: a systematic review and meta-analysis. Psychiatry Res. 2019;273:770–81. https://doi.org/10.1016/j.psychres.2018.12.041.
32. Holtzheimer PE, McDonald WM, Mufti M, Kelley ME, Quinn S, Corso G, Epstein CM. Accelerated repetitive transcranial magnetic stimulation for treatment-resistant depression. Depress Anxiety. 2010;27(10):960–3. https://doi.org/10.1002/da.20731.
33. Senda M, Johnson K, Taylor I, Jensen M, Flynn H, Kozel FA. Accelerated transcranial magnetic stimulation provided rapid improvement in depressive, anxiety, trauma, and pain symptoms in a woman experiencing postpartum depression. J ECT. 2023;39(3):e14–5. https://doi.org/10.1097/yct.0000000000000936.
34. Tovar-Perdomo S, McGirr A, Eynde FV d, Santos NRD, Berlim MT. High frequency repetitive transcranial magnetic stimulation treatment for major depression: dissociated effects on psychopathology and neurocognition. J Affect Disord. 2017;217:112–7. https://doi.org/10.1016/j.jad.2017.03.075.
35. George MS, Raman R, Benedek DM, Pelic CG, Grammer GG, Stokes KT, Schmidt M, Spiegel C, Dealmeida N, Beaver KL, Borckardt JJ, Sun X, Jain S, Stein MB. A two-site pilot randomized 3 day trial of high dose left prefrontal repetitive transcranial magnetic stimulation (rTMS) for suicidal inpatients. Brain Stimul. 2014;7(3):421–31. https://doi.org/10.1016/j.brs.2014.03.006.
36. Yiftach R, Alon A, Yechiel L, Abraham Z. Three-dimensional distribution of the electric field induced in the brain by transcranial magnetic stimulation using figure-8 and deep H-coils. J Clin Neurophysiol. 2007;24(1):31. https://doi.org/10.1097/WNP.0b013e31802fa393.
37. Paul BF, Sarah F, Zafiris JD. A comprehensive review of the effects of rTMS on motor cortical excitability and inhibition. Clin Neurophysiol. 2006;117(12):2584. https://doi.org/10.1016/j.clinph.2006.06.712.
38. Yiftach R, Abraham Z, Mark H. A coil design for transcranial magnetic stimulation of deep brain regions. J Clin Neurophysiol. 2002;19(4):361. https://doi.org/10.1097/00004691-200208000-00008.
39. Kent RD, Mark R. Suppressing the surface field during transcranial magnetic stimulation. IEEE Trans Biomed Eng. 2006;53(2):190. https://doi.org/10.1109/tbme.2005.862545.
40. Zhi-De D, Angel VP, Sarah HL. Coil design considerations for deep-brain transcranial magnetic stimulation (dTMS). Annu Int Conf IEEE Eng Med Biol Soc. 2009;2008. https://doi.org/10.1109/iembs.2008.4650502.

41. Bersani FS, Minichino A, Enticott PG, Mazzarini L, Khan N, Antonacci G, Raccah RN, Salviati M, Delle Chiaie R, Bersani G, Fitzgerald PB, Biondi M. Deep transcranial magnetic stimulation as a treatment for psychiatric disorders: a comprehensive review. Eur Psychiatry. 2012;28(1):30. https://doi.org/10.1016/j.eurpsy.2012.02.006.
42. Passa A-MD, Prokop-Millar S, Yaya H, Dabir M, McIntyre-Wood C, Fein A, MacKillop E, MacKillop J, Duarte D. Clinical efficacy of deep transcranial magnetic stimulation (dTMS) in psychiatric and cognitive disorders: a systematic review. J Psychiatr Res. 2024;175:287–315. https://doi.org/10.1016/j.jpsychires.2024.05.011.
43. Isserles M, Shalev AY, Roth Y, Peri T, Kutz I, Zlotnick E, Zangen A. Effectiveness of deep transcranial magnetic stimulation combined with a brief exposure procedure in post-traumatic stress disorder—a pilot study. Brain Stimul. 2013;6(3):377–83. https://doi.org/10.1016/j.brs.2012.07.008.
44. Moshe I, Aron T, Yiftach R, Alexander B, Daniel MB, Herbert W, David F, Laura V, Walter D, Joseph Z, Corey JK, Mahendra TB, Amit E, Mark SG, Igor F, Kyle L, Leah C, Sandeep V, Ahava S, et al. Deep transcranial magnetic stimulation combined with brief exposure for posttraumatic stress disorder: a prospective multisite randomized trial. Biol Psychiatry. 2021;90(10):721–8.
45. Isserles M, Tendler A, Roth Y, Bystritsky A, Blumberger DM, Ward H, Feifel D, Viner L, Duffy W, Zohar J, Keller CJ, Bhati MT, Etkin A, George MS, Filipcic I, Lapidus K, Casuto L, Vaishnavi S, Stein A, et al. Deep transcranial magnetic stimulation combined with brief exposure for posttraumatic stress disorder: a prospective multisite randomized trial. Biol Psychiatry. 2021;90(10):721–8. https://doi.org/10.1016/j.biopsych.2021.04.019.
46. Yechiel L, Moshe I, Frank P, Sarah HL, Alexander B, Guohua X, Aron T, Zafiris JD, Jaron LW, Pinhas D, Hisham MH, Irving MR, Oscar GM, Thomas ES, Eric H, Joshua AB, Mustafa MH, Uzi S, Ahava S, et al. Efficacy and safety of deep transcranial magnetic stimulation for major depression: a prospective multicenter randomized controlled trial. World Psychiatry. 2015;14(1):64. https://doi.org/10.1002/wps.20199.
47. Andre RB, Anna C, Adriano HM, Lais BR, Wagner FG, Zafiris JD, Andre FC. Repetitive transcranial magnetic stimulation for the acute treatment of major depressive episodes: a systematic review with network meta-analysis. JAMA Psychiatry. 2017;74(2):143–52.
48. Yu-Yung H, Li-Heng Y, Bredon S, Dian-Jeng L, Ping-Tao T, Ta-Chuan Y, Tien-Yu C, Chih-Sung L, Che-Sheng C. Efficacy and tolerability of deep transcranial magnetic stimulation for treatment-resistant depression: a systematic review and meta-analysis. Prog Neuro-Psychopharmacol Biol Psychiatry. 2019;99:109850. https://doi.org/10.1016/j.pnpbp.2019.109850.
49. Meenakshi BI, Nicole S, Eric MW. Priming stimulation enhances the depressant effect of low-frequency repetitive transcranial magnetic stimulation. J Neurosci. 2003;23(34):10867. https://doi.org/10.1523/jneurosci.23-34-10867.2003.
50. Paul BF, Kate H, Susan M, Sally H, Rebecca S, Greg B, Jayashri K, Zafiris JD. Priming stimulation enhances the effectiveness of low-frequency right prefrontal cortex transcranial magnetic stimulation in major depression. J Clin Psychopharmacol. 2008;28(1):52. https://doi.org/10.1097/jcp.0b013e3181603f7c.
51. Opie GM, Vosnakis E, Ridding MC, Ziemann U, Semmler JG. Priming theta burst stimulation enhances motor cortex plasticity in young but not old adults. Brain Stimul. 2017;10(2):298–304. https://doi.org/10.1016/j.brs.2017.01.003.
52. Paul BF, Kate EH, Ajeet S, Ranil G, Christopher S, Samir I, Phillip JH, Jeff Z, D. Equivalent beneficial effects of unilateral and bilateral prefrontal cortex transcranial magnetic stimulation in a large randomized trial in treatment-resistant major depression. Int J Neuropsychopharmacol. 2013;16(9):1975. https://doi.org/10.1017/s1461145713000369.
53. Andrew FL, Ian AC, David F, John WG, Mustafa H, Linda LC, Michael ET, Andrew DK, Noah SP, Mahendra TB, William JB, Robert HH, Yvette IS, Scott TA, Dan VI, John POR, William SG, Rakesh J, Karl SB, et al. Efficacy and safety of low-field synchronized transcranial magnetic stimulation (sTMS) for treatment of major depression. Brain Stimul. 2015;8(4):787. https://doi.org/10.1016/j.brs.2015.05.005.

54. Flavio F. Endogenous and exogenous electric fields as modifiers of brain activity: rational design of noninvasive brain stimulation with transcranial alternating current stimulation. Dialogues Clin Neurosci. 2014;16(1):93. https://doi.org/10.31887/DCNS.2014.16.1/ffroehlich.
55. Noah SP, Andrew FL, Ian AC, Joe M, John WG, Linda LC. Predictors of response to synchronized transcranial magnetic stimulation for major depressive disorder. Depress Anxiety. 2018;36(3):278. https://doi.org/10.1002/da.22862.
56. Noah SP, Emily EA, Mary EK, William B, Laurie W, Paul EH. Synchronized transcranial magnetic stimulation for posttraumatic stress disorder and comorbid major depression. Brain Stimul. 2019;12(5):1335. https://doi.org/10.1016/j.brs.2019.06.010.
57. Cole E, O'Sullivan SJ, Tik M, Williams NR. Accelerated theta burst stimulation: safety, efficacy, and future advancements. Biol Psychiatry. 2024;95(6):523–35. https://doi.org/10.1016/j.biopsych.2023.12.004.
58. Bation R, Magnin C, Poulet E, Mondino M, Brunelin J. Intermittent theta burst stimulation for negative symptoms of schizophrenia-A double-blind, sham-controlled pilot study. NPJ Schizophr. 2021;7(1):Article 10. https://doi.org/10.1038/s41537-021-00138-3.
59. Chang C-H, Liu C-Y, Chen S-J, Tsai H-C. Accelerated theta-burst transcranial magnetic stimulation of the bilateral dorsolateral prefrontal cortex in a patient with post-traumatic stress disorder and major depressive disorder: case report. Front Psychiatry. 2023;14:Article 1061088. https://doi.org/10.3389/fpsyt.2023.1061088.
60. Philip NS, Barredo J, Aiken E, Larson V, Jones RN, Shea MT, Greenberg BD, van't Wout-Frank, M. Theta-burst transcranial magnetic stimulation for posttraumatic stress disorder. Am J Psychiatry. 2019;176(11):939–48. https://doi.org/10.1176/appi.ajp.2019.18101160.
61. Yuan HL, Liu B, Li FZ, Jin YC, Zheng S, Ma ZJ, Wu ZY, Chen C, Zhang L, Gu YA, Gao X, Yang Q. Effects of intermittent theta-burst transcranial magnetic stimulation on post-traumatic stress disorder symptoms: A randomized controlled trial. Psychiatry Res. 2023;329:Article 115533. https://doi.org/10.1016/j.psychres.2023.115533.
62. Cirillo P, Gold AK, Nardi AE, Ornelas AC, Nierenberg AA, Camprodon J, Kinrys G. Transcranial magnetic stimulation in anxiety and trauma-related disorders: a systematic review and meta-analysis. Brain Behavior. 2019;9(6):Article e01284. https://doi.org/10.1002/brb3.1284.
63. Nursey J, Sbisa A, Knight H, Ralph N, Cowlishaw S, Forbes D, O'Donnell M, Hinton M, Cooper J, Hopwood M, McFarlane A, Herring S, Fitzgerald P. Exploring theta burst stimulation for post-traumatic stress disorder in Australian veterans-A pilot study. Mil Med. 2020;185(9–10):E1770–8. https://doi.org/10.1093/milmed/usaa149.
64. Vaithianathan T, Blair M, Soares V, Rybak YE, Palaniyappan L, Richardson JD, Burhan AM. Bilateral sequential theta burst stimulation in depressed veterans with service related posttraumatic stress disorder: a feasibility study. BMC Psychiatry. 2022;22(1):Article 81. https://doi.org/10.1186/s12888-022-03729-1.
65. Stohrmann P, Godbersen GM, Reed MB, Unterholzner J, Klobl M, Baldinger-Melich P, Vanicek T, Hahn A, Lanzenberger R, Kasper S, Kranz GS. Effects of bilateral sequential theta-burst stimulation on functional connectivity in treatment-resistant depression: first results [article]. J Affect Disord. 2023;324:660–9. https://doi.org/10.1016/j.jad.2022.12.088.
66. Cheng CM, Li CT, Tsai SJ. Current updates on newer forms of transcranial magnetic stimulation in major depression. Adv Exp Med Biol. 2021;1305:333–49. https://doi.org/10.1007/978-981-33-6044-0_18.
67. Lisanby SH. Update on magnetic seizure therapy: a novel form of convulsive therapy. J ECT. 2002;18(4):182–8. https://doi.org/10.1097/00124509-200212000-00003.
68. Daskalakis ZJ, Dimitrova J, McClintock SM, Sun Y, Voineskos D, Rajji TK, Goldbloom DS, Wong AHC, Knyahnytska Y, Mulsant BH, Downar J, Fitzgerald PB, Blumberger DM. Magnetic seizure therapy (MST) for major depressive disorder. Neuropsychopharmacology. 2020;45(2):276–82. https://doi.org/10.1038/s41386-019-0515-4.

69. Tang VM, Blumberger DM, Throop A, McClintock SM, Voineskos D, Downar J, Knyahnytska Y, Mulsant BH, Fitzgerald PB, Daskalakis ZJ. Continuation magnetic seizure therapy for treatment-resistant unipolar or bipolar depression. J Clin Psychiatry. 2021;82(6). doi:https://doi.org/10.4088/JCP.20m13677.
70. Zhang XY, Chen HD, Liang WN, Yang XH, Cai DB, Huang X, Huang XB, Liu CY, Zheng W. Adjunctive magnetic seizure therapy for schizophrenia: a systematic review. Front Psych. 2021;12:813590. https://doi.org/10.3389/fpsyt.2021.813590.
71. Chen M, Yang X, Liu C, Li J, Wang X, Yang C, Hu X, Li J, Zhao J, Li X, Xu Y, Liu S. Comparative efficacy and cognitive function of magnetic seizure therapy vs. electroconvulsive therapy for major depressive disorder: a systematic review and meta-analysis. Transl Psychiatry. 2021;11(1):437. https://doi.org/10.1038/s41398-021-01560-y.
72. Asthana M, Nueckel K, Mühlberger A, Neueder D, Polak T, Domschke K, Deckert J, Herrmann MJ. Effects of transcranial direct current stimulation on consolidation of fear memory. Front Psych. 2013;4:107. https://doi.org/10.3389/fpsyt.2013.00107.
73. Coffman BA, Clark VP, Parasuraman R. Battery powered thought: enhancement of attention, learning, and memory in healthy adults using transcranial direct current stimulation. NeuroImage. 2014;85(Pt 3):895–908. https://doi.org/10.1016/j.neuroimage.2013.07.083.
74. Mungee A, Burger M, Bajbouj M. No effect of cathodal transcranial direct current stimulation on fear memory in healthy human subjects. Brain Sci. 2016;6(4). https://doi.org/10.3390/brainsci6040055.
75. Nitsche MA, Cohen LG, Wassermann EM, Priori A, Lang N, Antal A, Paulus W, Hummel F, Boggio PS, Fregni F, Pascual-Leone A. Transcranial direct current stimulation: state of the art 2008. Brain Stimul. 2008;1(3):206–23. https://doi.org/10.1016/j.brs.2008.06.004.
76. Ahmadizadeh MJ, Rezaei M, Fitzgerald PB. Transcranial direct current stimulation (tDCS) for post-traumatic stress disorder (PTSD): a randomized, double-blinded, controlled trial. Brain Res Bull. 2019;153:273–8. https://doi.org/10.1016/j.brainresbull.2019.09.011.
77. Woods AJ, Antal A, Bikson M, Boggio PS, Brunoni AR, Celnik P, Cohen LG, Fregni F, Herrmann CS, Kappenman ES, Knotkova H, Liebetanz D, Miniussi C, Miranda PC, Paulus W, Priori A, Reato D, Stagg C, Wenderoth N, Nitsche MA. A technical guide to tDCS, and related non-invasive brain stimulation tools. Clin Neurophysiol. 2016;127(2):1031–48. https://doi.org/10.1016/j.clinph.2015.11.012.
78. Eyraud N, Poupin P, Legrand M, Caille A, Sauvaget A, Bulteau S, Gohier B, Harika-Germaneau G, Drapier D, Jaafari N, Bodic O, Brizard B, Gissot V, Belzung C, Courtine J-B, El-Hage W. Combining trauma script exposure with tDCS to alleviate symptoms of posttraumatic stress disorder: a two-arm randomized sham-controlled multicenter trial. Brain Stimul. 2024;17(3):591–3. https://doi.org/10.1016/j.brs.2024.04.018.
79. Bremner JD, Vermetten E, Schmahl C, Vaccarino V, Vythilingam M, Afzal N, Grillon C, Charney DS. Positron emission tomographic imaging of neural correlates of a fear acquisition and extinction paradigm in women with childhood sexual-abuse-related post-traumatic stress disorder. Psychol Med. 2005;35(6):791–806. https://doi.org/10.1017/s0033291704003290.
80. Wout MV t, Longo SM, Reddy MK, Philip NS, Bowker MT, Greenberg BD. Transcranial direct current stimulation may modulate extinction memory in posttraumatic stress disorder. Brain Behav. 2017;7(5):e00681. https://doi.org/10.1002/brb3.681.
81. Saunders N, Downham R, Turman B, Kropotov J, Clark R, Yumash R, Szatmary A. Working memory training with tDCS improves behavioral and neurophysiological symptoms in pilot group with post-traumatic stress disorder (PTSD) and with poor working memory. Neurocase. 2015;21(3):271–8. https://doi.org/10.1080/13554794.2014.890727.
82. Wout-Frank M v t, Shea MT, Larson VC, Greenberg BD, Philip NS. Combined transcranial direct current stimulation with virtual reality exposure for posttraumatic stress disorder: feasibility and pilot results. Brain Stimul. 2019;12(1):41–3. https://doi.org/10.1016/j.brs.2018.09.011.

83. Wout-Frank M v t, Arulpragasam AR, Faucher C, Aiken E, Shea MT, Jones RN, Greenberg BD, Philip NS. Virtual reality and transcranial direct current stimulation for posttraumatic stress disorder: a randomized clinical trial. JAMA Psychiatry. 2024;81(5):437–46. https://doi.org/10.1001/jamapsychiatry.2023.5661.
84. Hampstead BM, Mascaro N, Schlaefflin S, Bhaumik A, Laing J, Peltier S, Martis B. Variable symptomatic and neurophysiologic response to HD-tDCS in a case series with posttraumatic stress disorder. Int J Psychophysiol. 2020;154:93–100. https://doi.org/10.1016/j.ijpsycho.2019.10.017.
85. Milad MR, Quirk GJ. Fear extinction as a model for translational neuroscience: ten years of progress. Annu Rev Psychol. 2012;63:129–51. https://doi.org/10.1146/annurev.psych.121208.131631.
86. Quirk GJ, Garcia R, González-Lima F. Prefrontal mechanisms in extinction of conditioned fear. Biol Psychiatry. 2006;60(4):337–43. https://doi.org/10.1016/j.biopsych.2006.03.010.
87. VanElzakker MB, Dahlgren MK, Davis FC, Dubois S, Shin LM. From Pavlov to PTSD: the extinction of conditioned fear in rodents, humans, and anxiety disorders. Neurobiol Learn Mem. 2014;113:3–18. https://doi.org/10.1016/j.nlm.2013.11.014.
88. Fregni F, Marcolin MA, Myczkowski M, Amiaz R, Hasey G, Rumi DO, Rosa M, Rigonatti SP, Camprodon J, Walpoth M, Heaslip J, Grunhaus L, Hausmann A, Pascual-Leone A. Predictors of antidepressant response in clinical trials of transcranial magnetic stimulation. Int J Neuropsychopharmacol. 2006;9(6):641–54. https://doi.org/10.1017/s1461145705006280.
89. Bikson M, Grossman P, Thomas C, Zannou AL, Jiang J, Adnan T, Mourdoukoutas AP, Kronberg G, Truong D, Boggio P, Brunoni AR, Charvet L, Fregni F, Fritsch B, Gillick B, Hamilton RH, Hampstead BM, Jankord R, Kirton A, et al. Safety of transcranial direct current stimulation: evidence based update 2016. Brain Stimul. 2016;9(5):641–61. https://doi.org/10.1016/j.brs.2016.06.004.
90. Richardson J, Datta A, Dmochowski J, Parra LC, Fridriksson J. Feasibility of using high-definition transcranial direct current stimulation (HD-tDCS) to enhance treatment outcomes in persons with aphasia. NeuroRehabilitation. 2015;36(1):115–26. https://doi.org/10.3233/nre-141199.
91. Reckow J, Rahman-Filipiak A, Garcia S, Schlaefflin S, Calhoun O, DaSilva AF, Bikson M, Hampstead BM. Tolerability and blinding of 4x1 high-definition transcranial direct current stimulation (HD-tDCS) at two and three milliamps. Brain Stimul. 2018;11(5):991–7. https://doi.org/10.1016/j.brs.2018.04.022.
92. Etkin A, Wager TD. Functional neuroimaging of anxiety: a meta-analysis of emotional processing in PTSD, social anxiety disorder, and specific phobia. Am J Psychiatry. 2007;164(10):1476–88. https://doi.org/10.1176/appi.ajp.2007.07030504.
93. Simmons AN, Matthews SC. Neural circuitry of PTSD with or without mild traumatic brain injury: a meta-analysis. Neuropharmacology. 2012;62(2):598–606. https://doi.org/10.1016/j.neuropharm.2011.03.016.
94. Liu H, Wang X, Gong T, Xu S, Zhang J, Yan L, Zeng Y, Yi M, Qian Y. Neuromodulation treatments for post-traumatic stress disorder: a systematic review and network meta-analysis covering efficacy, acceptability, and follow-up effects. J Anxiety Disord. 2024;106:102912. https://doi.org/10.1016/j.janxdis.2024.102912.
95. Sauvaget A, Bulteau S, Guilleux A, Leboucher J, Pichot A, Valrivière P, Vanelle J-M, Sébille-Rivain V, Grall-Bronnec M. Both active and sham low-frequency rTMS single sessions over the right DLPFC decrease cue-induced cravings among pathological gamblers seeking treatment: A randomized, double-blind, sham-controlled crossover trial. J Behav Addict. 2018;7(1):126–36. https://doi.org/10.1556/2006.7.2018.14.
96. Pulopulos MM, Schmausser M, Smet SD, Vanderhasselt M-A, Baliyan S, Venero C, Baeken C, Raedt RD. The effect of HF-rTMS over the left DLPFC on stress regulation as measured by cortisol and heart rate variability. Horm Behav. 2020;124:104803. https://doi.org/10.1016/j.yhbeh.2020.104803.

97. Pulopulos MM, Allaert J, Vanderhasselt M-A, Sanchez-Lopez A, Witte SD, Baeken C, Raedt RD. Effects of HF-rTMS over the left and right DLPFC on proactive and reactive cognitive control. Soc Cogn Affect Neurosci. 2022;17(1):109–19. https://doi.org/10.1093/scan/nsaa082.
98. Nasiri F, Mashhadi A, Bigdeli I, Chamanabad AG, Ellard KK. Augmenting the unified protocol for transdiagnostic treatment of emotional disorders with transcranial direct current stimulation in individuals with generalized anxiety disorder and comorbid depression: a randomized controlled trial. J Affect Disord. 2020;262:405–13. https://doi.org/10.1016/j.jad.2019.11.064.
99. Siddiqi SH, Fox MD. Targeting symptom-specific networks with transcranial magnetic stimulation. Biol Psychiatry. 2024;95(6):502–9. https://doi.org/10.1016/j.biopsych.2023.11.011.
100. Callens L, Sienaert P. The role of electroconvulsive therapy in posttraumatic stress disorder. J ECT. 2024; https://doi.org/10.1097/yct.0000000000001063.
101. Margoob MA, Ali Z, Andrade C. Efficacy of ECT in chronic, severe, antidepressant- and CBT-refractory PTSD: an open, prospective study. Brain Stimul. 2010;3(1):28–35. https://doi.org/10.1016/j.brs.2009.04.005.
102. Ahmadi N, Moss L, Simon E, Nemeroff CB, Atre-Vaidya N. Efficacy and long-term clinical outcome of comorbid posttraumatic stress disorder and major depressive disorder after electroconvulsive therapy. Depress Anxiety. 2016;33(7):640–7. https://doi.org/10.1002/da.22451.
103. Naser A, Lori M, Peter H, Charles N, Nutan A-V. Clinical outcome of maintenance electroconvulsive therapy in comorbid posttraumatic stress disorder and major depressive disorder. J Psychiatr Res. 2018;105:132. https://doi.org/10.1016/j.jpsychires.2018.08.023.
104. Youssef NA, McCall WV, Andrade C. The role of ECT in posttraumatic stress disorder: a systematic review. Ann Clin Psychiatry. 2017;29:62–70
105. Zhong M, Liu Q, Li L, Tang VM, Wong AHC, Liu Y. Evaluating the effect of electroconvulsive therapy (ECT) on post-traumatic stress disorder (PTSD): A systematic review and meta-analysis of five studies. J Psychiatr Res. 2023;164:37–45. https://doi.org/10.1016/j.jpsychires.2023.05.080.
106. Marijn CWK, Indira T, Guido A v W, Jeroen A v W, Bryan AS, Guillén F. An electroconvulsive therapy procedure impairs reconsolidation of episodic memories in humans. Nat Neurosci. 2013;17(2):204. https://doi.org/10.1038/nn.3609.
107. Nagy AY, Sadiya D, Peter BR, Laryssa M, William VM. Treating posttraumatic stress disorder symptoms with low amplitude seizure therapy (LAP-ST) compared with standard right unilateral electroconvulsive therapy: a pilot double-blinded randomized clinical trial. J ECT. 2020;36(4):291. https://doi.org/10.1097/yct.0000000000000701.
108. Youssef NA. Posttraumatic stress disorder comorbidity in patients undergoing ECT for major depressive disorder. Ann Clin Psychiatry. 2023;35(4). https://doi.org/10.12788/acp.0115.

12 Sleep Disorders

Jiahui Deng, Yinjiao Li, Mingyue Zhang, and Wenrong Wen

Abstract

Sleep disorders comprise prevalent, intricate, and recurring set of conditions. Chronic sleep disturbances not only heighten the risk of cardiovascular and systemic diseases but also significantly increase susceptibility to mental health disorders including depression and anxiety, thereby placing substantial burdens on both families and society. Considering the adverse effects associated with pharmacological treatments and the protracted nature of psychotherapeutic interventions, neuromodulation technologies have gained increasing clinical acceptance for insomnia management. Their therapeutic appeal lies in demonstrated safety profiles, noninvasive application modalities, and promising clinical outcomes. Both transcranial magnetic stimulation (TMS) and transcranial direct current stimulation (tDCS) have shown efficacy in improving sleep architecture and mitigating insomnia severity through diverse stimulation protocols targeting specific neuroanatomical regions. This section systematically examines the recent advancements in TMS and tDCS applications for sleep disorders, providing evidence-based clinical recommendations while critically evaluating existing research findings and methodological limitations. These studies identified several promising treatment parameters: low-frequency TMS protocols, continuous theta-burst stimulation (cTBS), and cathodal tDCS directed at the right dorsolateral prefrontal cortex demonstrate particularly robust therapeutic potential. Emerging evidence also suggests the parietal cortex as a viable stimulation target, while combined neuromodulation approaches integrating TMS and tDCS may offer synergistic enhancement of treatment efficacy.

J. Deng (✉) · Y. Li · M. Zhang · W. Wen
Peking University Sixth Hospital, Peking University Institute of Mental Health, NHC Key Laboratory of Mental Health (Peking University), National Clinical Research Center for Mental Disorders (Peking University Sixth Hospital), Beijing, China
e-mail: jiahuideng2012@bjmu.edu.cn

W. Zheng, Y. Ning (eds.), *TMS and tDCS for Psychiatric Disorders*,
https://doi.org/10.1007/978-981-96-8504-2_12

Keywords

Sleep disorders · Insomnia · Transcranial magnetic stimulation · Transcranial direct current stimulation · Dorsolateral prefrontal cortex

Abbreviations

cTBS	Continuous theta-burst stimulation
DLPFC	Dorsolateral prefrontal cortex
DMN	Default mode network
dTMS	Deep transcranial magnetic stimulation
iTBS	Intermittent theta-burst stimulation
LF	Low-frequency
M1	Primary motor cortex
MST	Magnetic seizure therapy
PSQI	Pittsburgh sleep quality index
PTSD	Post-traumatic stress disorder
rMT	Resting motor threshold
rTMS	Repetitive transcranial magnetic stimulation
sTMS	Synchronized transcranial magnetic stimulation
TBS	Theta burst stimulation
tDCS	Transcranial direct current stimulation
TMS	Transcranial magnetic stimulation
TRD	Treatment-resistant depression

12.1 Introduction

Sleep occupies approximately one-third of the human lifespan. Sleep disorders are a set of common, complex, and recurrent diseases, which seriously impact psychosomatic health and serve as a risk factor for diabetes, obesity, depression, and anxiety [1, 2]. The diagnostic category for sleep disorders encompasses primary disorders such as insomnia, parasomnias, and obstructive sleep apnea, as well as secondary syndromes that disrupt sleep architecture through quantitative, qualitative, or cyclical abnormalities. Insomnia represents the most prevalent diagnostic entity within this spectrum, affecting 10–20% of the general population [3]. Clinical management challenges arise from multifactorial contributors including polypharmacy, environmental disruptions, sedentary lifestyles, excessive screen exposure, chronic stress, and comorbid affective disorders.

Pharmacological interventions for insomnia predominantly include benzodiazepines and Z-hypnotics (e.g., zolpidem and zopiclone). However, their clinical utility remains constrained by adverse effects including excessive sedation, parasomnias, withdrawal syndromes, and abuse liability [4]. Melatonin receptor agonists, typified by ramelteon, demonstrate efficacy in circadian rhythm sleep disorders when administered with precise timing, primarily through reduction of sleep onset latency [5]. Nevertheless, these agents show limited efficacy in addressing sleep maintenance impairment [6]. Psychotherapeutic approaches, particularly cognitive behavioral therapy, have established efficacy for both insomnia [7] and restless legs syndrome [8], while the insufficient access limits its utilization [9]. Complementary non-pharmacological interventions, including music therapy [10], acupuncture [11], essential oils [12], and light therapy [13], demonstrate variable therapeutic efficacy. These collective findings underscore the critical need for developing novel therapeutic modalities with enhanced efficacy and broader accessibility.

Noninvasive brain stimulation (NIBS) has emerged as a promising therapeutic approach for treating sleep disorders through modulation of neural plasticity and functional connectivity. The NIBS paradigm comprises two primary modalities—transcranial magnetic stimulation (TMS) and transcranial direct current stimulation (tDCS). Each employs multiple stimulation protocols specifically designed to address specific clinical objectives [14]. Compared to pharmacological interventions and psychotherapy, TMS demonstrates cost-efficient superiority in improving sleep architecture and reducing hyperarousal states, characterized by clinically significant efficacy, sustained therapeutic benefits, and absent risk of pharmacological dependence [15]. Furthermore, TMS has proven effective in alleviating cognitive deficits consequent to sleep deprivation [16], and emerging evidence suggests its therapeutic potential for restless legs syndrome management [17]. However, limited therapeutic efficacy has been observed in OSA patients receiving TMS, [18] with current evidence base requiring Systematic integration for other sleep disorder indications [19]. Similarly, while tDCS shows promise for sleep quality improvement across varied clinical contexts [20], suboptimal stimulation parameters may produce inconsistent outcomes in narcolepsy and insomnia patients [21, 22]. This underscores the critical need for developing standardized stimulation protocols capable of effectively modulating arousal process and sleep-wake homeostasis [23]. Given the inconsistencies regarding the efficacy and stimulation parameters of NIBS, further research is necessary to establish the optimal stimulation strategy. This chapter presents a comprehensive review and clinical recommendations of the application of TMS and tDCS in the treatment of sleep disorders, especially insomnia disorders. The content will cover standard repetitive transcranial magnetic stimulation (rTMS), accelerated rTMS (aTMS), deep TMS (dTMS), priming TMS (pTMS), synchronized TMS (sTMS), theta burst stimulation (TBS), magnetic seizure therapy (MST) and tDCS. We further provide evidence-based clinical recommendations while identifying existing research limitations and proposing future investigative directions to optimize therapeutic outcomes.

12.2 TMS

12.2.1 rTMS

12.2.1.1 Unilateral rTMS

Efficacy

A systematic review synthesized evidence from 28 clinical trials investigating the effects of rTMS on sleep quality and sleep disturbances across various neuropsychiatric conditions [24]. Meta-analysis revealed consistent improvements in both objective polysomnographic metrics (including rapid eye movement sleep latency and slow-wave sleep duration) and subjective sleep quality. The majority of included studies targeted the dorsolateral prefrontal cortex (DLPFC), with left prefrontal predominance, employing low-frequency protocols (≤1 Hz) that demonstrated sustained therapeutic efficacy. Notably, rTMS-induced benefits persisted for up to 30 days post-intervention, with longitudinal investigations reporting significantly reduced relapse rates and symptom recurrence at 3-month follow-up compared to pharmacotherapy or psychotherapy cohorts. Emerging evidence further supports dose-dependent therapeutic effects, where higher stimulation intensity (≥110% motor threshold) correlates with enhanced clinical outcomes in insomnia management [25].

While numerous studies consistently report subjective sleep quality improvement following rTMS, objective polysomnographic measures show less consistent improvements. Only a restricted subset of investigations have implemented sham-controlled designs to adequately control for placebo effects. Among these rigorously designed trials, significant placebo responses have been observed in certain cohorts, whereas negligible effects emerged in others. This methodological heterogeneity may reflect variations in stimulation parameters, with most investigations utilizing low-frequency (LF) rTMS protocols (≤1 Hz) directed at either DLPFC or parietal cortical regions [26].

Emerging evidence underscores the therapeutic potential of rTMS in managing comorbid sleep disorders associated with psychiatric conditions. Two seminal clinical trials specifically evaluated divergent rTMS protocols in patients presenting with insomnia comorbid with generalized anxiety disorder or major depressive disorder. Both studies reported improvements in all subjective and objective sleep parameters. However, the reliability of these findings must be considered alongside limitations, including an unblinded study design. Additionally, the results of both studies may be influenced by the fact that participants with psychiatric conditions were permitted to take variable doses of short-acting benzodiazepines and agomelatine [27].

Safety

rTMS exhibits an exceptional tolerability profile. The therapeutic profile of rTMS is characterized by a low incidence of adverse events, with transient cephalalgia and cervical myalgia representing the most frequently reported minor complaints.

Moderate-intensity reactions typically involve scalp dysesthesia and periorbital muscle contractions, while severe neurological complications such as generalized tonic-clonic seizures or hypomanic episodes remain exceptionally rare. Cardiac pacemaker recipients and individuals with seizure disorders warrant particular caution, necessitating rigorous pretreatment screening protocols. Absolute contraindications include intracranial metallic implants (e.g., cochlear prostheses, aneurysm clips) and deep brain stimulation hardware, which may pose significant safety risks through electromagnetic interference [28]. These evidence-based safety parameters enable rTMS implementation across diverse clinical populations, with risk mitigation strategies effectively managing potential treatment-related complications.

Treatment Regimen

From a neurophysiological perspective, rTMS protocols are classified into high-frequency (>5 Hz) and low-frequency (≤1 Hz) stimulation paradigms, with conventional clinical applications favoring high-frequency protocols at approximately 10 Hz. A review published in 2021 on the effectiveness of rTMS for the treatment of insomnia revealed considerable variability in study characteristics and outcomes. The majority of studies employed low-frequency rTMS targeting the DLPFC or parietal cortex. Other target sites included the vertex, the raphe nuclei, the medial aspects of the bilateral frontal, occipital, and temporal cortices, as well as specific acupuncture points [28]. Technical parameters demonstrated considerable variability: stimulation intensity ranged from 80% to 130% of motor threshold (MT), with intersession pulse counts varying between 1100 and 2400. Each session lasted from 10 to 90 min, with the majority lasting for 20 min. Session duration spanned 10–90 min (modal value: 20 min), and treatment regimens most commonly involved daily sessions (5 days/week) over 2–4 weeks. Outcome assessments were predominantly conducted immediately post-intervention, with only seven studies (21.9%) extending follow-up to 1–22 weeks. Notably, the functional topography of neural networks modulates rTMS effects across different cortical regions. This neurobiological phenomenon suggests that identical stimulation frequencies may elicit divergent neurophysiological responses depending on target localization within cortico-limbic-thalamic circuits [29].

Clinical Recommendations

Clinical Recommendation: Unclear (Table 12.1). The evidence base remains insufficient to establish standardized treatment protocols for rTMS in sleep disorders. This deficit primarily stems from the nascent stage of research in this domain, compounded by methodological limitations across existing studies. Critical shortcomings include: the majority lack sham-controlled designs, a considerable number of recent studies have small sample sizes, and inconsistent objective outcomes despite subjective improvement reports. These discrepancies underscore the necessity for rigorous, multimodal evaluation frameworks in future investigations.

Table 12.1 TMS in sleep disorder

TMS	Clinical recommendations				Levels + references
	Target	Frequency	Length (min)	Duration (sessions)	
Unilateral rTMS	Right DLPFC	LF (1 Hz)	20	10–20	[60–62]
Bilateral rTMS	Right DLPFC and left DLPFC	LF (1 Hz)	30–40	10–15	[31, 59]
Accelerated rTMS	/	/	/	/	Unclear
Deep TMS	/	/	/	/	Unclear
Priming TMS	/	/	/	/	Unclear
Synchronised TMS	/	/	/	/	Unclear
iTBS	/	/	/	/	Unclear
Accelerated iTBS	Right DLPFC and left DLPFC	cTBS and iTBS	20	10	[57]
cTBS	Right DLPFC	cTBS	30	10	[64]
Bilateral TBS	/	/	/	/	Unclear
MST	/	/	/	/	Unclear

12.2.1.2 Bilateral rTMS

Efficacy

Current empirical evidence remains scarce regarding bilateral rTMS for applications in sleep disorders, with most investigations omitting sham-controlled or control groups. Notably, a study investigated the effects of sequential bilateral LF rTMS over DLPFC on patients with primary insomnia [30]. They found that sequential bilateral LF rTMS over DLPFC alleviated primary insomnia, including reduced the amplitude of MEP, and elevated the serum concentrations of brain-derived neurotrophic factor and gamma-aminobutyric acid. Furthermore, correlations among these effects were demonstrated. Another open-label pilot study utilizing sequential bifrontal LF-rTMS reported subjective sleep quality enhancements as per PSQI scores [31]. These preliminary findings highlight the therapeutic potential of bilateral LF-rTMS while underscoring the critical need for larger-scale, sham-controlled trials to validate these neurophysiological and biochemical correlates.

There was one study that found bilateral rTMS ameliorated sleep disorder and HPA axis dysfunction in patients with major depression [32]. To date, no randomized controlled trials (RCTs) have directly compared the neuromodulation efficacy of unilateral versus bilateral TMS modalities in sleep disorders, representing a critical knowledge gap requiring urgent investigation through double-blind RCTs.

Safety

Among the extant limited studies employing bilateral rTMS protocols, sequential bilateral LF-rTMS demonstrated excellent tolerability profiles.

Treatment Regimen

A study utilized sequential bilateral LF-rTMS (1 Hz) with 10-s active-duty cycles separated by 2-s interstimulus intervals, delivered at 80% resting motor threshold (rMT). Each 30-min session comprised 15 min of left DLPFC stimulation followed by 15 min targeting the right DLPFC, delivering a total of 1500 pulses per session. This twice-daily protocol (5 days/week) was maintained for 14 days. In two additional studies mentioned previously, one involved sequential bilateral LF rTMS stimulation over a 3-week period, with each session lasting 40 min. Studies also revealed two additional treatment paradigms: (1) a 3-week sequential bilateral LF-rTMS regimen (40-min sessions; left→right DLPFC sequence), and (2) a dual-frequency approach combining 10 Hz excitatory left DLPFC stimulation with concurrent 1 Hz inhibitory right DLPFC modulation.

Clinical Recommendations

Clinical Recommendation: Unclear (Table 12.1). Insufficient evidence to establish clinical efficacy. While current evidence demonstrates clinically promising outcomes for bilateral LF-rTMS targeting DLPFC, the absence of sham-controlled comparison substantially limits interpretational validity. This methodological limitation is compounded by the lack of RCT data and small sample sizes in existing studies. To address these critical gaps, future investigations should prioritize large-scale, double-blind sham-controlled RCTs and extended follow-up periods. Such rigorous evaluation is essential to differentiate true therapeutic effects from nonspecific placebo responses in insomnia management.

12.2.1.3 Accelerated rTMS

Accelerated TMS (aTMS) an emerging therapeutic modality characterized by ≥2 daily treatment sessions, demonstrates significant potential for optimizing neuromodulation outcomes. This accelerated paradigm demonstrates particular promise in treating neuropsychiatric disorders, including depression, schizophrenia, and others. However, critical knowledge gaps persist regarding its therapeutic potential for primary insomnia and insomnia comorbid with affective disorders.

Clinical Recommendation Unclear (Table 12.1). Current data do not support definitive conclusions regarding aTMS efficacy in insomnia management.

12.2.2 Deep TMS

Deep TMS (dTMS) an FDA-approved intervention for treatment-resistant depression (TRD), has received regulatory approval in Europe for a broader spectrum of neuropsychiatric indications. The H1-coil variant is authorized for treating unipolar depression, bipolar depression, negative symptoms of schizophrenia, and post-traumatic stress disorder (PTSD), while other H-coil configurations (e.g., H7) demonstrate utility in managing Alzheimer's disease (prevalence: 5.8 million in the United States), chronic pain syndromes, substance use disorders, obsessive-compulsive disorder, autism spectrum disorder, Parkinson's disease, stroke rehabilitation, and multiple sclerosis [33]. Emerging evidence suggests dTMS may improve sleep architecture in comorbid insomnia populations.

There are few studies that examined the improvement of insomnia symptoms by dTMS. One study explored the impact of dTMS on comorbid insomnia in patients with TRD using a retrospective analysis, which suggested dTMS alleviated insomnia symptoms in patients with TRD [34]. Further research incorporating a prospective study design in a multicenter setting is warranted to replicate these findings. Clinical Recommendation: Unclear (Table 12.1). Current data do not support definitive clinical adoption pending robust validation studies.

12.2.3 Priming TMS

Emerging evidence from RCTs in diverse clinical populations supports the therapeutic efficacy of priming rTMS (prTMS) across neuropsychiatric and neurological domains, demonstrating enhanced cortical plasticity and network connectivity compared to conventional protocols. However, no clinical investigations to date have evaluated prTMS's efficacy in insomnia, representing a critical gap in translating these neurophysiological advancements to sleep disorders. While prTMS shows promise for refractory neuropsychiatric conditions, current data insufficiently support its clinical adoption for insomnia pending robust validation through sham-controlled multicenter trials. Clinical Recommendation: Unclear (Table 12.1).

12.2.4 Synchronised TMS

Synchronized Transcranial Magnetic Stimulation (sTMS), an emerging therapeutic noninvasive brain stimulation, employs three-axis rotational neodymium magnet arrays to generate low-energy, sinusoidal magnetic fields synchronized with endogenous α-band oscillations. Preliminary investigations demonstrate sTMS-induced neurophysiological modulation, including enhanced α-rhythm synchronization and reduced theta power in healthy volunteers. While current evidence supports sTMS efficacy in major depressive disorder (MDD) and substance use disorder, its therapeutic potential for sleep disorders remains uninvestigated. Neurobiological mechanisms underlying sTMS may involve entrainment of thalamocortical rhythms and normalization of functional connectivity, though rigorous clinical validation is required. Current data is insufficient to recommend sTMS for sleep disorders. Multicenter RCTs are needed to evaluate therapeutic efficacy. Clinical Recommendation: Unclear (Table 12.1).

12.2.5 TBS

12.2.5.1 iTBS

TBS, an innovative neuromodulation paradigm, employs theta-frequency (50 Hz) magnetic pulse trains to induce cortical plasticity through frequency-specific synaptic plasticity mechanisms. Intermittent TBS (iTBS) has demonstrated an ability to enhance motor cortical excitability, whereas continuous TBS (cTBS) induces prolonged cortical inhibition.[35] Emerging evidence positions iTBS as a clinically viable therapeutic modality for MDD. A prior study has shown that iTBS-induced reduction of DMN functional connectivity and insular cortex activation, correlating with improved insomnia severity index scores.[36] These neurophysiological effects suggest iTBS may target maladaptive thalamocortical rhythms implicated in insomnia pathophysiology. Clinical Recommendation: Unclear (Table 12.1). Current evidence is insufficient to recommend routine clinical application pending validation through large-scale RCTs.

12.2.5.2 Accelerated iTBS

Accelerated iTBS protocols, characterized by multiple daily sessions over consecutive days, demonstrate significant advantages in treatment efficiency by reducing total treatment duration (≤5 days) and cumulative pulse delivery (≤3000 pulses) compared to conventional schedules. Beyond their established rapid-acting antidepressant effect, these protocols show promise in mitigating acute suicidality in TRD. Emerging clinical evidence from RCT. Notably, there has been a single reported instance where accelerated iTBS targeted at the bilateral DLPFC may ameliorate neuropsychiatric sequelae of long-COVID syndrome, including anxiety and insomnia after 10 daily sessions [37]. Clinical Recommendation: Unclear (Table 12.1). While accelerated iTBS shows compelling antidepressant and neuropsychiatric benefits, current evidence insufficiently supports routine use for insomnia pending validation through large-scale RCTs.

12.2.5.3 cTBS

Efficacy

The standardized cTBS protocol delivers 200 contiguous burst trains (600 pulses total) within a 40-s period, inducing LTD-like neuroplastic changes characterized by sustained cortical excitability reduction [38]. This brief intervention (2–3 min per session) demonstrates enhanced participant tolerability compared to conventional rTMS protocols requiring 20–30 min per session [35], while maintaining comparable neuromodulatory efficacy. cTBS has garnered attention for its capacity to suppress corticospinal excitability for up to 60 min following brief stimulation sessions [35, 39, 40]. Recent evidence extends its therapeutic scope, demonstrating cTBS-mediated enhancement of NREM stage 2 sleep architecture through increased slow-wave activity and decreased sleep onset latency in healthy volunteers [41]. A RCT evaluated 10 daily cTBS sessions (5 days/week for 2 weeks) targeting the DLPFC. The intervention group exhibited significant insomnia symptom amelioration, evidenced by reduction in ISI scores and improved polysomnographic parameters including increased total sleep time and slow-wave sleep [42]. These findings corroborate cTBS's potential as a time-efficient intervention for sleep disorders.

Safety

Current evidence supports the safety profile of cTBS when targeting the right DLPFC for insomnia treatment. One study found that targeting the right DLPFC with cTBS effectively reduced insomnia severity in patients, without any adverse effects. Participants exhibited excellent tolerability, with no occurrences of seizures, neurological deficits, or complaints related to cognitive impairment [42]. Chung et al. reviewed studies on the effects of TBS on clinical symptom improvement in patients with depression. Their findings revealed that cTBS has significant efficacy in alleviating symptoms, with no reported adverse effects [38]. A meta-analysis has reported that seizures associated with TBS have occurred only once to date, corresponding to a crude seizure risk of 0.02% per session. The overall crude risk of mild adverse events was estimated to be 1.1%, which is comparable to that of high-frequency rTMS protocols [43]. Notably, the reported seizure occurred at an intensity of 100% of the rMT on primary motor cortex (M1), whereas most studies adhere to the original TBS protocol using 80% of the of the active motor threshold [44].

Treatment Regimen

TBS protocols exhibit neuroanatomical and temporal specificity across different sleep disorders, with therapeutic parameters (intervention targets, session frequency, and treatment duration) tailored to underlying pathophysiology. In patients with insomnia, a standardized 2-week protocol (10 daily sessions of cTBS targeting the right DLPFC) demonstrated significant reductions in insomnia severity. Follow-up evaluations at 1 month and 6 months posttreatment also demonstrated markedly lower insomnia severity compared to the sham stimulation group [38]. cTBS interventions for other sleep disorders should be tailored based on brain regions closely associated with the condition. For example, targeting the M1 may be a suitable choice for patients with restless legs syndrome or obstructive sleep apnea. In cases of sleep deprivation, stimulation of the occipital cortex may be an effective strategy. However, further scientific research is required to validate these approaches [19].

Clinical Recommendations

Previous studies have suggested that primary insomnia may be associated with hyperactivity of the DLPFC. Therefore, cTBS intervention targeting this region to suppress its hyperactivity may be a suitable approach for patients with insomnia. The standard pattern of cTBS was employed and consisted of bursts of 3 pulses at 50 Hz with bursts repeated at a frequency of 5 Hz. Each session encompassed a total of 1800 pulses, delivered in intervals of 15 minutes between sets of 600 pulses.

Given the current evidence base being limited to an investigation, rigorous validation through multicenter RCTs becomes imperative to establish both the efficacy and safety profiles of these intervention strategies. Clinical Recommendation: Unclear (Table 12.1).

12.2.5.4 Bilateral TBS

Two protocols of TBS with opposing effects on cortical excitability have been proposed; iTBS and cTBS with excitatory and inhibitory consequences, respectively. Bilateral TBS, integrating left prefrontal iTBS (targeting demonstrated hypoactivity in the left DLPFC) with right prefrontal cTBS (addressing hyperactivity in the right DLPFC), has emerged as a promising candidate for TRD. No studies have explored the effect of bilateral TBS intervention on insomnia disorders. Clinical Recommendation: Unclear (Table 12.1).

12.2.6 MST

Magnetic Seizure Therapy (MST), an emerging neuroregulatory technique in psychiatric, employs time-varying magnetic fields to induce therapeutic seizures through cortical and subcortical network activation. While sharing mechanistic similarities with electroconvulsive therapy, including glutamate-mediated excitotoxicity and neuroplasticity induction, MST demonstrates reduced cognitive side effects. Emerging evidence from MST trials reveals promising efficacy in TRD. Additional research is necessary to ascertain the long-term effects of MST and to establish its role within the spectrum of psychiatric interventions. Current data insufficiently support routine clinical adoption pending completion of sham-controlled multicenter trials. Clinical Recommendation: Unclear (Table 12.1).

12.3 tDCS

12.3.1 Efficacy

tDCS is a noninvasive technique that employs a constant, low-intensity direct current (ranging from 1 to 2 mA) to modulate cortical neuronal activity [45]. Anodal tDCS is generally believed to enhance the functionality of the underlying cortical areas, whereas cathodal tDCS exerts a suppressive effect [46]. Emerging clinical evidence demonstrates tDCS's therapeutic potential in sleep disorders. They found that sleep disturbances comorbid with organic lesions can be significantly improved after tDCS by increasing slow-wave sleep, sleep quality, and modulating synaptic connectivity [47]. However, contradictory findings exist in PTSD-related insomnia. A trial targeting DLPFC with 1.5 mA cathode tDCS for 15 sessions in patients with insomnia caused by post-traumatic encephalopathy showed no significant therapeutic effect [48].

12.3.2 Safety

tDCS demonstrates significant therapeutic potential due to its cost-effectiveness, clinical portability, and exceptional safety profile. While transient side effects such as itching, a burning sensation, or headaches are prevalent, these adverse events exhibit an absence of long-term sequelae [49, 50]. A systematic meta-analysis of 64 clinical trials (N = 2262) revealed high tolerability rates [50]. Notably, conventional tDCS protocols (≤40 min, ≤4 milliamperes, ≤7.2 Coulombs) demonstrate an unprecedented safety record across 33,200 stimulation sessions in 1000+ subjects [51]. No serious adverse events or irreversible neurological injuries have been documented, even in populations with comorbid neurologic conditions.

12.3.3 Treatment Regimen

DLPFC emerges as the predominant cortical target in tDCS for sleep disorders. tDCS targeting the bilateral DLPFC significantly reduced insomnia severity in patients with depression comorbid with insomnia. The intervention protocol involved 20 sessions of 2-mA direct current stimulation administered for 30 minutes over a 4-week period (excluding weekends), followed by four weekly maintenance sessions [52]. However, contradictory findings emerge in post-traumatic brain injury and insomnia. They found that cathode tDCS over the right DLPFC (1.5 mA, 15-min sessions, 5 days/week for 3 weeks), the anode electrode on the left shoulder, failed to demonstrate significant sleep quality improvements [48]. Overall, further research and exploration are required to investigate the utility of tDCS in treating sleep disorders.

12.3.4 Clinical Recommendations

Currently, tDCS treatment for sleep disorders should be tailored to the specific brain regions associated with different sleep conditions, involving individualized selection of intervention targets, intensity, frequency, and other parameters. For insomnia, the DLPFC is a commonly targeted region, with current intensities typically ranging from 1 to 2 mA and intervention durations spanning 15–30 min, 5 days/week for 4 weeks. Optimal protocols require individualized parameter selection based on neuroanatomical targeting and pathophysiological mechanisms. Clinical Recommendation: Unclear (Table 12.2).

Table 12.2 tDCS in sleep disorder

tDCS	Clinical recommendations					Levels +references
	Anode position	Cathode position	Intensity (mA)	Length (min)	Duration (sessions)	
tDCS	Left DLPFC	Right DLPFC	1-2 mA	15–30	10–20	[58]
TMS combined tDCS	Left DLPFC	Right DLPFC	2 mA	20	20	[63]
		Right DLPFC	1 Hz	30	20	

12.4 TMS Combined tDCS

12.4.1 Efficacy

Combined neuromodulation strategies integrating rTMS and tDCS represent an innovative therapeutic paradigm for cortical excitability modulation and neurobehavioral enhancement in both healthy populations and neuropsychiatric patients [53]. When performed simultaneously, these two stimulation techniques induce synergistic neuroplastic changes, as evidenced by improved sleep architecture and cognitive performance in patients with Alzheimer's disease [54]. Current evidence for combined rTMS-tDCS in chronic insomnia disorder remains sparse, with only one RCT conducted to date. This study demonstrated that combined anodal on right DLPFC and cathode on the right DLPFC tDCS (2 mA, 20-minute sessions) with 1 Hz rTMS (1200 pulses/session) targeting the right DLPFC achieved superior therapeutic outcomes versus monotherapy [52]. Mechanistic studies suggest that "Facilitatory preconditioning" with anodal TDCS caused a subsequent period of 1 Hz rTMS to reduce corticospinal excitability to below baseline levels for >20 min. Conversely, "inhibitory preconditioning" with cathodal tDCS resulted in 1 Hz rTMS increasing corticospinal excitability for at least 20 min [55]. In conclusion, the combination of rTMS and tDCS holds promise as a therapeutic option for insomnia and further research with larger sample sizes is necessary.

12.4.2 Safety

Combined neuromodulation protocols integrating rTMS and tDCS exhibited similar side effects to those observed with single interventions, including skin redness, transient headache, mild pruritus, and discomfort at the stimulated site [52]. Most of these side effects resolved within a few days without the need for additional medications. Given the well-tolerated and safe profiles of rTMS and tDCS, respectively, no serious adverse events such as seizures or mania were reported in the combination therapy [56]. However, due to the limited sample size, it remains uncertain whether the simultaneous administration of rTMS and tDCS increases the risk of other side effects or serious adverse events in the treatment of insomnia. Further clinical trials investigating different combination strategies and interventions for patients with insomnia of various ages are required to verify the safety of this combined approach. Clinical Recommendation: Unclear (Table 12.2).

12.5 Conclusion

In conclusion, rTMS and tDCS represent promising non-pharmacological intervention paradigms for sleep disorders. Current evidence demonstrates efficacy of LF-TMS, cTBS, and cathodic tDCS targeting the right DLPFC in alleviating insomnia symptoms. These interventions exhibit not only therapeutic superiority over sham controls but also exceptional tolerability profiles. Additionally, the parietal cortex has emerged as a potential candidate for stimulation, and the synergistic application of TMS and tDCS may present an even more potent intervention strategy. In light of its considerable potential, future research necessitates the conduct of large-scale, multicenter RCTs to investigate novel protocols, targets, and modes of intervention. Moreover, there is a pressing need to further elucidate the therapeutic mechanisms of these interventions through the use of advanced neuroimaging techniques such as magnetoencephalography, electroencephalography, and magnetic resonance imaging. As such, development of standardized clinical pathways for neuromodulation therapies necessitates collaborative efforts between neurologists, psychiatrists, and biomedical engineers to establish evidence-based treatment algorithms.

Acknowledgments None.

Disclosure/Conflicts of Interest The authors declare no conflicts of interest in conducting this study or preparing the manuscript.

Financial Support None.

References

1. Ge L, Guyatt G, Tian J, et al. Insomnia and risk of mortality from all-cause, cardiovascular disease, and cancer: systematic review and meta-analysis of prospective cohort studies. Sleep Med Rev. 2019;48:101215.
2. Li L, Gan Y, Zhou X, et al. Insomnia and the risk of hypertension: a meta-analysis of prospective cohort studies. Sleep Med Rev. 2021;56:101403.
3. Pavlova MK, Latreille V. Sleep disorders. Am J Med. 2019;132(3):292–9.

4. De Crescenzo F, Dalò GL, Ostinelli EG, et al. Comparative effects of pharmacological interventions for the acute and long-term management of insomnia disorder in adults: a systematic review and network meta-analysis. Lancet. 2022;400(10347):170–84.
5. Tan X, van Egmond L, Partinen M, et al. A narrative review of interventions for improving sleep and reducing circadian disruption in medical inpatients. Sleep Med. 2019;59:42–50.
6. Sateia MJ, Buysse DJ, Krystal AD, et al. Clinical practice guideline for the pharmacologic treatment of chronic insomnia in adults: an American Academy of Sleep Medicine Clinical practice guideline. J Clin Sleep Med. 2017;13(2):307–49.
7. Song ML, Park KM, Motamedi GK, et al. Cognitive behavioral therapy for insomnia in restless legs syndrome patients. Sleep Med. 2020;74:227–34.
8. Starling CM, Greenberg D, Lewin D, et al. Voice-activated cognitive behavioral therapy for insomnia: a randomized clinical trial. JAMA Netw Open. 2024;7(9):e2435011.
9. Koffel E, Bramoweth AD, Ulmer CS. Increasing access to and utilization of cognitive behavioral therapy for insomnia (CBT-I): a narrative review. J Gen Intern Med. 2018;33(6):955–62.
10. Feng F, Zhang Y, Hou J, et al. Can music improve sleep quality in adults with primary insomnia? A systematic review and network meta-analysis. Int J Nurs Stud. 2018;77:189–96.
11. Yin X, Gou M, Xu J, et al. Efficacy and safety of acupuncture treatment on primary insomnia: a randomized controlled trial. Sleep Med[J]. 2017;37:193–200.
12. Feyzabadi Z, Rezaeitalab F, Badiee S, et al. Efficacy of violet oil, a traditional Iranian formula, in patients with chronic insomnia: a randomized, double-blind, placebo-controlled study. J Ethnopharmacol. 2018;214:22–8.
13. Chambe J, Reynaud E, Maruani J, et al. Light therapy in insomnia disorder: a systematic review and meta-analysis. J Sleep Res. 2023;32(6):e13895.
14. Hyde J, Carr H, Kelley N, et al. Efficacy of neurostimulation across mental disorders: systematic review and meta-analysis of 208 randomized controlled trials. Mol Psychiatry. 2022;27(6):2709–19.
15. Wu H, Lv J, Liu M, et al. The long-term effect of repetitive transcranial magnetic stimulation in the treatment of intractable insomnia. Sleep Med. 2021;85:309–12.
16. Luber B, Steffener J, Tucker A, et al. Extended remediation of sleep deprived-induced working memory deficits using fMRI-guided transcranial magnetic stimulation. Sleep. 2013;36(6):857–71.
17. Lanza G, Cantone M, Aricò D, et al. Clinical and electrophysiological impact of repetitive low-frequency transcranial magnetic stimulation on the sensory-motor network in patients with restless legs syndrome. Ther Adv Neurol Disord. 2018;11:1756286418759973.
18. Opie GM, Catcheside PG, Usmani ZA, et al. Motor cortex plasticity induced by theta burst stimulation is impaired in patients with obstructive sleep apnoea. Eur J Neurosci. 2013;37(11):1844–52.
19. Lanza G, Fisicaro F, Cantone M, et al. Repetitive transcranial magnetic stimulation in primary sleep disorders. Sleep Med Rev. 2023;67:101735.
20. Zhou Q, Yu C, Yu H, et al. The effects of repeated transcranial direct current stimulation on sleep quality and depression symptoms in patients with major depression and insomnia. Sleep Med. 2020;70:17–26.
21. Hohenester M, Langguth B, Wetter TC, et al. Single sessions of transcranial direct current stimulation and transcranial random noise stimulation exert no effect on sleepiness in patients with narcolepsy and idiopathic hypersomnia. Front Psych. 2023;14:1288976.
22. Qiao MX, Yu H, Li T. Non-invasive neurostimulation to improve sleep quality and depressive symptoms in patients with major depressive disorder: A meta-analysis of randomized controlled trials. J Psychiatr Res. 2024;176:282–92.
23. Frase L, Selhausen P, Krone L, et al. Differential effects of bifrontal tDCS on arousal and sleep duration in insomnia patients and healthy controls. Brain Stimul. 2019;12(3):674–83.
24. Sun N, He Y, Wang Z, et al. The effect of repetitive transcranial magnetic stimulation for insomnia: a systematic review and meta-analysis. Sleep Med. 2021;77:226–37.

25. Herrero Babiloni A, Brazeau D, De Koninck BP, et al. The utility of non-invasive brain stimulation in relieving insomnia symptoms and sleep disturbances across different sleep disorders: a topical review. Curr Sleep Med Rep. 2023;9(2):124–32.
26. Oroz R, Kung S, Croarkin PE, et al. Transcranial magnetic stimulation therapeutic applications on sleep and insomnia: a review. Sleep Sci Pract. 2021;5(1):3.
27. Krone LB, Fehér KD, Rivero T, et al. Brain stimulation techniques as novel treatment options for insomnia: a systematic review. J Sleep Res. 2023;32(6):e13927.
28. Ma H, Lin J, He J, et al. Effectiveness of TES and rTMS for the treatment of insomnia: Meta-analysis and meta-regression of randomized sham-controlled trials. Front Psychiatry. 2021:12.
29. Zheng H, Zhou Q, Yang JJ, et al. Altered functional connectivity of the default mode and frontal control networks in patients with insomnia. CNS Neurosci Ther. 2023;29(8):2318–26.
30. Feng J, Zhang Q, Zhang C, et al. The Effect of sequential bilateral low-frequency rTMS over dorsolateral prefrontal cortex on serum level of BDNF and GABA in patients with primary insomnia. Brain Behav. 2019;9(2):e01206.
31. Holbert RC, Carr BR, Bussing R. An open label pilot trial of sequential bifrontal low frequency r-TMS in the treatment of primary insomnia. Psychiatry Res. 2023;324:115194.
32. Chen X, Jiang F, Yang Q, et al. Bilateral repetitive transcranial magnetic stimulation ameliorated sleep disorder and hypothalamic–pituitary–adrenal axis dysfunction in subjects with major depression. Front Psych. 2022:13.
33. Tendler A, Barnea Ygael N, Roth Y, et al. Deep transcranial magnetic stimulation (dTMS) – beyond depression. Expert Rev Med Devices. 2016;13(10):987–1000.
34. Chopra A, Singal P, Kodya S. Impact of deep transcranial magnetic stimulation on insomnia outcomes in patients with treatment-resistant depression: a retrospective study. J Clin Sleep Med. 2024;20(5):813–5.
35. Huang YZ, Edwards MJ, Rounis E, et al. Theta burst stimulation of the human motor cortex. Neuron. 2005;45(2):201–6.
36. Qi S, Zhang Y, Li X, et al. Improved functional organization in patients with primary insomnia after individually-targeted transcranial magnetic stimulation. Front Neurosci-Switz. 2022;16
37. Chang CH, Chen SJ, Chen YC, et al. A 30-year-old woman with an 8-week history of anxiety, depression, insomnia, and mild cognitive impairment following COVID-19 who responded to accelerated bilateral theta-burst transcranial magnetic stimulation over the prefrontal cortex. Am J Case Rep. 2023:24.
38. Chung SW, Hoy KE, Fitzgerald PB. Theta-burst stimulation: a new form of TMS treatment for depression? Depress Anxiety. 2015;32(3):182–92.
39. Volz LJ, Benali A, Mix A, et al. Dose-dependence of changes in cortical protein expression induced with repeated transcranial magnetic theta-burst stimulation in the rat. Brain Stimul. 2013;6(4):598–606.
40. Cai DB, Qin XD, Qin ZJ, et al. Adjunctive continuous theta burst stimulation for major depressive disorder or bipolar depression: a meta-analysis of randomized controlled studies. J Affect Disord. 2024;346:266–72.
41. Mensen A, Gorban C, Niklaus M, et al. The effects of theta-burst stimulation on sleep and vigilance in humans. Front Hum Neurosci. 2014;8:420.
42. Zhu X, Tabarak S, Que J, et al. Efficiency and safety of continuous theta burst stimulation for primary insomnia: a randomized clinical trial. Sleep Med. 2024;124:77–83.
43. Oberman L, Edwards D, Eldaief M, et al. Safety of theta burst transcranial magnetic stimulation: a systematic review of the literature. J Clin Neurophysiol. 2011;28(1):67–74.
44. Oberman LM, Pascual-Leone A. Report of seizure induced by continuous theta burst stimulation. Brain Stimul. 2009;2(4):246–7.
45. Chase HW, Boudewyn MA, Carter CS, et al. Transcranial direct current stimulation: a roadmap for research, from mechanism of action to clinical implementation. Mol Psychiatry. 2020;25(2):397–407.
46. Nitsche MA, Paulus W. Transcranial direct current stimulation–update 2011. Restor Neurol Neurosci. 2011;29(6):463–92.

47. Ebajemito JK, Furlan L, Nissen C, et al. Application of transcranial direct current stimulation in neurorehabilitation: the modulatory effect of sleep. Front Neurol. 2016;7:54.
48. Bakhshayesh Eghbali B, Ramezani S, Sedaghat Herfeh S, et al. notTranscranial direct current stimulation improves sleep quality in patients with insomnia after traumatic brain injury. Brain Inj. 2023;37(1):63–73.
49. Moffa AH, Brunoni AR, Fregni F, et al. Safety and acceptability of transcranial direct current stimulation for the acute treatment of major depressive episodes: analysis of individual patient data. J Affect Disord. 2017;221:1–5.
50. Aparicio LVM, Guarienti F, Razza LB, et al. A systematic review on the acceptability and tolerability of transcranial direct current stimulation treatment in neuropsychiatry trials. Brain Stimul. 2016;9(5):671–81.
51. Godinho MM, Junqueira DR, Castro ML, et al. Safety of transcranial direct current stimulation: evidence based update 2016. Brain Stimul. 2017;10(5):983–5.
52. Zhou Q, Liu Z, Yu C, et al. Effect of combined treatment with transcranial direct current stimulation and repetitive transcranial magnetic stimulation compared to monotherapy for the treatment of chronic insomnia: a randomised, double-blind, parallel-group, controlled trial. BMC Med. 2024;22(1):538.
53. Li X, Liu J, Wei S, et al. Cognitive enhancing effect of rTMS combined with tDCS in patients with major depressive disorder: a double-blind, randomized, sham-controlled study. BMC Med. 2024;22(1):253.
54. Hu Y, Jia Y, Sun Y, et al. Efficacy and safety of simultaneous rTMS-tDCS over bilateral angular gyrus on neuropsychiatric symptoms in patients with moderate Alzheimer's disease: a prospective, randomized, sham-controlled pilot study. Brain Stimul. 2022;15(6):1530–7.
55. Siebner HR, Lang N, Rizzo V, et al. Preconditioning of low-frequency repetitive transcranial magnetic stimulation with transcranial direct current stimulation: evidence for homeostatic plasticity in the human motor cortex. J Neurosci. 2004;24(13):3379–85.
56. Wang H, Wang K, Xue Q, et al. Transcranial alternating current stimulation for treating depression: a randomized controlled trial. Brain. 2022;145(1):83–91.
57. Chang CH, Chen SJ, Chen YC, Tsai HC. A 30-year-old woman with an 8-week history of anxiety, depression, insomnia, and mild cognitive impairment following COVID-19 who responded to accelerated bilateral theta-burst transcranial magnetic stimulation over the prefrontal cortex. Am J Case Rep. 2023;24
58. Ebajemito JK, Furlan L, Nissen C, Sterr A. Application of transcranial direct current stimulation in neurorehabilitation: the modulatory effect of sleep. Front Neurol. 2016;7:54.
59. Feng J, Zhang Q, Zhang C, Wen Z, Zhou X. The Effect of sequential bilateral low-frequency rTMS over dorsolateral prefrontal cortex on serum level of BDNF and GABA in patients with primary insomnia. Brain Behav. 2019;9(2):e01206.
60. Herrero Babiloni A, Brazeau D, De Koninck BP, Lavigne GJ, De Beaumont L. The utility of non-invasive brain stimulation in relieving insomnia symptoms and sleep disturbances across different sleep disorders: a topical review. Curr Sleep Med Rep. 2023;9(2):124–32.
61. Oroz R, Kung S, Croarkin PE, Cheung J. Transcranial magnetic stimulation therapeutic applications on sleep and insomnia: a review. Sleep Sci Pract. 2021;5(1):3.
62. Sun N, He Y, Wang Z, Zou W, Liu X. The effect of repetitive transcranial magnetic stimulation for insomnia: a systematic review and meta-analysis. Sleep Med. 2021;77:226–37.
63. Zhou Q, Liu Z, Yu C, Wang Q, Zhuang W, Tang Y, Zheng T, Yu H, Zhou D. Effect of combined treatment with transcranial direct current stimulation and repetitive transcranial magnetic stimulation compared to monotherapy for the treatment of chronic insomnia: a randomised, double-blind, parallel-group, controlled trial. BMC Med. 2024;22(1):538.
64. Zhu X, Tabarak S, Que J, Yan W, Lin X, Liu X, Chen W, Shi J, Deng J, Lu L. Efficiency and safety of continuous theta burst stimulation for primary insomnia: a randomized clinical trial. Sleep Med. 2024;124:77–83.

Neurocognitive Disorders 13

Chaomeng Liu, Yan Li, Li Ren, Nan Ding, and Qinge Zhang

Abstract

Neurocognitive disorders, including mild cognitive impairment (MCI) and Alzheimer's disease (AD), are characterized by a decline in cognitive functions from previously higher levels. The complex and multifactorial etiologies of these disorders pose substantial obstacles to cognitive enhancement. Noninvasive neurostimulation methods, such as transcranial magnetic stimulation (TMS) and transcranial direct current stimulation (tDCS), have emerged as promising therapeutic strategies. Despite numerous clinical trials, results remain inconsistent, and a consensus is still lacking. This chapter provides a thorough review of the latest developments in TMS and tDCS interventions for neurocognitive disorders, aiming to inform the advancement of novel treatment approaches.

Keywords

Neurocognitive disorders · Repetitive transcranial magnetic stimulation · Mild cognitive impairment · Alzheimer's disease · Transcranial direct current stimulation

C. Liu · Y. Li · L. Ren · N. Ding · Q. Zhang (✉)
Beijing Key Laboratory of Mental Disorders, National Clinical Research Center for Mental Disorders & National Center for Mental Disorders, Beijing Anding Hospital, Capital Medical University, Beijing, China

W. Zheng, Y. Ning (eds.), *TMS and tDCS for Psychiatric Disorders*,
https://doi.org/10.1007/978-981-96-8504-2_13

Abbreviations

AD	Alzheimer's Disease
ADAS-cog	Alzheimer's Disease Assessment Scale—cognitive subscale
AM	Associative Memory
aMCI	amnestic Mild Cognitive Impairment
AVLT	Auditory Verbal Learning Test
DLPFC	Dorsolateral Prefrontal Cortex
DSM-5	The fifth edition of the Diagnostic and Statistical Manual of Mental Disorders
ECT	Electroconvulsive Therapy
FDA	Food and Drug Administration
GMV	Grey Matter Volume
H-coil	Hesed coil
HD-tDCS	High-Definition transcranial Direct Current Stimulation
iTBS	Intermittent Theta Burst Stimulation
MCI	Mild Cognitive Impairment
MDD	Major Depressive Disorder
MoCA	Montreal Cognitive Assessment
MST	Magnetic Seizure Therapy
NR	No Record
pTMS	priming Transcranial Magnetic Stimulation
RCT	Randomized Controlled Trials
RMT	Resting Motor Threshold
sTMS	Synchronized Transcranial Magnetic Stimulation
tDCS	Transcranial Direct Current Stimulation
TBS	Theta Burst Stimulation
TMS	Transcranial Magnetic Stimulation

13.1 Introduction

Neurocognitive disorders—encompassing delirium, mild cognitive impairment (MCI), and dementia—are characterized by a decline in cognitive function from a previously higher baseline [1]. These disorders present a wide array of clinical manifestations and etiologies, with common causes including Alzheimer's disease (AD), cerebrovascular disease, Lewy body disease, frontotemporal degeneration, traumatic brain injury, infections, and alcohol abuse [2]. Defining the cognitive domains affected is crucial when discussing neurocognitive disorders. While various classifications have been proposed, consensus remains elusive [1]. The Neurocognitive Work Group has outlined six key cognitive domains: complex attention, executive function, learning and memory, language, perceptual-motor function, and social cognition [1]. The fifth edition of the Diagnostic and Statistical Manual of Mental Disorders (DSM-5) establishes criteria to distinguish specific etiological subtypes of mild and major neurocognitive disorders [3]. This chapter primarily addresses MCI and AD, the two neurological conditions most commonly studied with transcranial magnetic stimulation (TMS).

Individuals with MCI face an elevated risk of progressing to dementia, with annual conversion rates ranging from 8% to 18% in community settings and 10% to 15% in clinical environments [4]. Current US Food and Drug Administration (FDA)-approved medications for AD primarily offer temporary relief of cognitive and behavioral symptoms; however, the recent approvals of aducanumab and lecanemab show promise in delaying disease progression [5, 6]. Nonpharmacological interventions such as risk reduction, cognitive training, psychosocial therapies, and nutraceuticals remain under investigation and require further validation [7]. Continued research is essential to identify novel therapies capable of improving cognitive function or delaying progression in MCI and AD.

Recent studies suggest that TMS holds promise for enhancing cognitive function [8–10]; however, much of the existing literature is hindered by methodological inconsistencies. Treatment protocols vary significantly across studies, including differences in the location, intensity, and frequency of magnetic stimulation in clinical trials. Additionally, outcome measures differ, with some studies assessing global cognition while others focus on specific cognitive functions, making it challenging to draw definitive and coherent conclusions from the available evidence.

In contrast to TMS, transcranial direct current stimulation (tDCS) has been more widely adopted due to its convenience, superior safety profile, fewer side effects, and lower cost. tDCS is believed to modulate the resting membrane potential of neurons, thereby influencing neuronal excitability and altering spontaneous neural discharge [11]. Anodal stimulation typically enhances neuronal excitability, while cathodal stimulation reduces it [12]. The effects of tDCS last from minutes to hours poststimulation, with repeated sessions potentially extending effects from days to months. It has been proposed that repeated tDCS sessions promote neuroplasticity by facilitating the remodeling of neural circuits, thereby enhancing cognitive functions such as learning and memory [13]. A prior study demonstrated that anodal tDCS alleviated age-related cognitive deficits [14]. This chapter specifically examines the effectiveness and safety of TMS and tDCS in improving cognitive function in AD and MCI populations, focusing on randomized controlled trials (RCTs) comparing TMS with sham stimulation, given the current level of clinical evidence.

13.2 TMS

13.2.1 TMS

13.2.1.1 Conventional rTMS

Efficacy

A review of the literature reveals that for patients with MCI, the left dorsolateral prefrontal cortex (DLPFC) is the most frequently targeted site for unilateral stimulation, followed by the right DLPFC and left precuneus [15]. A recent study from Mexico assessed the effects of 5 Hz rTMS at 100% resting motor threshold (RMT) on cognitive function in patients with MCI, using sham stimulation as a control. After 10 weeks of treatment, delivering a total of 1500 pulses, rTMS was found to significantly improve Montreal Cognitive Assessment (MoCA) scores compared to sham stimulation [16]. The left DLPFC is considered crucial due to its positive correlation with attention, memory, and executive functions [17, 18], which could provide valuable insights into cognitive processing areas and enhance intervention models by targeting this region. Conversely, an earlier study investigated the impact of rTMS at 10 Hz and 90% motor threshold (MT) on memory in patients with amnestic MCI, targeting the right DLPFC. Results indicated that, compared to sham stimulation, rTMS improved performance on the Auditory Verbal Learning Test (AVLT) across multiple recall stages [19]. Despite promising findings, recent research on unilateral stimulation for patients with MCI has been limited by small sample sizes, requiring further validation to strengthen the reliability of conclusions.

In the case of AD, numerous meta-analyses support the efficacy of rTMS in enhancing overall cognitive function [15, 20, 21]. Notably, the study by Teselink et al. demonstrated significant cognitive improvements in the rTMS group targeting the left DLPFC, compared to sham stimulation [21]. There is considerable variation in TMS protocols across studies, but the left DLPFC remains the most common site, with high-frequency stimulation (over 5 Hz) applied using 1000 to 1500 pulses per session, typically at 90% to 100% RMT, across 10 to 20 sessions. The FDA's current approval for TMS in major depressive disorder (MDD) specifies the left DLPFC as the stimulation site, using either high-frequency stimulation at 10 to 20 Hz (1800 to 3000 pulses per session) or intermittent theta burst stimulation (iTBS) at 120% RMT (600 pulses per session) over 30 sessions [15]. In contrast, treatment parameters for AD generally involve fewer pulses per session, lower intensity as a percentage of RMT, and shorter overall treatment durations. Although TMS has been predominantly studied in depression, its application is increasingly being explored for other neurological and psychiatric conditions.

Safety

TMS has been generally regarded as safe and well-tolerated, with a low incidence of adverse events consistent with the known side effects of the procedure. Seizures, though the most serious adverse event, occur infrequently, with an estimated risk of less than 1 in 30,000 [22]. More common side effects include transient headaches, scalp discomfort, and muscle twitches during stimulation [22]. In patients with cognitive impairments, age is a critical safety consideration due to age-related physiological changes, the presence of comorbid medical and neurological conditions, implants or devices, and polypharmacy, all of which can influence TMS response. However, when standard safety protocols are followed, TMS remains safe and well-tolerated, even in older adults with depression [23]. Adherence to up-to-date safety guidelines [22], which include thorough participant screening, maintaining stimulation parameters within safe thresholds, and ensuring that trained technicians and clinicians oversee the procedure, effectively mitigates the risk of seizures [24, 25].

Treatment Regimen

There is significant variability in the TMS parameters used across studies involving MCI or AD. The left DLPFC is the most frequently targeted site, typically with high-frequency stimulation (frequencies greater than 5 Hz). Standard protocols often involve 1000 to 1500 pulses per session at 90% to 100% of RMT, over a course of 10 to 20 sessions. While many studies focus on the left DLPFC and its effect on cognitive functions, they often overlook potential mood-related effects [15]. As stimulation of the left DLPFC is associated with antidepressant outcomes, improvements in cognitive function may result directly from stimulation or indirectly due to the alleviation of depressive symptoms. This highlights the importance of controlling for depression in studies to accurately assess the cognitive benefits of TMS [21].

Cognitive processes are linked to specific brain regions, necessitating exploration beyond the left DLPFC. Stimulation at a single site may affect brain connectivity, potentially influencing other regions. High-frequency stimulation, known for its excitatory effects, is beneficial for conditions such as depression and dementia, whereas low-frequency stimulation, considered inhibitory, has been employed in the treatment of anxiety and depression [26]. Future rTMS studies for dementia should consider using stimulation at 120% of RMT, along with increasing the number of pulses per session and the total number of sessions. Recognizing that higher stimulation parameters are used in other clinical and research settings will be instrumental in designing future rTMS research for dementia.

Clinical Recommendations

For both MCI and AD, the left DLPFC is the most commonly targeted site for high-frequency stimulation, defined as frequencies exceeding 5 Hz, with 1000–1500 pulses per session at 90% to 100% of the RMT, across 10 to 20 treatment sessions. However, no guidelines or consensus offer specific clinical recommendations (Table 13.1), so current research findings should be approached cautiously.

13.2.1.2 Accelerated rTMS

Standard rTMS treatment protocols can be challenging for patients with cognitive impairment due to their length. To address this, "compression" regimens such as enhanced, accelerated, or interval TMS/TBS have been explored for their safety and feasibility. These regimens involve multiple daily treatments over a few days to achieve the total number of sessions typically prescribed in standard rTMS protocols Fitzgerald et al. [27]. The accelerated rTMS program has been widely utilized in depression treatment, demonstrating good efficacy and safety Cheng et al. [28]. However, further research is required in the context of cognitive impairments.

13.2.2 Deep TMS

13.2.2.1 Efficacy

Compared to traditional focal coils, the Hesed coil (H-coil) is designed to target broader brain regions, thanks to its reduced electric field decay over distance [29]. The H-coil also enables concurrent stimulation of multiple brain areas. H-coil rTMS has been employed in the treatment of MDD [30], Parkinson's disease [31], language disorders [32], and motor deficits in patients with stroke [33]. However, the

Table 13.1 Optimization strategies of TMS in neurocognitive disorders

TMS	Clinical recommendations				Levels + references
	Target	Frequency	Length (min)	Duration (sessions)	
Unilateral rTMS	Unclear	Unclear	Unclear	Unclear	Both[a]: [15]
Bilateral rTMS	Unclear	Unclear	Unclear	Unclear	Both[a]: [15]
Accelerated rTMS	Unclear	Unclear	Unclear	Unclear	Both[a]: NR
Deep TMS	Unclear	Unclear	Unclear	Unclear	MCI: NR AD: [19]
Priming TMS	Unclear	Unclear	Unclear	Unclear	Both[a]: NR
Synchronized TMS	Unclear	Unclear	Unclear	Unclear	Both[a]: NR
iTBS	Unclear	Unclear	Unclear	Unclear	MCI: [34] AD: [35]
Accelerated iTBS	Unclear	Unclear	Unclear	Unclear	Both[a]: NR
Continuation TBS	Unclear	Unclear	Unclear	Unclear	Both[a]: NR
Bilateral TBS	Unclear	Unclear	Unclear	Unclear	Both[a]: NR
MST	Unclear	Unclear	Unclear	Unclear	Both[a]: NR

[a]Both including MCI and AD

Abbreviations: *DLPFC* dorsolateral prefrontal cortex, *RMT* resting motor threshold, *iTBS* intermittent theta burst stimulation, *MST* magnetic seizure therapy, *rTMS* repetitive transcranial magnetic stimulation, *TBS* theta burst stimulation, *TMS* transcranial magnetic stimulation, *NR* no record

potential efficacy of the H-coil for AD remains largely unexplored. The H2-coil, in particular, is capable of stimulating the medial prefrontal, lateral frontal, and temporoparietal regions simultaneously.

Preliminary results from a small open-label case series suggest that H2-coil rTMS may show promise in the AD context [36]. A small placebo-controlled study investigated the feasibility, safety, and efficacy of deep rTMS targeting bilateral fronto-temporo-parietal regions using the H2-coil in patients with AD [37]. Thirty patients, with a mean age of 70.9 years, were randomly assigned to receive either real or sham 10 Hz rTMS with the H2-coil. Participants received three sessions per week for 4 weeks, followed by one session per week for an additional 4 weeks during the maintenance phase. The primary outcome measure was the improvement in the Alzheimer's Disease Assessment Scale—cognitive subscale (ADAS-cog) scores at 4 and 8 weeks compared to baseline. Patients receiving real rTMS showed a trend toward improved ADAS-cog scores, with a mean decrease of −1.01 points per time point [37]. However, this trend was not sustained 2 months posttreatment. Additionally, no significant changes were observed in the Mini-Mental State Examination (MMSE) scores over time. These preliminary findings indicate that rTMS with the H-coil is feasible for patients with probable AD and may offer transient cognitive improvements.

13.2.2.2 Safety

In the placebo-controlled pilot study conducted by Leocani et al. [37], no participants reported serious adverse effects directly attributable to the treatment. One participant in the sham group experienced an acute myocardial infarction 2 weeks after the initiation of treatment, which was deemed unrelated to their participation in the study. This participant also reported a transient headache following one rTMS session. Overall, the procedure was well-tolerated, with the exception of one participant in the active treatment group who was unable to tolerate the intensity of 120% RMT in 11 out of 16 rTMS sessions, prompting a reduction to 95–110% RMT.

13.2.2.3 Treatment Regimen

The H-coil used in deep TMS differs from conventional TMS coils and is incorporated into a helmet designed for secure fitting on the user. The H-coil consists of two layers, each containing four elliptically shaped windings arranged in a stacked configuration. The major and minor axes of these windings range from 70 to 130 mm and 55 to 105 mm, respectively [38]. The magnetic field generated by the H-coil attenuates at a slower rate, enabling the stimulation of neurons over a larger and deeper area. The subdural penetration depth and stimulation volume of the H-coil are approximately 3 cm and 40.3 cm^3, respectively [38].

The application of iTBS for MCI or AD is still in its early stages, necessitating further extensive research to establish optimal treatment parameters. In the study by Leocani et al. [37], the right and left prefrontal cortices, along with the temporal-parietal regions, were selected as the target areas for stimulation. Each rTMS session involved the administration of 840 stimuli at a frequency of 10 Hz, organized into 42 trains of 20 stimuli with inter-train intervals of 22 s, delivered at an intensity of 120% of the RMT. Participants were randomly assigned a blank-coded magnetic card that activated either the real or sham mode on the H-coil controller, with the decoding information retained by administrative personnel who were not involved in the study, ensuring the blinding of both patients and operators administering rTMS. The treatment protocol consisted of 16 rTMS sessions, including an initial 4-week period of intensive treatment with three sessions per week, followed by a 4-week maintenance phase with one session per week. Patient and caregiver interviews were conducted after each session to monitor potential side effects.

13.2.2.4 Clinical Recommendations

Given the absence of large-scale RCTs in this area, continuous monitoring of future developments is essential. As a result, specific treatment recommendations remain to be established (Table 13.1). In comparison to traditional rTMS, deep rTMS offers the potential to deliver therapeutic effects to deeper brain regions, suggesting promising prospects for both research and clinical application.

13.2.3 Priming TMS

Priming TMS (pTMS) is a modified low-frequency TMS protocol that begins with a brief, low-intensity high-frequency TMS session, followed by low-frequency rTMS targeting the right DLPFC Iyer et al. [39]. For example, 6-Hz priming at 90% RMT enhances the cortical suppression effects of 1-Hz rTMS in the motor cortex. Fitzgerald and colleagues demonstrated in an RCT that this priming approach resulted in improved antidepressant outcomes compared to 1-Hz right-sided rTMS alone Fitzgerald et al. [40]. It is anticipated that pTMS will be explored further in future studies involving individuals with cognitive impairments.

13.2.4 Synchronized TMS

Synchronized TMS (sTMS) utilizes rotating spherical neodymium magnets positioned along the scalp midline to administer low-field sinusoidal waveform stimulation synchronized with an individual's alpha EEG frequency Leuchter et al. [41]. By aligning with the brain's natural resonance at the individual's alpha frequency, sTMS is believed to achieve antidepressant efficacy with lower energy levels than conventional high-frequency rTMS, potentially reducing adverse effects Fröhlich [42]. Given that cognitive impairment often coexists with depression, alleviating depressive symptoms may, to some extent, improve cognitive function Grahek et al. [43]. Consequently, sTMS could offer superior therapeutic outcomes for patients with cognitive impairment comorbid with depression.

13.2.5 iTBS

13.2.5.1 Efficacy

Completing a traditional rTMS regimen, typically requiring daily sessions over 4–6 weeks, may present significant logistical challenges for individuals with cognitive impairments. However, recent innovations in rTMS delivery, particularly the use of iTBS within an accelerated treatment framework, offer potential solutions to these challenges. iTBS involves delivering 3-pulse bursts at 50 Hz every 200 ms, with each session lasting less than 3 min, as opposed to the 20–30 min required for conventional 10 Hz rTMS sessions [44]. The shorter duration of iTBS sessions makes it feasible to administer multiple sessions within a single day, facilitating an accelerated treatment protocol. This approach reduces the overall treatment duration while maintaining the same number of sessions and pulses as conventional rTMS, which typically spans 4–6 weeks. Accelerated iTBS treatment, completed within 3 to 5 days [45], may thus offer a more viable and accessible therapeutic option for individuals with MCI. Notably, accelerated iTBS has received FDA approval for treating refractory major depression, demonstrating advantages such as enhanced patient adherence and a more rapid therapeutic response compared to conventional rTMS.

Given the limited research on accelerated iTBS for MCI, a crucial first step in advancing this promising and minimally burdensome neuromodulation therapy is to systematically assess its safety and carefully evaluate factors that may influence its implementation. A recent Phase I open-label trial involving 24 elderly individuals diagnosed with amnestic MCI (aMCI) explored the feasibility of accelerated iTBS targeting the left dorsolateral prefrontal cortex. The protocol consisted of eight stimulation sessions, each delivering 600 pulses of iTBS per day over 3 consecutive days [34]. The study's findings confirm the safety, feasibility, and acceptability of accelerated iTBS in individuals with aMCI, and also provide evidence of target engagement, as demonstrated by enhanced cognitive function following treatment. These promising results offer valuable insights for future trials aimed at optimizing treatment parameters.

For AD, recent research has not only explored the efficacy of iTBS in improving cognitive function but also sought to elucidate the underlying mechanisms [35, 46, 47, 48]. However, the long-term effects of iTBS on cognitive decline and brain structure in patients with AD remain poorly understood. A recent study assessed the potential of repeated administration of accelerated iTBS every 3 months to mitigate cognitive decline in patients with AD [35]. This randomized, assessor-blinded, controlled trial involved 42 patients with AD who received iTBS to the left DLPFC over 14-day periods, spaced 13 weeks apart. Assessment tools included the MoCA, a comprehensive neuropsychological battery, and hippocampal grey matter volume (GMV) measurements. The results indicated that the iTBS group maintained MoCA scores while showing reduced hippocampal atrophy, a finding that was significantly correlated with changes in the global degeneration scale. These results suggest that DLPFC-iTBS could serve as a viable and easily implementable non-pharmacological intervention to help mitigate the progressive cognitive decline and deterioration in quality of life associated with AD. Deficits in associative memory (AM) are one of the earliest and most prominent symptoms of AD. In a randomized, double-blind, sham-controlled trial, iTBS was applied to the left DLPFC of patients with AD over a 14-day period [47]. The active stimulation group demonstrated significantly greater improvements in AM compared to the sham group by the second week, with these enhancements persisting until the tenth week. Growing evidence suggests that neural network dysfunction may directly contribute to cognitive impairment in AD, making it a promising treatment target. A recent study further investigated the impact of iTBS on functional connectivity and cognition in mild to moderate AD [46]. In this double-blind, randomized, sham-controlled trial, 58 participants received either active or sham iTBS targeting the left and right dorsolateral prefrontal cortex and posterior parietal cortex. Neurobiological, cognitive, and behavioral assessments were performed at the beginning and end of treatment. The results showed that active iTBS enhanced resting-state gamma connectivity and improved delayed recall in an episodic memory task. Baseline gamma connectivity and its changes were predictive of better delayed recall following active treatment. These findings underscore the need for further research to refine iTBS protocols for AD, with the potential to optimize therapeutic outcomes.

13.2.5.2 Safety

In studies examining iTBS for individuals with MCI or AD, no serious adverse events, including seizures, were reported. A small number of participants in the active group experienced minor discomfort, such as scalp sensations or sleep disturbances, though all reported events were deemed tolerable.

13.2.5.3 Treatment Regimen

Despite the promising findings, there are limited clinical studies investigating iTBS for MCI or AD, and further exploration of treatment parameters is warranted. A recent study showed enhanced fluid cognition following 3 days of accelerated iTBS, which involved a total of 14,400 pulses [34]. This result indicates successful target engagement in less than one-third of the treatment duration typically required in earlier iTBS studies for AD. One of the key advantages of accelerated iTBS is the substantial reduction in both the frequency and duration of treatment sessions, a factor particularly relevant for older adults with MCI, for whom time commitment is often a significant barrier to participation. The majority of participants in this study expressed a preference for completing multiple sessions per day in a condensed timeframe, rather than daily sessions spread over several weeks. Although the optimal intersession interval for same-day rTMS sessions has not been systematically explored, with existing protocols varying from 10 min to 2 h, it is proposed that a 10–15-min interval between sessions is both safe and well-received. This approach was further supported by the majority of participants reporting high motivation, interest, and commitment to the treatment, along with a low rate of discontinuation and a willingness to continue with future sessions.

13.2.5.4 Clinical Recommendations

Given the limited number of studies, making robust recommendations is challenging. For patients with MCI, a reasonable stimulation parameter might involve 50 Hz iTBS with a triple pulse every 200 ms, repeated every 10 s, for a duration of 8 s, totaling 190 s and 600 pulses per day over 3 days [34]. To ensure participant comfort, the interval between each stimulation session on the same day should be 10–15 min or longer. For patients with AD, a recent RCT suggested that a suitable treatment protocol may involve administering three cycles of left DLPFC iTBS per day, with a 15-min interval between cycles (totaling 1800 pulses per day), over 14 consecutive days, at 70% of the RMT [35]. However, neither MCI nor AD have expert consensus or treatment guidelines recommending the clinical use of iTBs (Table 13.1). We anticipate that ongoing research will clarify this issue.

13.2.6 Magnetic Seizure Therapy

Magnetic seizure therapy (MST), which uses magnetic fields to induce a generalized seizure, targets more superficial brain areas compared to electroconvulsive therapy (ECT), potentially sparing deeper structures like the hippocampus Lisanby [49]. A recent systematic review found no significant difference in antidepressant response between MST and ECT, although MST showed better outcomes compared to sham therapy Mutz et al. [50]. However, the application of MST in psychiatry and neurology remains limited, with small sample sizes. Future studies will explore this therapeutic approach further.

13.3 tDCS for Neurocognitive Disorder

Transcranial Direct Current Stimulation (tDCS) has emerged as a promising noninvasive brain stimulation technique for treating neurocognitive disorders, including AD, MCI, and other dementias [51, 52]. By delivering a low-intensity electrical current to targeted brain regions, tDCS modulates cortical excitability and aims to enhance cognitive function. This chapter examines both conventional tDCS and high-definition tDCS (HD-tDCS), focusing on their efficacy, safety, treatment regimens, and clinical recommendations for individuals with neurocognitive disorders.

13.3.1 Conventional tDCS

13.3.1.1 Efficacy

The efficacy of tDCS for neurocognitive disorders has been explored in numerous clinical trials [53–56]. Evidence suggests that tDCS improves cognitive function by modulating neuroplasticity and enhancing synaptic efficiency [57–59]. A meta-analysis of 23 comparisons from 13 qualified studies supports the conclusion that tDCS protocols have positive effects in ameliorating AD-induced cognitive dysfunction, particularly in memory and global cognitive domains [60]. Studies targeting the DLPFC have reported improvements in memory, attention, and executive functions in patients with neurocognitive disorders [61–63]. However, while some RCTs have demonstrated significant cognitive benefits, others have reported mixed or minimal effects, indicating variability in patient response. Meta-analyses suggest that this variability may stem from factors such as stimulation parameters—including current intensity and density, electrode size and location, stimulation duration, and the number of sessions—as well as inter-individual differences in cranial and brain anatomy. These anatomical variations, such as skull thickness, subcutaneous fat, gyral patterns, local tissue heterogeneities, and neuron orientation, all influence the current distribution, particularly in patients with neurocognitive disorders who experience varying degrees of brain atrophy [63]. Overall, while tDCS shows promise as an adjunctive therapy to improve cognitive function, further robust, large-scale studies are necessary to establish its consistent efficacy across diverse patient populations.

13.3.1.2 Safety

The safety profile of tDCS has been well-established in clinical studies involving patients with neurocognitive disorders. In general, tDCS is considered safe and well-tolerated, with only minor side effects reported. Common adverse events include itching, tingling, and mild discomfort at the stimulation site, which typically resolves quickly after treatment [64, 65]. In patients with neurocognitive disorders, these side effects are similar to those experienced by healthy individuals, suggesting that elderly or cognitively impaired populations tolerate tDCS effectively. Serious adverse effects are rare, and no significant cognitive or neurological deterioration has been linked to tDCS use [66]. However, caution is warranted for patients with implanted medical devices, epilepsy, or other neurological conditions. The long-term safety of repeated sessions remains an area of ongoing research, and future studies should continue to assess potential cumulative effects over extended periods.

13.3.1.3 Treatment Regimen

The treatment regimen for tDCS in neurocognitive disorders typically involves specific parameters to achieve optimal results. The DLPFC is the most commonly targeted brain region due to its central role in executive functions and working memory. Stimulation intensity generally ranges from 1 to 2 mA, depending on patient tolerance and treatment goals [67, 68]. Sessions are typically administered 3 to 5 times per week over several weeks, with each session lasting approximately 20 to 30 min [67, 68]. The total number of treatment sessions may range from 10 to 20, though some studies have explored longer-term maintenance protocols over several months [60, 67]. The number of treatment sequences may range from 10 to 20 sessions, although some studies have explored maintenance protocols over several months [51, 67]. The ideal treatment regimen depends on factors such as disease severity, patient response, and therapeutic objectives. Personalizing treatment parameters is crucial to optimizing cognitive benefits while minimizing adverse effects.

13.3.1.4 Clinical Recommendations

Clinical recommendations for the use of tDCS in neurocognitive disorders emphasize the importance of individualized treatment plans [68–70]. Given the variability in response, clinicians are encouraged to adjust parameters such as stimulation intensity, frequency, and target area based on each patient's specific needs and characteristics. tDCS should be considered an adjunctive therapy alongside conventional treatments like pharmacotherapy and cognitive rehabilitation to enhance overall therapeutic outcomes. It is also essential to conduct a thorough assessment of the patient's medical history, including contraindications such as epilepsy or implanted medical devices. Due to the experimental nature of tDCS in this patient population, its use should be confined to controlled clinical settings, where patients can be closely monitored for both safety and efficacy. Future guidelines should be revised as more data from large-scale, randomized trials become available.

13.3.2 HD-tDCS

13.3.2.1 Efficacy

The optimal number of tDCS sessions for MCI remains under investigation, influenced by factors such as the tDCS protocol, MCI severity, and individual response [69]. Session durations range from 5 to 30 min, with longer sessions generally yielding more sustained effects, although no additional benefits are observed beyond

30 min [71]. A key concern with tDCS is its lack of focality, which has led to the development of HD-tDCS. This technique employs five smaller electrodes arranged in a circular configuration to improve focality and target the surrounding cortex, with the central and peripheral electrodes set at opposite polarities [72].

Research on HD-tDCS for MCI or AD is limited. One study assessed whether HD-tDCS could improve cognitive function in patients with MCI and its potential link to brain activity changes, measured by resting-state functional magnetic resonance imaging [73]. Forty-three patients with MCI were randomly assigned to either 10 sessions of HD-tDCS or sham treatment targeting the left DLPFC. Posttreatment analysis revealed no significant changes in MMSE or MoCA scores. However, HD-tDCS enhanced regional homogeneity in several brain regions, suggesting possible mechanisms for its clinical effects. Similarly, a recent study found no beneficial effects of HD-tDCS on spatial navigation ability in patients with MCI [74]. In contrast, a Norwegian study evaluated the effects of personalized HD-tDCS on cognitive function in 19 patients with AD [75]. Participants were randomly assigned to either active or sham treatment, with the optimal montage for each patient targeting the left DLPFC using MRI-based computational modeling. The treatment involved 2 mA sessions for 20 min, conducted twice over 2 days. Cognitive tests and MRI data were analyzed before and after treatment. The active HD-tDCS group showed significant improvements in delayed memory and MMSE scores compared to the sham group, with five participants improving in delayed memory, four remaining stable, and one declining. These findings suggest that HD-tDCS may enhance delayed memory in AD, warranting further investigation with larger sample sizes.

13.3.2.2 Safety

Due to the limited clinical research available, no significant adverse reactions have been reported thus far. However, some researchers suggest that more targeted interventions may enhance safety [76]. By reducing stimulation to adjacent brain regions, this approach could potentially minimize side effects, allowing for increased intensity or repetition of stimulation to improve efficacy. Kuo et al. demonstrated that when the HD-tDCS 4 × 1 montage is centered on the primary motor cortex, the motor-evoked potential amplitude is more pronounced and sustained compared to bipolar stimulation [77]. Moreover, HD-tDCS is associated with less discomfort and improved applicability for elderly populations [78].

13.3.2.3 Treatment Regimen

In Rasmussen et al.'s study, the researchers utilized MRI scans to examine the computational model of HD-tDCS-induced electric fields in individual patients with AD [75]. The montage was optimized to deliver the highest net anodic electric field to the left DLPFC. An accelerated HD-tDCS protocol (2 mA, 3 sessions of 20 min each) was applied over 2 days, yielding promising therapeutic outcomes. Given the variability in brain atrophy among patients with AD, precise targeting is critical to ensure effective stimulation, as current distribution is influenced by cerebrospinal fluid and the extent of atrophy. Further studies with larger sample sizes are needed to produce more robust results.

Table 13.2 Optimization strategies of tDCS in neurocognitive disorder

tDCS	Clinical recommendations					Levels + references
	Anode position	Cathode position	Intensity (mA)	Length (min)	Duration (sessions)	
Conventional tDCS	Unclear	Unclear	Unclear	Unclear	Unclear	MCI: [79] AD: [51]
HD-tDCS	Unclear	Unclear	Unclear	Unclear	Unclear	MCI: [73] AD: [75]

Abbreviations: *HD-tDCS* high-definition transcranial direct current stimulation, *tDCS* transcranial direct current stimulation

13.3.2.4 Clinical Recommendations

The lack of large-scale RCTs underscores the need for ongoing monitoring of emerging developments in this field. At present, the literature does not support definitive recommendations for improving cognitive function in patients with AD through HD-tDCS (Table 13.2). Compared to conventional tDCS, HD-tDCS offers improved facility and the potential for more personalized, symptom-specific stimulation, suggesting promising prospects for future research.

13.4 TMS vs. tDCS

13.4.1 Efficacy

Cognitive impairment can arise in patients with various neurological disorders, including schizophrenia, depression, dementia, Parkinson's disease, stroke, traumatic brain injury, and multiple sclerosis. A previous meta-analysis assessed the efficacy of TMS and tDCS in improving multiple cognitive domains [80]. The analysis included 82 studies with 2784 participants, revealing that both TMS and tDCS had small but statistically significant effects on working memory, with age acting as a positive moderator for TMS effects. Furthermore, tDCS outperformed sham treatment in enhancing attention and vigilance. These effects were consistent across different neurological conditions. However, no significant changes were found for the other five cognitive domains examined. The study concluded that both TMS and tDCS produce modest trans-diagnostic effects on working memory, with tDCS also improving attention and vigilance. Although effect sizes were small, even modest cognitive improvements can enhance daily functioning.

13.4.2 Safety

To date, no direct comparative clinical studies have evaluated the safety profiles of TMS and tDCS for neurocognitive disorders. However, existing literature suggests that both modalities demonstrate favorable safety profiles and patient tolerance in conditions like MCI and AD, with no reports of serious adverse events, including seizures or shock. Despite this, given the variability in patient characteristics and treatment parameters, vigilant monitoring for potential adverse reactions remains essential.

13.5 Conclusion

This study offers a comprehensive overview of recent advancements in TMS and tDCS interventions for neurocognitive disorders. Despite the development of several new TMS paradigms, rTMS remains the dominant approach. Evidence suggests that rTMS significantly enhances global cognitive function, as measured by tools such as the MMSE, MoCA, and ADAS-Cog, when compared to sham stimulation. While improvements in specific cognitive domains, including memory, working memory, and executive function, have been observed, these may reflect the relatively limited number of studies focusing on these areas. Future research should move beyond broad cognitive assessments and utilize more sensitive measures that target specific cognitive domains. This approach could help identify the neuropsychological functions most responsive to improvement. Additionally, research may reveal that TMS applied to specific brain regions can lead to more pronounced effects in certain cognitive domains. For example, stimulation of the frontal regions may enhance executive function and working memory more than new learning [15]. This would facilitate a more personalized approach, where TMS interventions are tailored to address the individual cognitive impairments of each patient. The safety and tolerability of TMS are well-established, particularly when appropriate safety protocols are followed, even in older adults with depression [23]. Adherence to contemporary safety guidelines [22], including thorough participant screening, maintaining stimulation parameters within safety thresholds, and using qualified technicians and clinicians, can significantly mitigate the risk of seizures [25].

Extensive research suggests that tDCS may be a valuable tool for neuromodulating cognitive function and may potentially support successful cognitive aging [81]. However, these findings need replication and clinical application. Further studies are necessary to enhance our understanding of the underlying mechanisms, therapeutic outcomes, and influencing factors, as well as to optimize stimulation parameters. Future research should focus on determining the ideal stimulation parameters, including dosage, session duration, intersession intervals, optimal stimulation sites, and the most effective combinations with pharmacological agents and other techniques that may influence disease progression. Despite the positive feedback from clinical trials, tDCS remains an evolving technique, requiring further investigation to establish its long-term safety. Additionally, healthcare professionals must receive adequate training in tDCS and adhere strictly to safety guidelines when implementing the technique.

Acknowledgments Acknowledgments are extended to all contributors who have facilitated the completion of this academic manuscript. We thank Bullet Edits Limited for the linguistic editing and proofreading of the manuscript. Special recognition is given to the geriatric team at Beijing Anding Hospital for their substantial support and to Dr. Zheng Wei for his valuable input.

Disclosure/Conflicts of Interest The authors declare no conflicts of interest in conducting this study or preparing the manuscript.

Financial Support This study was funded by the National Natural Science Foundation of China (82401791); Capital's Funds for Health Improvement and Research (CFH 2022-3-2124); High-Level Public Health Technical Personnel Construction Training Program (Discipline Backbone-02-38); Beijing Municipal Science and Technology Project (Z141107002514121). The funders had no role in study design, data collection and analysis, decision to publish, or preparation of the manuscript.

References

1. Sachdev PS, Blacker D, Blazer DG, et al. Classifying neurocognitive disorders: the DSM-5 approach. Nat Rev Neurol. 2014;10(11):634–42. https://doi.org/10.1038/nrneurol.2014.181.
2. Sachdev P, Andrews G, Hobbs MJ, et al. Neurocognitive disorders: cluster 1 of the proposed meta-structure for DSM-V and ICD-11. Psychol Med. 2009;39(12):2001–12. https://doi.org/10.1017/s0033291709990262.
3. American Psychological Association. Diagnostic and statistical manual of mental disorders. 5th ed. Washington, DC: American Psychiatric Pub; 2013.
4. Petersen RC, Lopez O, Armstrong MJ, et al. Practice guideline update summary: mild cognitive impairment: report of the guideline development, dissemination, and implementation subcommittee of the American Academy of Neurology. Neurology. 2018;90(3):126–35. https://doi.org/10.1212/wnl.0000000000004826.
5. Esang M, Gupta M. Aducanumab as a novel treatment for Alzheimer's disease: a decade of hope, controversies, and the future. Cureus. 2021;13(8):e17591. https://doi.org/10.7759/cureus.17591.
6. van Dyck CH, Swanson CJ, Aisen P, et al. Lecanemab in early Alzheimer's disease. N Engl J Med. 2023;388(1):9–21. https://doi.org/10.1056/NEJMoa2212948.
7. Arvanitakis Z, Shah RC, Bennett DA. Diagnosis and management of dementia: review. JAMA. 2019;322(16):1589–99. https://doi.org/10.1001/jama.2019.4782.
8. Birba A, Ibáñez A, Sedeño L, et al. Non-invasive brain stimulation: a new strategy in mild cognitive impairment? Front Aging Neurosci. 2017;9:16. https://doi.org/10.3389/fnagi.2017.00016.
9. Cheng CPW, Wong CSM, Lee KK, et al. Effects of repetitive transcranial magnetic stimulation on improvement of cognition in elderly patients with cognitive impairment: a systematic review and meta-analysis. Int J Geriatr Psychiatry. 2018;33(1):e1–e13. https://doi.org/10.1002/gps.4726.
10. Dong X, Yan L, Huang L, et al. Repetitive transcranial magnetic stimulation for the treatment of Alzheimer's disease: a systematic review and meta-analysis of randomized controlled trials. PLoS One. 2018;13(10):e0205704. https://doi.org/10.1371/journal.pone.0205704.
11. Fileccia E, Di Stasi V, Poda R, et al. Effects on cognition of 20-day anodal transcranial direct current stimulation over the left dorsolateral prefrontal cortex in patients affected by mild cognitive impairment: a case-control study. Neurol Sci. 2019;40(9):1865–72. https://doi.org/10.1007/s10072-019-03903-6.
12. Ciullo V, Spalletta G, Caltagirone C, et al. Transcranial direct current stimulation and cognition in neuropsychiatric disorders: systematic review of the evidence and future directions. Neuroscientist. 2021;27(3):285–309. https://doi.org/10.1177/1073858420936167.
13. Goette WF, Goette HE. A meta-analysis of the accuracy of embedded performance validity indicators from the repeatable battery for the assessment of neuropsychological status. Clin Neuropsychol. 2019;33(6):1044–68. https://doi.org/10.1080/13854046.2018.1538429.

14. Meinzer M, Lindenberg R, Phan MT, et al. Transcranial direct current stimulation in mild cognitive impairment: behavioral effects and neural mechanisms. Alzheimers Dement. 2015;11(9):1032–40. https://doi.org/10.1016/j.jalz.2014.07.159.
15. Pagali SR, Kumar R, LeMahieu AM, et al. Efficacy and safety of transcranial magnetic stimulation on cognition in mild cognitive impairment, Alzheimer's disease, Alzheimer's disease-related dementias, and other cognitive disorders: a systematic review and meta-analysis. Int Psychogeriatr. 2024:1–49. https://doi.org/10.1017/s1041610224000085.
16. Gy RR, Jv RL, Ricardo Garcell J, López Hidalgo M, Aguilar Fabré L, Trejo Cruz G, et al. Effect of transcranial magnetic stimulation as an enhancer of cognitive stimulation sessions on mild cognitive impairment: preliminary results. Psychiatry Res. 2021;304:114151. https://doi.org/10.1016/j.psychres.2021.114151.
17. Dubreuil-Vall L, Chau P, Ruffini G, et al. tDCS to the left DLPFC modulates cognitive and physiological correlates of executive function in a state-dependent manner. Brain Stimul. 2019;12(6):1456–63. https://doi.org/10.1016/j.brs.2019.06.006.
18. Turriziani P, Smirni D, Zappalà G, et al. Enhancing memory performance with rTMS in healthy subjects and individuals with mild cognitive impairment: the role of the right dorsolateral prefrontal cortex. Front Hum Neurosci. 2012;6:62. https://doi.org/10.3389/fnhum.2012.00062.
19. Cui H, Ren R, Lin G, et al. Repetitive transcranial magnetic stimulation induced hypoconnectivity within the default mode network yields cognitive improvements in amnestic mild cognitive impairment: a randomized controlled study. J Alzheimers Dis. 2019;69(4):1137–51. https://doi.org/10.3233/jad-181296.
20. Lin Y, Jiang WJ, Shan PY, et al. The role of repetitive transcranial magnetic stimulation (rTMS) in the treatment of cognitive impairment in patients with Alzheimer's disease: a systematic review and meta-analysis. J Neurol Sci. 2019;398:184–91. https://doi.org/10.1016/j.jns.2019.01.038.
21. Teselink J, Bawa KK, Koo GK, et al. Efficacy of non-invasive brain stimulation on global cognition and neuropsychiatric symptoms in Alzheimer's disease and mild cognitive impairment: a meta-analysis and systematic review. Ageing Res Rev. 2021;72:101499. https://doi.org/10.1016/j.arr.2021.101499.
22. Rossi S, Antal A, Bestmann S, et al. Safety and recommendations for TMS use in healthy subjects and patient populations, with updates on training, ethical and regulatory issues: expert guidelines. Clin Neurophysiol. 2021;132(1):269–306. https://doi.org/10.1016/j.clinph.2020.10.003.
23. Iriarte IG, George MS. Transcranial magnetic stimulation (TMS) in the elderly. Curr Psychiatry Rep. 2018;20(1):6. https://doi.org/10.1007/s11920-018-0866-2.
24. Pandis D, Scarmeas N. Seizures in Alzheimer disease: clinical and epidemiological data. Epilepsy Curr. 2012;12(5):184–7. https://doi.org/10.5698/1535-7511-12.5.184.
25. Targa Dias Anastacio H, Matosin N, Ooi L. Neuronal hyperexcitability in Alzheimer's disease: what are the drivers behind this aberrant phenotype? Transl Psychiatry. 2022;12(1):257. https://doi.org/10.1038/s41398-022-02024-7.
26. Antal A, Luber B, Brem AK, et al. Non-invasive brain stimulation and neuroenhancement. Clin Neurophysiol Pract. 2022;7:146–65. https://doi.org/10.1016/j.cnp.2022.05.002.
27. Fitzgerald PB, Chen L, Richardson K, et al. A pilot investigation of an intensive theta burst stimulation protocol for patients with treatment resistant depression. Brain Stimul. 2020;13(1):137–44. https://doi.org/10.1016/j.brs.2019.08.013.
28. Cheng CM, Li CT, Tsai SJ. Current updates on newer forms of transcranial magnetic stimulation in major depression. Adv Exp Med Biol. 2021;1305:333–49. https://doi.org/10.1007/978-981-33-6044-0_18.
29. Zangen A, Roth Y, Voller B, et al. Transcranial magnetic stimulation of deep brain regions: evidence for efficacy of the H-coil. Clin Neurophysiol. 2005;116(4):775–9. https://doi.org/10.1016/j.clinph.2004.11.008.

30. Harel EV, Rabany L, Deutsch L, et al. H-coil repetitive transcranial magnetic stimulation for treatment resistant major depressive disorder: an 18-week continuation safety and feasibility study. World J Biol Psychiatry. 2014;15(4):298–306. https://doi.org/10.3109/15622975.2011.639802.
31. Spagnolo F, Volonté MA, Fichera M, et al. Excitatory deep repetitive transcranial magnetic stimulation with H-coil as add-on treatment of motor symptoms in Parkinson's disease: an open label, pilot study. Brain Stimul. 2014;7(2):297–300. https://doi.org/10.1016/j.brs.2013.10.007.
32. Chieffo R, Ferrari F, Battista P, et al. Excitatory deep transcranial magnetic stimulation with H-coil over the right homologous Broca's region improves naming in chronic post-stroke aphasia. Neurorehabil Neural Repair. 2014;28(3):291–8. https://doi.org/10.1177/1545968313508471.
33. Chieffo R, De Prezzo S, Houdayer E, et al. Deep repetitive transcranial magnetic stimulation with H-coil on lower limb motor function in chronic stroke: a pilot study. Arch Phys Med Rehabil. 2014;95(6):1141–7. https://doi.org/10.1016/j.apmr.2014.02.019.
34. Aghamoosa S, Lopez J, Rbeiz K, et al. A phase I trial of accelerated intermittent theta burst rTMS for amnestic MCI. J Neurol Neurosurg Psychiatry. 2024;95(11):1036–45. https://doi.org/10.1136/jnnp-2023-332680.
35. Wu X, Yan Y, Hu P, et al. Effects of a periodic intermittent theta burst stimulation in Alzheimer's disease. Gen Psychiatr. 2024;37(1):e101106. https://doi.org/10.1136/gpsych-2023-101106.
36. Avirame K, Stehberg J, Todder D. Benefits of deep transcranial magnetic stimulation in Alzheimer disease: case series. J ECT. 2016;32(2):127–33. https://doi.org/10.1097/yct.0000000000000286.
37. Leocani L, Dalla Costa G, Coppi E, et al. Repetitive transcranial magnetic stimulation with H-coil in Alzheimer's disease: a double-blind, placebo-controlled pilot study. Front Neurol. 2020;11:614351. https://doi.org/10.3389/fneur.2020.614351.
38. Tzirini M, Roth Y, Harmelech T, et al. Detailed measurements and simulations of electric field distribution of two TMS coils cleared for obsessive compulsive disorder in the brain and in specific regions associated with OCD. PLoS One. 2022;17(8):e0263145. https://doi.org/10.1371/journal.pone.0263145.
39. Iyer MB, Schleper N, Wassermann EM. Priming stimulation enhances the depressant effect of low-frequency repetitive transcranial magnetic stimulation. J Neurosci. 2003;23(34):10867–72. https://doi.org/10.1523/jneurosci.23-34-10867.2003.
40. Fitzgerald PB, Hoy K, McQueen S, et al. Priming stimulation enhances the effectiveness of low-frequency right prefrontal cortex transcranial magnetic stimulation in major depression. J Clin Psychopharmacol. 2008;28(1):52–8. https://doi.org/10.1097/jcp.0b013e3181603f7c.
41. Leuchter AF, Cook IA, Feifel D, et al. Efficacy and safety of low-field synchronized transcranial magnetic stimulation (sTMS) for treatment of major depression. Brain Stimul. 2015;8(4):787–94. https://doi.org/10.1016/j.brs.2015.05.005.
42. Fröhlich F. Endogenous and exogenous electric fields as modifiers of brain activity: rational design of noninvasive brain stimulation with transcranial alternating current stimulation. Dialogues Clin Neurosci. 2014;16(1):93–102. https://doi.org/10.31887/DCNS.2014.16.1/ffroehlich.
43. Grahek I, Shenhav A, Musslick S, et al. Motivation and cognitive control in depression. Neurosci Biobehav Rev. 2019;102:371–81. https://doi.org/10.1016/j.neubiorev.2019.04.011.
44. Liu C, Li L, Pan W, et al. Executive function deficits in patients with the first episode of late-life depression before and after SSRI treatment: a pilot fMRI study. Int J Geriatr Psychiatry. 2024;39(5):e6095. https://doi.org/10.1002/gps.6095.
45. Caulfield KA, Fleischmann HH, George MS, et al. A transdiagnostic review of safety, efficacy, and parameter space in accelerated transcranial magnetic stimulation. J Psychiatr Res. 2022;152:384–96. https://doi.org/10.1016/j.jpsychires.2022.06.038.

46. Hoy KE, Emonson MRL, Bailey NW, et al. Gamma connectivity predicts response to intermittent theta burst stimulation in Alzheimer's disease: a randomized controlled trial. Neurobiol Aging. 2023;132:13–23. https://doi.org/10.1016/j.neurobiolaging.2023.08.006.
47. Wu X, Ji GJ, Geng Z, et al. Accelerated intermittent theta-burst stimulation broadly ameliorates symptoms and cognition in Alzheimer's disease: a randomized controlled trial. Brain Stimul. 2022;15(1):35–45. https://doi.org/10.1016/j.brs.2021.11.007.
48. Xiao G, Wu Y, Yan Y, et al. Optimized magnetic stimulation induced hypoconnectivity within the executive control network yields cognition improvements in Alzheimer's patients. Front Aging Neurosci. 2022;14:847223. https://doi.org/10.3389/fnagi.2022.847223.
49. Lisanby SH. Update on magnetic seizure therapy: a novel form of convulsive therapy. J ect. 2002;18(4):182–8. https://doi.org/10.1097/00124509-200212000-00003.
50. Mutz J, Vipulananthan V, Carter B, et al. Comparative efficacy and acceptability of non-surgical brain stimulation for the acute treatment of major depressive episodes in adults: systematic review and network meta-analysis. BMJ. 2019;364:l1079. https://doi.org/10.1136/bmj.l1079.
51. Benussi A, Dell'Era V, Cosseddu M, et al. Transcranial stimulation in frontotemporal dementia: a randomized, double-blind, sham-controlled trial. Alzheimers Dement (N Y). 2020;6(1):e12033. https://doi.org/10.1002/trc2.12033.
52. Saxena V, Pal A. Role of transcranial direct current stimulation in the management of Alzheimer's disease: a meta-analysis of effects, adherence and adverse effects. Clin Psychopharmacol Neurosci. 2021;19(4):589–99. https://doi.org/10.9758/cpn.2021.19.4.589.
53. Cespón J, Rodella C, Miniussi C, et al. Behavioural and electrophysiological modulations induced by transcranial direct current stimulation in healthy elderly and Alzheimer's disease patients: a pilot study. Clin Neurophysiol. 2019;130(11):2038–52. https://doi.org/10.1016/j.clinph.2019.08.016.
54. Gangemi A, Colombo B, Fabio RA. Effects of short- and long-term neurostimulation (tDCS) on Alzheimer's disease patients: two randomized studies. Aging Clin Exp Res. 2021;33(2):383–90. https://doi.org/10.1007/s40520-020-01546-8.
55. Khedr EM, Salama RH, Abdel Hameed M, et al. therapeutic role of transcranial direct current stimulation in Alzheimer disease patients: double-blind, placebo-controlled clinical trial. Neurorehabil Neural Repair. 2019;33(5):384–94. https://doi.org/10.1177/1545968319840285.
56. Liu C, Pan W, Jia L, et al. Efficacy and safety of repetitive transcranial magnetic stimulation for peripartum depression: a meta-analysis of randomized controlled trials. Psychiatry Res. 2020;294:113543. https://doi.org/10.1016/j.psychres.2020.113543.
57. Das S, Holland P, Frens MA, et al. Impact of transcranial direct current stimulation (tDCS) on neuronal functions. Front Neurosci. 2016;10:550. https://doi.org/10.3389/fnins.2016.00550.
58. Qi S, Cao L, Wang Q, et al. The physiological mechanisms of transcranial direct current stimulation to enhance motor performance: a narrative review. Biology (Basel). 2024;13(10). https://doi.org/10.3390/biology13100790.
59. Rroji O, van Kuyck K, Nuttin B, et al. Anodal tDCS over the primary motor cortex facilitates long-term memory formation reflecting use-dependent plasticity. PLoS One. 2015;10(5):e0127270. https://doi.org/10.1371/journal.pone.0127270.
60. Majdi A, van Boekholdt L, Sadigh-Eteghad S, et al. A systematic review and meta-analysis of transcranial direct-current stimulation effects on cognitive function in patients with Alzheimer's disease. Mol Psychiatry. 2022;27(4):2000–9. https://doi.org/10.1038/s41380-022-01444-7.
61. Cheng CP, Chan SS, Mak AD, et al. Would transcranial direct current stimulation (tDCS) enhance the effects of working memory training in older adults with mild neurocognitive disorder due to Alzheimer's disease: study protocol for a randomized controlled trial. Trials. 2015;16:479. https://doi.org/10.1186/s13063-015-0999-0.
62. Fernandes SM, Mendes AJ, Rodrigues PFS, et al. Efficacy and safety of repetitive transcranial magnetic stimulation and transcranial direct current stimulation in memory deficits in patients with Alzheimer's disease: meta-analysis and systematic review. Int J Clin Health Psychol. 2024;24(2):100452. https://doi.org/10.1016/j.ijchp.2024.100452.

63. Galli G, Vadillo MA, Sirota M, et al. A systematic review and meta-analysis of the effects of transcranial direct current stimulation (tDCS) on episodic memory. Brain Stimul. 2019;12(2):231–41. https://doi.org/10.1016/j.brs.2018.11.008.
64. Godinho MM, Junqueira DR, Castro ML, et al. Safety of transcranial direct current stimulation: evidence based update 2016. Brain Stimul. 2017;10(5):983–5. https://doi.org/10.1016/j.brs.2017.07.001.
65. Rossi S, Hallett M, Rossini PM, et al. Safety, ethical considerations, and application guidelines for the use of transcranial magnetic stimulation in clinical practice and research. Clin Neurophysiol. 2009;120(12):2008–39. https://doi.org/10.1016/j.clinph.2009.08.016.
66. Brunoni AR, Amadera J, Berbel B, et al. A systematic review on reporting and assessment of adverse effects associated with transcranial direct current stimulation. Int J Neuropsychopharmacol. 2011;14(8):1133–45. https://doi.org/10.1017/s1461145710001690.
67. Im JJ, Jeong H, Bikson M, et al. Effects of 6-month at-home transcranial direct current stimulation on cognition and cerebral glucose metabolism in Alzheimer's disease. Brain Stimul. 2019;12(5):1222–8. https://doi.org/10.1016/j.brs.2019.06.003.
68. Bystad M, Grønli O, Rasmussen ID, et al. Transcranial direct current stimulation as a memory enhancer in patients with Alzheimer's disease: a randomized, placebo-controlled trial. Alzheimers Res Ther. 2016;8(1):13. https://doi.org/10.1186/s13195-016-0180-3.
69. Fregni F, El-Hagrassy MM, Pacheco-Barrios K, et al. Evidence-based guidelines and secondary meta-analysis for the use of transcranial direct current stimulation in neurological and psychiatric disorders. Int J Neuropsychopharmacol. 2021;24(4):256–313. https://doi.org/10.1093/ijnp/pyaa051.
70. Lefaucheur JP, Antal A, Ayache SS, et al. Evidence-based guidelines on the therapeutic use of transcranial direct current stimulation (tDCS). Clin Neurophysiol. 2017;128(1):56–92. https://doi.org/10.1016/j.clinph.2016.10.087.
71. Woods AJ, Antal A, Bikson M, et al. A technical guide to tDCS, and related non-invasive brain stimulation tools. Clin Neurophysiol. 2016;127(2):1031–48. https://doi.org/10.1016/j.clinph.2015.11.012.
72. Bikson M, Esmaeilpour Z, Adair D, et al. Transcranial electrical stimulation nomenclature. Brain Stimul. 2019;12(6):1349–66. https://doi.org/10.1016/j.brs.2019.07.010.
73. He F, Li Y, Li C, et al. Repeated anodal high-definition transcranial direct current stimulation over the left dorsolateral prefrontal cortex in mild cognitive impairment patients increased regional homogeneity in multiple brain regions. PLoS One. 2021;16(8):e0256100. https://doi.org/10.1371/journal.pone.0256100.
74. Iordan AD, Ryan S, Tyszkowski T, et al. High-definition transcranial direct current stimulation enhances network segregation during spatial navigation in mild cognitive impairment. Cereb Cortex. 2022;32(22):5230–41. https://doi.org/10.1093/cercor/bhac010.
75. Rasmussen ID, Boayue NM, Mittner M, et al. High-definition transcranial direct current stimulation improves delayed memory in Alzheimer's disease patients: a pilot study using computational modeling to optimize electrode position. J Alzheimers Dis. 2021;83(2):753–69. https://doi.org/10.3233/jad-210378.
76. Lang N, Siebner HR, Ward NS, et al. How does transcranial DC stimulation of the primary motor cortex alter regional neuronal activity in the human brain? Eur J Neurosci. 2005;22(2):495–504. https://doi.org/10.1111/j.1460-9568.2005.04233.x.
77. Kuo HI, Bikson M, Datta A, et al. Comparing cortical plasticity induced by conventional and high-definition 4 × 1 ring tDCS: a neurophysiological study. Brain Stimul. 2013;6(4):644–8. https://doi.org/10.1016/j.brs.2012.09.010.
78. Boayue NM, Csifcsák G, Kreis IV, et al. The interplay between executive control, behavioural variability and mind wandering: Insights from a high-definition transcranial direct-current stimulation study. Eur J Neurosci. 2021;53(5):1498–516. https://doi.org/10.1111/ejn.15049.

79. Palimariciuc M, Oprea DC, Cristofor AC, et al. The effects of transcranial direct current stimulation in patients with mild cognitive impairment. Neurol Int. 2023;15(4):1423–42. https://doi.org/10.3390/neurolint15040092.
80. Begemann MJ, Brand BA, Ćurčić-Blake B, et al. Efficacy of non-invasive brain stimulation on cognitive functioning in brain disorders: a meta-analysis. Psychol Med. 2020;50(15):2465–86. https://doi.org/10.1017/s0033291720003670.
81. Pezzoli S, Giorgio J, Martersteck A, et al. Successful cognitive aging is associated with thicker anterior cingulate cortex and lower tau deposition compared to typical aging. Alzheimers Dement. 2024;20(1):341–55. https://doi.org/10.1002/alz.13438.

MIX
Papier aus verantwortungsvollen Quellen
Paper from responsible sources
FSC® C105338

If you have any concerns about our products,
you can contact us on
ProductSafety@springernature.com

In case Publisher is established outside the EU,
the EU authorized representative is:
Springer Nature Customer Service Center GmbH
Europaplatz 3, 69115 Heidelberg, Germany

Printed by Libri Plureos GmbH
in Hamburg, Germany